West Southwest

West Southwest

Vertebrate Life in Southern California

Gregory K. Pregill

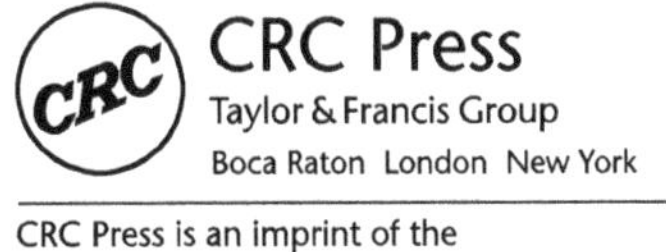

CRC Press is an imprint of the
Taylor & Francis Group, an **informa** business

CRC Press
Taylor & Francis Group
6000 Broken Sound Parkway NW, Suite 300
Boca Raton, FL 33487-2742

Printed on acid-free paper

International Standard Book Number-13: 978-1-1384-9696-5 (Paperback)
International Standard Book Number-13: 978-1-1385-8504-1 (Hardback)

Library of Congress Cataloging-in-Publication Data

Names: Pregill, Gregory K., author.
Title: West southwest : vertebrate life in southern California / Gregory K. Pregill.
Description: Boca Raton : Taylor & Francis, 2018. | Includes bibliographical references.
Identifiers: LCCN 2018000549| ISBN 9781138496965 (paperback : alk. paper) | ISBN 9781138585041 (hardback : alk. paper)
Subjects: LCSH: Vertebrates--California, Southern.
Classification: LCC QL606.52.U6 P74 2018 | DDC 596.09794/9--dc23
LC record available at https://lccn.loc.gov/2018000549

Visit the Taylor & Francis Web site at
http://www.taylorandfrancis.com

and the CRC Press Web site at
http://www.crcpress.com

Contents

SECTION I Fundamentals

SECTION II Settings

SECTION III Vertebrates

Introduction

> We are all mere proofreaders of Nature's book ... and we are apt to insert our commas according to our day's belief; but we cannot alter the true manuscript.
>
> **—Elliott Coues**

Early in the days of the *padres* and *conquistadores*, there was an upper *Alta* California and a lower *Baja* California, both controlled by Spain and later by Mexico. By the middle of the nineteenth century, the United States took control of Alta California, the *alta* prefix was dropped, and California became part of the Union. California is an almost imaginary land—a vast, fertile empire of mountains, valleys, and wilderness that fronts the Pacific Ocean with more than 3,400 miles of coastline.

Its continuity notwithstanding, California is composed of three geographic regions from north to south, each a reflection of latitude and terrain. This book is about Southern California, a place that is peerless in so many ways and unrivaled in so many others. In some places, the transition from southern to central California is fairly abrupt, as defined for instance by the Tehachapi Mountains in southern Kern County that separate the Central Valley from the Mojave Desert. Along the coast, Southern California extends to southern San Luis Obispo County north of Point Conception where the climate and plant life transition to that of the Central Coast. I consider all of the Mojave Desert, including Death Valley, as part of Southern California, so here the region extends farther north, well into Inyo County before ceding to the Great Basin. So defined, Southern California encompasses around 55,000 mi^2, an area about the size of Illinois, Iowa, or Wisconsin.

Southern California is recognized as a biological "hot spot" because of its diversity of so many unique species of plants and animals. True enough, but it is not an island, and its natural history has been influenced by adjacent areas to the north, a good part of the Southwest, and of course, Baja California, Mexico to the south of it. Indeed, the northern 200 miles or so of the Baja Peninsula is geologically and biologically nearly indistinguishable from much of Southern California. The international boundary between the two is a political one and not at all a natural divide. In that spirit, I dispense with the formality of referring to the peninsula as Baja California, Mexico, and simply call it Baja, a comfortable name familiar to practically everybody living in the West.

I was born in Southern California and grew up here as a boy. Many years later I had the opportunity to return as an eager vertebrate biologist, first as a museum curator and then as a professor at the University of San Diego (USD). At USD I taught a course in vertebrate natural history for more than two decades. With its varied habitats from the coast to the valleys and foothills, the mountains and the deserts, there is no place more ideal to introduce students to vertebrate life than Southern California. Over the course of the semester, my students and I would spend one afternoon a week in the field, and by the end of the course we had completed a west-to-east transect from the coast to the deserts. My purpose was straightforward enough; I wanted the students to become familiar with the amphibians, snakes, lizards, birds, and mammals—what they are, how they are related, why they are here, and how they survive. Because I also taught a course in the evolution of vertebrate structure (known in some circles as comparative anatomy), I infused natural history with some basic principles of vertebrate design. The objective was to illustrate how anatomical form is both an opportunity and a constraint in allowing, or not, a vertebrate do the things that it does. In the end, students come to appreciate, or at least realize, that field biology requires a lot of patience. Nature often seems reluctant to disclose its ways, past or present, and there are no shortcuts to learning them. Only persistence and paying attention will yield the opportunities for discovery.

For most students, the course was their first occasion to use a field guide for species identification. Field guides are a marvelous resource and the authors of such books deserve high praise. Good field guides become smudged, dog-eared, and well-worn additions to our personal libraries, as they are intended. Yet as I discuss later in this book in the chapter on birds, field guides are most effective only *after* one has acquired some basic familiarity with the plants or animals in question. Too, because of their scope and purpose, field guides cannot offer much in the way of evolutionary history, which is essential if we are to place organisms into context with the environments they occupy. To provide that perspective my course was heavily supplemented with handouts and outside readings gleaned from all manner of sources. This book is an attempt to share that experience.

The book is divided into three sections. Section I begins by examining the land and its climate. The ancient forces of mountain building and plate tectonics gave Southern California its renowned landscape, along with the San Andreas Fault. Here and elsewhere, reference is made to events that happened millions of years ago, so a geologic time table is provided (Chapter 1) as a reference to the various periods and epochs. Climate is something easy to take for granted, and it is often confused with weather. Weather is simply the local expression of climate at any particular time. Climate, like the Mediterranean one we live in, is a product of latitude, altitude, and in our case, adjacency to a cold ocean. Rainfall is strongly seasonal for most of Southern California and that has implications for its vertebrate inhabitants.

Next is a recounting of some of the relevant history of Southern California's Spanish settlement that opened the way for the earliest naturalists of the seventeenth and eighteenth centuries. These were adventurous, passionate men who were the first to discover California's extraordinary diversity of animals and plants. They began the tradition of biological exploration, collection, and interpretation that led eventually to the establishment of several of our region's outstanding natural history museums.

Section I concludes with a discussion of what natural history is and its relevance for our time. Readers unfamiliar with phylogenetics, the study of evolutionary relationships among species (part of the first of what I call the three pillars of natural history), will find here an introductory discussion of the topic sufficient to make sense of the evolutionary "trees" presented elsewhere in the book.

Section II is an excursion through the biotic communities of Southern California with an emphasis on the types of plants that help define them and the vertebrates that live there. An animal (or plant) returned to the laboratory for study becomes a specimen, an absolute necessity for learning the structural details of its kind. Alive in the field, it seizes our attention, compels us to look for context, and reminds us to continually evaluate our assumptions.

Section III takes up the vertebrates of Southern California themselves. In systematic fashion, major groups are put into historical and ecological perspective before commenting on the individual families, genera, and species—their general attributes and where they occur. For ease of reference all the species in our region are also listed in respective tables. This book is about the land vertebrates including those that secondarily have taken up life in the sea. Southern California is home to 30 species of amphibians, 80 species of reptiles (turtles, lizards, and snakes), 125 species of mammals plus four others that were eliminated from Southern California in historic times. Approximately 550 species of birds have been recorded in Southern California, too numerous to consider individually, but as a group are discussed as the unique biological phenomena that they are. The last chapter of Section III goes back in time to around 12,000 years ago near the beginning of our current epoch, the Holocene. Imagine what Southern California looked like before the extinction of all the large mammals that once roamed the area.

I want this book to be useful and informative to most anybody interested in nature and Southern California's unique place in the environment of the Southwest. To that end I've tried to keep technical terms to a minimum or define them where needed. But a jargon-free narrative is almost impossible to accomplish completely. Where the unexplained or unfamiliar word pops up, my hope is that you can continue with minimal distraction.

Acknowledgments

It is a pleasure to acknowledge the colleagues who shared their insights and comments on previous drafts of the manuscript. David Archibald, Michael Simpson, and David Steadman each made helpful contributions toward improving clarity and content.

Others who provided information, hospitality, referrals, assistance in the field, or other courtesies were Analissa Berta, the California State University Desert Studies Center, Brian Crother, Robert Fisher, Rob Fulton, Jim Hanken, the library staff at the Museum of Comparative Zoology, Harvard University, Mary Sue Lowery, Mike Mayer, Ray Munson, Bill Presch, Kristen Quarles and Tom Jorstadt at the Department of Paleobiology, Smithsonian Institution, Jay Savage, and George Zug.

To David Archibald, Don Genero, Mitch Leslie, Kurt Spanos, Jason Wallace, and David Wyatt, I extend hearty thanks for the photographs.

My wife Melissa was involved in many phases of this project. Her wit, sensibilities and good cheer made this all the more fun to do.

Section I

Fundamentals

1 The Land

Southern California's stunning landscape varies from beaches, coastal bluffs, mesas, wide inland valleys and foothills to mountains that rise from base to summit higher than any other in the lower 48 states. There are the high deserts of the Mojave and the below-sea-level expanse of the Salton Trough. Yet it is the mountains that all other Southern California landforms owe their character. The Transverse Ranges trend east to west, whereas the Peninsular Ranges are oriented north to south (Figure 1.1). The two ranges are separated from each other by the San Gorgonio Pass that extends from Banning to Palm Springs along I-10. The two ranges are a geological consequence of Southern California straddling two crustal plates that are creating tectonic deformations at a pace unmatched anywhere else in North America. Their origin and uplift are fairly recent events, geologically speaking, and are much younger than the rocks from which they are composed.

California's geological history extends back 1.8 billion years, yet little is known of that beginning because only remnants remain. The traceable record starts around 800 million years ago when a large chunk of the ancient North American continent split away and a new ocean basin was formed. Sediment filled the coastal margin and became sedimentary rock that in some places accumulated to several miles thick. Rocks in the Inyo Mountains, for example, and even portions of the Grand Canyon date to this time. Southern California was under shallow water, well west of the ancient, meandering coastline that ran from the Mojave Desert into Nevada and southern Utah, even to the vicinity of the Grand Canyon as sea levels rose and fell.

During the early Mesozoic, about 240 million years ago, things started to change. The enormous batholith of basement rock that would eventually form the Sierra Nevada began to accumulate, as did the rocks that would form the Peninsular Ranges. The tectonic history of the Sierra Nevada is deep and complex and serves as the evidentiary source for tracking the assembly of California so compellingly described by John McPhee. This mass of rock that began emerging just 3 million years ago would eventually extend north and south for 400 mi (644 km) nearly vertical on its east side, sloping on its west, and rising to an elevation of 14,495 ft (4,400 m) at Mt. Whitney, the highest point in the continental United States. The Sierra Nevada uplift is woven into the story of Southern California's own tectonic development, and both are products of the events that created the San Andreas Fault system (Table 1.1).

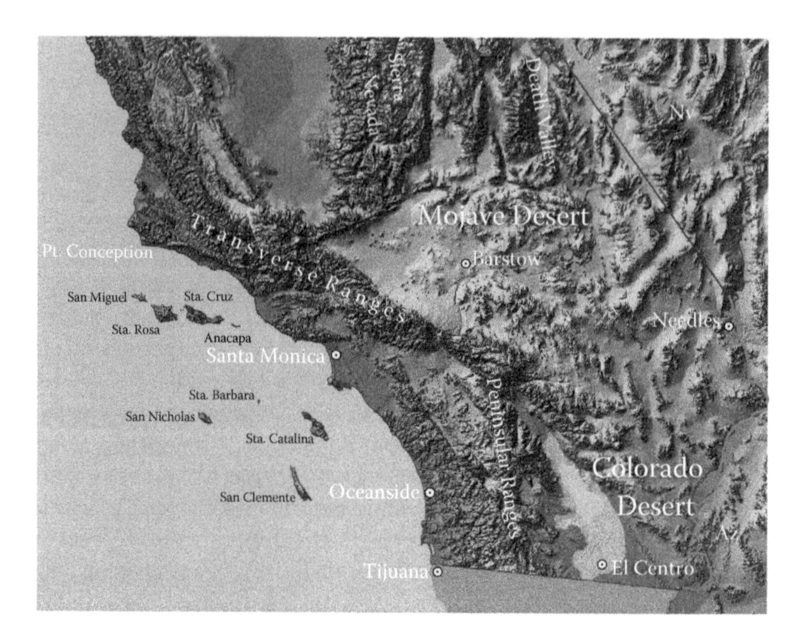

FIGURE 1.1 The Transverse and Peninsular mountain ranges are the topographic axes of Southern California. The foothills and valleys slope to the coast from their western flanks. To their east are the deserts (Mojave and Colorado) and the Salton Trough. (Base map adapted from www.nationalatlas.gov.)

TABLE 1.1
Geologic Time

Era	Period	Epoch	Age (millions of years)	
Cenozoic	Quaternary	Holocene	0.01	Historic period
		Pleistocene	1.8	Ice ages
		Pliocene	5.3	
		Miocene	23.8	
	Tertiary	Oligocene	33.7	
		Eocene	54.8	
		Paleocene	65.5	Major radiation of mammals and birds
Mesozoic	Cretaceous		145	Flowering plants; mass extinctions
	Jurassic		206	Dinosaurs diverse
	Triassic		251	Origin of mammals
Paleozoic	Permian		299	Radiation of reptiles; mass extinctions
	Carboniferous		360	
	Devonian		416	First land vertebrates
	Silurian		444	Early vascular plants
	Ordovician		485	
	Cambrian		541	

1.1 SAN ANDREAS FAULT

On Easter Sunday 2010, just past 3:30 p.m., the shaking started and continued for the next minute and a half. It was like trying to stand in a rubber raft on the open seas. Southern Californians expect earthquakes now and then, and this one was a ringer. The epicenter was 110 mi (177 km) from my house, near the northern Baja California community of Guadeloupe Victoria. The Sierra El Mayor earthquake, also called the 2010 Easter Quake, occurred along the Laguna Salada Fault at a depth of 10 km. It was the strongest quake in Southern California (magnitude of 7.2) since the Landers Quake in 1992. Four lives were lost, 100 more injured, and structural damage was serious in the border cities of Mexicali and Calexico. The quake resulted in 1 m of right-lateral surface faulting.

The Laguna Salada Fault is one of scores of faults that riddle California—all components of the San Andreas Fault system that defines the boundary between two of the Earth's crustal plates. The 2010 Easter Quake was yet another manifestation of tension being released when these plates grind past one another. Westernmost California sits atop the Pacific Plate, which is moving to the northwest, and the rest of the state (and the continent) rides on the North American Plate moving west. The plate boundary is not a tidy, uniform suture like a zipper. It's more of an aging Velcro™ fastener with a lot of play in it.

The Earth's crust, or lithosphere, is a rigid outer shell that covers the semimolten, plastic upper mantle. The crust can be a few or hundreds of kilometers thick, broken into a dozen gigantic plates. The continents ride on top. Plates are pushed apart from the mid-ocean spreading center, or rise, that circles the globe in a zigzag pattern, sort of like the stitching on a baseball as more than one structural geologist has described it. New ocean seafloor is created at these spreading centers when liquid basalt emerges and rapidly solidifies in the cold ocean. As sea floor is born, the plates on either side of the spreading center are pushed away, a few centimeters per year. New plate material added at the spreading center means that an equivalent amount has to be returned to the mantle somewhere else on the globe or uplifted. Recycling occurs at continental margins where plates meet head on and one overrides the other. Because oceanic lithosphere is denser than the more buoyant continents, it dives beneath them in a process called *subduction*. Subduction creates heat and pressure as the lithosphere melts, which in turn often results in an arc of volcanoes above the subduction zone. Thus, all plates have a leading edge on the side away from the spreading axis and a trailing edge where it diverges from the spreading axis. On the North American Plate, the western coast is the leading edge, and the east coast is trailing. Plates can also slide past one another, but in these cases, crust is conserved. Such sliding plate edges are known as *transform faults*, like the San Andreas.

The 800-mi-long San Andreas Fault runs in a fairly straight line from Tomales Bay (north of San Francisco) southwest to the southern end of the Great Central Valley near Maricopa. The southern end of the fault is likewise straight from the northern end of the Coachella Valley south through the Salton Trough. Between these two regions, however, the fault runs more or less east to west.

FIGURE 1.2 Hundreds of geologic faults are etched into Southern California's crust. The San Andreas Fault Zone, the largest in the state, is a plate boundary that extends from the Salton Trough to San Francisco Bay. Land to the west sits atop the Pacific Plate, which is sliding to the northwest. To the east of the fault, the rest of the continent is riding on the North American Plate in a more westerly direction. The Garlock, San Jacinto, and Elsinore Fault Zones are the next largest in Southern California. FZ, fault zone. (Base map adapted from www.nationalatlas.gov.)

Therefore, the entire fault is not a straight line. Instead it has a bend to it before continuing toward the Gulf of California (Figure 1.2). The bend crosses I-5 below Ft. Tejon between Lebec and Gorman, travels southeast toward Palmdale along the north side of the San Gabriel Mountains, east to Wrightwood, and crosses I-15 just northwest of Devore. It continues on this path along the north edge of San Bernardino all the way to Desert Hot Springs. Here it resumes its more southerly bearing down into the Gulf of California. Land to the west of the main fault zone is moving northwest with respect to the eastern side. In other words, looking across the fault from either side, land is moving horizontally (strike-slip) to the right, or in a right-lateral motion.

The fault system is a few miles wide in some places and more than 50 mi wide in others. The shorter associated faults run roughly parallel to the San Andreas northwest to southeast. In Southern California, for example, the 130-mi-long San Jacinto Fault extends through San Bernardino, Riverside, San Diego, and Imperial counties. Parallel to it on the west is the Elsinore Fault.

Since it first formed, total land displacement along the entire San Andreas Fault has been about 740 mi (1200 km). Try to visualize northern and central coastal California at the initiation of the fault's formation around 30 million years ago. All of the west coast from San Francisco south would be below the Mexican border. The fault system did not arise in a single, continuous event, but rather as separate events in the north and in the south, each by different mechanisms.

Tens of millions of years ago, an enormous piece of oceanic lithospheric, the Farallon Plate, sat between the Pacific plate and the North American Plate where it was being subducted beneath western North America. The Pacific Plate was far to the west, separated from the Farallon Plate by a spreading center. The spreading center moved east as the Farallon Plate was being consumed at the trench along the western edge of the North American Plate. That spreading center reached the western margin of North America near Los Angeles and Santa Barbara about 30 million years ago, when then remnants of the Farallon Plate were subducted, and the Pacific Plate made its initial contact with North America. When the spreading center met the continent, it was cut off from receiving new magma and the trench filled with sediment from the continent. The subduction trench and spreading center both ceased to function and became a transform fault when the trench deposits accreted to the northward moving Pacific Plate: The San Andreas was born. It was short at first but lengthened as the triple junction formed by the Pacific, North American, and diminishing Farallon plates moved north. By 3 million years ago, the triple junction was off San Francisco, the Pacific Plate rotated about 11° eastward, bumping up against the continent, and the Sierra Nevada began to emerge. The triple junction is now located at Cape Mendocino. Fragments of the Farallon Plate remain today; for example, the Juan de Fuca Plate off the Pacific Northwest whose subduction has given rise to the Cascade Range and its renowned volcanoes like Mt. Ranier, Mt. St. Helens, Mt. Hood, Mt. Shasta, and Lassen Peak.

As the Sierra was building, the Coast Ranges emerged from the depths of the defunct trench as island arcs that would eventually coalesce. Unlike the granitic Sierra Nevada, the Coast Ranges consist of a heterogeneous assemblage of terrestrial and marine rock and sediment called the Franciscan Mélange.

1.2 TRANSVERSE RANGES

The north-south trending Sierra Nevada, the spine of California, is paralleled by the smaller Coast Ranges to the west, and in between stretches, the long, sediment-filled Central Valley, almost as flat as the surface of the sea (its miniscule elevation shifts are read by laser). Defying California's north-south linearity are the Transverse Ranges that extend from Point Conception eastward for about 100 mi (160 km) following the bend in the San Andreas Fault. The San Gabriel Mountains are the dominant feature among the other topographic components but all are part of the same system: the Santa Ynez Mountains, San Rafael Mountains, Sierra Madre Mountains, Tecuya Ridge, Topatopa Mountains, Santa Susana Mountains, Santa Monica Mountains, and San Bernardino Mountains. Other features include the Simi Hills, Puente Hills, Chino Hills, and Little San

Bernardino Mountains. Off shore, the Transverse Ranges extend to the islands of San Miguel, Santa Rosa, Santa Cruz, and Anacapa. With a drop in sea level, such as occurred during the last glacial maximum, these islands would coalesce and contribute to a long peninsula that extended west from the Santa Monica Mountains. The Santa Barbara Channel then becomes a bay.

The San Gabriel Mountains are a steep, commanding landmark that separates greater Los Angeles from the Central Valley to the north of it. The highest point, Mt. San Antonio (Old Baldy), reaches 10,064 ft (3,060 m)—the highest point in Southern California is San Gorgonio Peak in the San Bernardino Mountains at 11,502 ft (3,500 m), (Figure 1.3). The San Gabriel Mountains are composed of a conglomerate of rocks and sediments quite unlike the granitic Sierra Nevada, or the San Bernardino Mountains separated from the San Gabriel's by the Cajon Pass. Densely wooded and steep, with grades as severe as 85%, the eroding canyons have for millennia deposited mud, boulders, and debris on to the lower slopes and out onto the Los Angeles basin. The conjoined alluvium stretches for 50 mi along the front of the San Gabriel's. Although the mountain continues to erode, it is also rising about 1 cm per year. Uplift and folding since its creation have provided access to underground petroleum reserves around Santa Monica and elsewhere adjacent to the range. The Ventura oil fields, for example, drill into an astonishingly young (geologically) bulge or fold called the Ventura anticline, an impressive feature extending east to west for 15 mi. Inclined sedimentary rocks on the sides of the fold were deposited only about 200,000 years ago, so the anticline itself must have initiated after that. The crest is still rising but slowing, and also eroding.

Geologists have known for a long time that the Transverse Ranges were formed by compression, but attempts at a plausible mechanism were sketchy and

FIGURE 1.3 San Gorgonio Peak in the San Bernardino Mountains, viewed here from the southeast, is the highest point in Southern California at 11,502 ft (ca. 3,500 m).

unsatisfactory until plate tectonics entered the discussion in the early 1970s. The origin of the Transverse Ranges is directly linked to the evolution of the southern end of the San Andreas Fault system. All of Southern California west of the San Andreas, and all of Baja California to the south, comprise a unified block that rides northwest on the Pacific Plate with the rest of the west coast north to San Francisco. Five and a half million years ago, San Diego, Los Angeles, and Santa Barbara were 300 mi south of their present latitude. Around 13 million years ago, the Gulf of California began to appear as a narrow basin when the spreading center known as the East Pacific Rise, which begins in the Southern Hemisphere near Tierra del Fuego, encroached from the south creating a rift between mainland Mexico and Baja California. By 5.5 million years ago, active widening of the gulf was underway. The rift continued north and formed the southern section of the San Andreas Fault.

Southern portions of the Baja California peninsula may have split off separately from mainland Mexico only later to reunite as a single unit. In any case, as the Gulf of California opened, the entire block rafted north attached to the Pacific Plate only to encounter the massively deep, granite batholiths (basement) of the Sierra Nevada and the San Bernardino Mountains. The only option was for the block to shift directions to the west where it compressed the terrain higher and higher into the Transverse Ranges of today. In doing so, it created the big bend in the San Andreas Fault and a potentially confusing orientation that greets the newcomer and often its residents.

When we think of "going up the coast" our natural inclination is to assume north. Not here. To go up the coast to Point Conception is to head mostly west. Traveling from San Bernardino to Santa Barbara, for instance, takes you 137 mi (220 km) west, but only 27 mi (60 km) north. Heading north from Los Angeles leads to the interior on a bearing to Antelope Valley near Lancaster, 95 mi (153 km) east of Vandenberg. Consider also that Reno, Nevada, lies west of Southern California from about Santa Barbara south. Southern California's orientation is also the reason that many coastal communities enjoy south-facing beaches.

There were further consequences. The Tehachapi Mountains also had their origin at this time. Like the Transverse Ranges, they do not run north and south, but rather southwest to northeast and so separate the Central Valley from the Mojave Desert. Traversing along the southern base of the Tehachapi's is the Garlock Fault, the second largest fault in California (Figure 1.2). It forms a junction with the San Andreas Fault at the west end of Antelope Valley, and then extends northeast for 160 mi toward the Great Basin. The crust in the Great Basin in Utah and Nevada is thin and stretching. Apparently, the Garlock Fault developed in response to strain between this left-lateral crustal stretching within the North American Plate, and the right-lateral slip of the Mojave block. Land on either side of the fault is moving in a left-lateral motion. Combined with the northwest movement of that part of California that sits on the Pacific Plate, the Mojave region is moving east with respect to the rest of the state. Crustal stretching of the Great Basin may also have contributed to the Great Bend in the San Andreas.

1.3 PENINSULAR RANGES

Southern California's Peninsular Range is part of a 900-mi string of mountains that extends to the southern end of the Baja, California peninsula. The northernmost components are the San Jacinto Mountains in Riverside County, rising to 10,800 ft (ca. 3,300 m). Moving south, the Peninsular Ranges include the Santa Rosa Mountains, Santa Ana Mountains, Mount Palomar, and the Laguna Mountains in San Diego County. Continuing into Baja California are the Sierra Juarez, Sierra San Pedro Martir, Sierra de la Giganta, and the Sierra de la Laguna. The Peninsula Range of mountains had their origin in the Jurassic, possibly as part of the same granitic batholith that created the Sierra Nevada (i.e., the Salinian Block). Magmatic activity reached its peak in the Cretaceous when the greatest volume of magma rose and cooled to form the plutonic rocks now exposed in the Peninsular Ranges. Plutons are rocks created from magma that cooled miles below the earth surface. At about 95 million years ago, the rapid subduction of the steeply dipping Farallon Plate beneath North America had uplifted the mountains to their highest extent. Thereafter erosion carried off much of the overlying metamorphic and volcanic cover to expose the plutonic rocks buried beneath, resulting in the boulder-studded landscape we see today.

In Southern California, a companion feature of the Peninsular Ranges is the Salton Trough on its eastern flank. The trough is a tectonic depression that has been stretching since the Miocene. Essentially, it is the northern end of the Gulf of California, which continues to widen by about 6 cm per year. The only reason that the trough is not underwater presently is because the Colorado River delta created a cross-wise dam that prevents gulf water from flowing north.

West of the mountain crests lie the topographic features of post-Cretaceous erosion and sediment deposition. Storm tracts come in from the west, and the mountains trap the low, moisture-laden clouds so they drop most of their precipitation before cresting and moving east. Thus, the western slopes are more deeply eroded than the eastern side, resulting in the foothills, valleys, and canyons.

1.4 MARINE TERRACES

Many coastal areas in Southern California are wide, flat mesas known as *marine terraces*. For the past 1 million years, the coastal San Diego/Orange County region has been raising an average of 5.5 in. (14 cm) per 1,000 years. Marine terraces are thus uplifted sea floor that begin as wave-cut platforms near shore. When sea levels drop and shorelines recede, the platform is covered by sediment eroding from the adjacent cliffs. Hence the platform is often blanketed with a layer of nonmarine sediment. The steep cliffs between successive platforms are old sea cliffs.

The terraces facing the ocean comprise some of Southern California's most desirable (i.e., expensive) real estate, despite the fact that these terraces are inherently unstable. Composed of soft, thinly bedded sediments, the cliffs are constantly being undercut by the sea. The slope steepens, and the cliff faces eventually collapse. Marine terraces are further dissected into westerly directed canyons carved out by rainwater runoff.

2 The Climate

Warm, dry summers, mild winters, and rain falling only during the cooler months of the year characterize Southern California. Mediterranean climates like ours occur between about 30° and 40° latitude on the western margins of continents—not only here but also the Mediterranean, southern Chile, southwest Australia, and the western cape region of Africa. Their climatic similarities are striking and so too is the structure of their scrub vegetation. Mediterranean climates essentially result from (1) a descending air mass drying out the atmosphere and the land below, and (2) the adjacent cold water circulating offshore.

Solar radiation is absorbed by the land and oceans, which in turn generate latitudinal gradients in temperature, precipitation, and air circulation—what we call *climate.* Because the Earth is a sphere, heat from the Sun is not evenly distributed over its surface. Solar radiation is most intense at the equator because the Sun is more directly overhead. At the poles, sunlight strikes the Earth's surface obliquely and is broadcast over a wider area. High latitudes are therefore cooler. Seasonal variation in climate results from (1) the declination of the earth with respect to its axis, a tilt of 23.5° and (2) the annual orbit of the earth around the Sun. Winters in the Northern Hemisphere result from the orbital position that puts the tilting axis away from the Sun. Concomitantly, exposure of the Southern Hemisphere to the Sun is at its greatest during winter (their summer) in the Northern Hemisphere.

The Earth's major biomes (e.g., tropics, deserts, boreal forests, chaparral) result from the distribution of heat and rainfall along latitudinal gradients from the equator to the poles. The desert and arctic tundra are both characterized by low precipitation but differ in mean temperatures, which is a function of latitude. Tropical forests and high-latitude boreal forests both receive significant rain, but like deserts and tundra, differ in temperature.

Air circulation globally is initiated by solar radiation at the equator. As air at the sea surface warms, it expands and rises carrying water vapor with it. Air will continue rising and expanding but eventually reaches an altitude where it begins to cool, causing the water vapor to condense and return to Earth as rain. Thus, tropical latitudes are wet. However, the now dry upper-level dry air masses circulate to higher latitudes where they descend at around 30° on either side of the equator (e.g., Southern California). As the cool, dry air descends, it heats up and absorbs moisture from the atmosphere and the ground. Some of this descending cool air returns to the equator, completing a giant circle known as a *Hadley cell.* Without this returning surface air, tropical regions would be considerably warmer than they are. Back at about 30°, the rest of the air mass that did not recirculate south moves north instead as another giant circulating cell of air. At around 60° latitude, that air mass rises and once again releases moisture as it cools, although not as much as in the tropics. As at 30°, this giant mass of air recirculates, with some of it moving poleward. There it descends, absorbs moisture, and coupled with a low incidence of solar radiation produces a dry, bitterly cold polar climate.

2.1 GLOBAL WINDS

The Earth's clockwise rotation has a profound influence on climate because it dictates the direction and flow of the large, recirculating global air masses, as well as the ocean currents. Because the Earth is widest at the equator, objects on the ground travel farther and faster than they would at higher latitudes as the Earth makes its 24-h 360° rotation. Viewing the equator from a northern perspective, objects (wind) will tend to veer to the right because ground speed is not as fast. Viewed from the south of the equator, objects will tend to move to the left. This phenomenon, known as the *Coriolis effect*, means that Hadley cells do not move in a strict north–south direction but are deflected by the angular momentum of the Earth's rotation. Winds from Hadley cells returning to the equator from 30° (the "Horse latitudes") blow from east to west in the Northern Hemisphere, and west to east in the Southern Hemisphere. These directional winds constitute the northeast and southeast *trade winds*, respectively. Cells moving toward the poles from the Horse latitudes deflect from west to east and are therefore called *westerlies*. Note that these names refer to the source of the wind, not its direction.

Ocean surface currents are also the product of the global wind patterns. Toward the equator, trade winds move currents to the west, and the westerlies move currents to the east. The result is a giant circulation pattern of surface currents adjacent to continental margins. The huge north Pacific gyre, for example, circulates clockwise, and as it moves back down the west coast of North America, it brings cold water from high latitudes that flows southward off the California coast. South of Point Conception, the current is slightly warmer because the configuration of the coastline accommodates subtropical water circulating up from lower latitudes.

2.2 WINTER

The combination of the clockwise circulation of the Pacific current coupled with the drying effects of the descending Hadley cell produce the pronounced seasonality of Southern California's rains. During California's summer months, an enormous expanse of high-pressure air dominates the Pacific and overlaps the continents, further guaranteeing that those months will be dry. The Pacific high weakens during the cool winter months allowing storm tracks to move down the California coast and bring seasonal rains, primarily from November to March. In fact, these 5 months account for 85% of the annual total for most of Southern California west of the Transverse and Peninsular Ranges; 60% falls between December and February (Figure 2.1). Farther east into desert communities like Barstow and Blythe, winter rain accounts for only about 50% of the annual total, and summer months (June–August) see a little more than 20% because of monsoonal rain drifting west from Arizona (Figure 2.1).

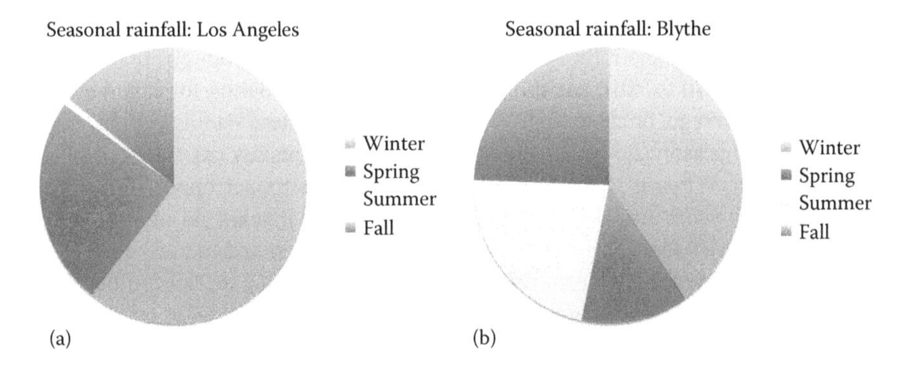

FIGURE 2.1 Seasonal distribution of rain in Southern California. (a) Los Angeles is typical of the region from the mountains and high deserts to the coast in that much of the annual accumulation, around 60%, falls during the winter months (December–February). Add the average accumulations for November (late fall) and March (early spring) and the total jumps to 85%. Those 5 months define the rainy season for most of Southern California, regardless of the total annual accumulations for any particular locality. By contrast, low desert communities, like (b) Blythe along the Colorado River, receive around 22% of their rain during the summer (June–August) and an additional 10% in September. The one constant for all of Southern California is that June is the driest month.

Winter storms that reach Southern California originate in the Gulf of Alaska and nearby areas of the north Pacific. They slide down the west coast in an east-southeast direction, wrap around the bight of California at Point Conception and approach land from the west or southwest. Winter storms are of such size that most all areas of Southern California will receive measurable rain. However, the amount that a particular location will receive from a single winter storm, and thus for the season, is a function of latitude, proximity to the coast, and topography. More northerly coastal localities receive more rain than those to the south. Los Angeles, for instance receives around 14.8 in. (390 mm), whereas San Diego, 100 mi to the south, experiences about 10.2 in. (260 mm).

A second rainfall gradient unfolds from west to east. When cold, saturated air pushes up the foothill and mountain slopes, more rain is received in these areas than along the coast. Pasadena at 860 ft (206 m) abuts the San Gabriel Mountains and averages around 21 in. (540 mm) annually, but it is only 9 mi (14.5 km) from downtown Los Angeles (350 ft [105 km]). Further east, San Bernardino receives about 16 in. (410 mm), around 1.2 in. (28 mm) more than Los Angeles, but 5 in. (130 mm) less than Pasadena. By the time a storm crests the mountains of Southern California, most of the rain has been released, and a *rainshadow* effect deprives the deserts of most moisture from a typical winter storm. Lytle Creek ranger station in the San Gabriel Mountains averages more than 36 in. (ca. 1 m), but the high desert city of Victorville, about 22 mi (35 km) northeast, receives less than 6 in. (150 mm). Indeed, none of the desert areas of Southern California receive on average more than about 5 in. (130 mm) per year, with some spots like El Centro averaging only about 2.7 in. (70 mm).

Rainfall averages reflect only long-term trends. Any single year will most likely under- or overshoot the statistical mean. Consider, for example, that in only 3 years from 1970 to 2010 has the annual rainfall accumulation for Los Angeles equaled the average, or close to it. When the tropical eastern Pacific is cool going into the rainy season (*La Nina*), Southern California typically experiences below normal years. When it is warm (*El Nino*), there is a stronger opportunity for a wet year. Wet years are less frequent than below-normal years (14 of 40 for Los Angeles), but they tend to be pretty soggy when they occur and therefore increase the statistical mean. Long-term drought such as California experienced between 2011 and 2016 will obviously depress annual averages.

As winter storms vacate California and move east, winds in Southern California gradually shift their approach from the northeast to the east over several days. The change in direction is a consequence of the huge mass of air that follows storms eastward and then becomes trapped in the Great Basin between the Rocky Mountains and the Sierra Nevada. High pressure squeezes the air to the ground, compressing and warming it at the rate of about 5° for every 1,000 ft (ca. 300 m) of descent. The Sierra Nevada has few low passes, so the only place for this warm, dry air to go is downhill toward the deserts of Southern California and over the low mountain passes into the foothills and inland valleys. These are the Santa Ana winds that can be ferocious, as residents well know. Occasionally, east winds, but not Santa Ana winds, can result in freezing temperatures at night because the air is so clear and dry. Generally, however, Southern California is spared prolonged freezes because a Canadian air mass is required to produce them; mid-continental cold air from high latitudes typically moves south and east of the continental divide, thereby sparing Southern California from its effects. Once in a while, persistent high pressure will force it south into the intermountain region with bitterly cold results.

2.3 SPRING AND SUMMER

When the gradual transition to spring arrives, inland areas begin a sustained warming that brings haze and conditions that prevent Santa Ana winds through the summer and early fall. By early May, the California strata crosses the Catalina Embayment and curls counterclockwise striking the coast at a right angle. This marine layer deepens to as much as 6,000 ft (1,820 m), blanketing the coast and inland valleys through June (Figure 2.2). Rarely does it burn off along the coast for more than an hour, much to the surprise and disappointment of sun-seeking, beach-goers from out of state. By mid-July, the marine layer begins to wane, and warm clear days are sustained into early October. During these months the immense pool of air (Hadley cell) sinking over Southern California gradually compresses and warms. Near the coast, the air at sea level remains cool because of the ocean, and it cools slightly more to about 2,000 ft (ca. 600 m) where it meets the warmer descending air mass. The result is the summer temperature inversion characteristic of Southern California lowlands; the air is markedly warmer above

FIGURE 2.2 The marine layer familiar to Southern Californians, a late spring–early summer phenomenon, often creeps many miles inland from the coast, isolating hilltops like islands. It usually retreats by late morning.

about 2,000 ft (ca. 600 m) than it is below. The near-surface air is denser, has more moisture content, and is therefore hazy. Dry summers are further assured because there is no vertical movement of air above 2,000 ft (ca. 600 m).

During July and August, sometimes into September, the only possibility of rain comes from "Sonoran storms" when massive thunderheads develop in the afternoons over the deserts and mountains. The activity is the product of warm moist air originating in the Gulf of Mexico that crosses the Sierra Madre, moves northwest over Sonora, Mexico, and into southern Arizona. In this part of the Southwest, the summer monsoons account for the majority of annual rainfall. Occasionally they push westward into Southern California or drift up from Baja California and the Sea of Cortez, bringing lightning and thunder along with brief but intense periods of rain to the deserts and high mountains, especially to the Peninsular Ranges and their eastern slopes. Once in a while, thunderstorms breach the mountains to dampen the foothills and slopes on the western side. Another, albeit infrequent, source of late summer showers to Southern California in August and September are disturbances in the tropical Pacific off Mexico called *chubascos*. More often they bring warm, steamy air to the region rather than rain.

2.4 TOPOGRAPHY AND TEMPERATURE

We have already noted how the mountain and foothills will influence winter rainfall accumulations throughout the region. So, too, are summertime temperatures in Southern California locally influenced by topography. The cooling sea breezes of summer mean that the coastal zones are mild and comfortable for people during the warmest months of the year. Gentle breezes blowing inland will follow surface contours that direct them through areas of low elevation to unobstructed inland valleys and foothills. Major river valleys are especially important for escorting sea breezes inland, for example the Los Angeles, San Gabriel, and Santa Ana Rivers in the Los Angeles Basin. Thus, distance from the coast is less a factor in the distribution of sea breeze than is accessibility. For instance, the Santa Monica Mountains and Simi Hills box in the San Fernando Valley, and the only passageway is the Glendale Narrows. The San Fernando Valley receives about the same amount of sea breeze as the upper Santa Ana basin, which is three times farther from the coast. Once sea breezes reach the San Gorgonio Pass, the high temperatures of the Coachella Valley cancel their cooling effect. Similarly, the comparatively low Cajon Pass funnels cool air toward its summit, but it is then overwhelmed by heat from the upper Mojave.

To the south, sea breezes follow canyons and low passes of the Peninsular Range that are at right angles to the coast. The San Diego River Valley (Mission Valley) as well as the Sweetwater River Valley south of it are major avenues for cool air to travel a considerable distance inland. By contrast, the Perris–Temecula Valley is occluded by the Santa Ana Mountains to the west of it, and hot summer temperatures there are inevitable.

2.5 CLIMATE ZONES

Southern California's climate zones specify temperature regimes across large areas and are determined mainly by the relative contribution of maritime versus continental influence. The latter becomes increasingly dominant inland. Broadly, these zones are the Maritime Fringe, Intermediate Valleys, Transitional, Mountain, High Desert, and Low Desert. Climate zones may be parsed even further depending on local topography and the creation of thermal belts, especially west of the deserts. Thermal belts are essentially the slopes of the hills, where winter temperatures are not as cold as the crests above or the floors below and where summer temperatures are similarly mitigated. Recognition of thermal belts and regional microclimates are especially useful to horticulturists (*Sunset* Magazine's *Western Garden Guide* lists 14 of them for Southern California).

2.5.1 Maritime Fringe

Along the coastal strip the warmest months are usually below 75°F (23.8°C) and the coolest are above 50°F (10°C). July and August are mild, whereas average January temperatures are typically warmer than anywhere else in Southern California. Such highly equable climate is of course a function of proximity to the ocean, which has an immense capacity to store heat. The trade off, if there is one, is that the sea surface infuses the overlying air with moisture that results in fog and cloud cover. Heat transfer from the ocean puts moisture into the atmosphere. Moist sea air must be imported from warmer latitudes by light winds from the south-southwest, but light enough so that turbulence will not lift it into a stratus cloud. In winter, this air has such a low dew point that contact with the sea surface is sufficient to result in fog without nighttime chilling. Winter sea air heats slowly on cool windless nights. When there is little difference in temperature between the sea and inland areas, sinking air chills to its dew point and produces ground fog in the basins and valleys. Fog belts are widest in the Lompoc/Santa Maria area, in which fog is concentrated for about 3 months beginning in the late summer. The inland incursion of fog weakens from Point Arguello south to the Santa Monica Mountains. It increases again in the Los Angeles basin, mainly in the cooler months and for less than 60 days. South of Newport, coastal cliffs keep fog from moving very far inland except through corridors. San Diego experiences significant fog for only about 30 days, primarily in the winter. Realize, though, that May and especially June are the cloudiest months because the California stratus (marine layer) parks itself along the coast and is a phenomenon separate from winter fog.

2.5.2 Intermediate Valleys

These are the Southern California lowlands adjacent to the Maritime Fringe to the west, and the mountain ranges to the north and east. Winters are usually mild and summers warm, warmer still farther inland. Average January temperatures are above the 50° isotherm, which varies in elevation from about 1,000 to 2,500 ft (ca. 300–760 m). This is not to say the overnight temperatures cannot dip into the twenties. Cold air sinks, and valley floors therefore can become cold air basins during the winter. In summer, temperatures from July through September will exceed 90°F (32.2°C) to more than 100°F (37.7°C) for half the days.

Mean annual rainfall is from 15 to 20 in. (390–510 mm) with those sites abutting mountains receiving the most. Streams from mountain runoff were an important source of irrigation water during early settlement, and much of the intermediate valleys later became the choicest regions for growing citrus. Up until the 1960s, citrus groves depicted the quintessential image of Southern California—expansive orchards on clear winter days under stunning blue sky and mountains in the distance. In the decades since, the grasslands and orchards of the inland valleys were obliterated by urbanization. Still, the image, or some vestige of it, lingers on—note the indulging panoramas of the Rose Bowl and snow-capped San Gabriel Mountains in January during telecasts of the annual football game.

2.5.3 Transitional

This zone comprises the foothills, the often steep and rugged slopes between the intermediate valleys and the mountain summits of the Peninsular and Transverse Ranges. Elevations are from around 2,000 to 4,500 ft (ca. 600–1,370 m). The foothill zone is separated from the intermediate valleys by the 50° winter isotherm. The climate overall has a stronger continental influence than a maritime one. The average January temperature usually is below 50°F (10°C). The transitional zone is dryer than the mountains, but not as cold and the differential between the warmest and coldest months is less than 30°.

2.5.4 Mountains

The Mountain climate zone differs primarily from the Transitional in that the mean temperature between the warmest and coldest months exceeds 30°. Although 100°F (37.7°C) is not uncommon in the summer, mean summer temperatures do not often exceed 75°F (23.8°C). Mountain climates receive the most rain, typically 25 in. (650 mm) or more. Amounts vary depending on latitude and orientation of the locality. Mount Wilson in the San Gabriel Mountains at 5,700 ft (1,730 m) receives almost 34 in. (850 mm), whereas Mt. St. San Jacinto gets slightly less than 26 in. (650 mm) but is 10,804 ft (3,280 m) in elevation. Mt. Palomar at 5,563 ft (1,690 m) receives slightly more than 28 in. (725 mm). Mountain climates experience snow annually above 4,000 ft (1,215 m) with the heaviest and most persistent accumulations occurring above 6,000 ft (1,820 m). Summer thunderstorms are also a feature of the ridgelines of the Transverse and Peninsular Ranges. They usually bring abundant lightning but little rain. Lightning-induced fires are indeed a threat from late July to September.

2.5.5 Deserts

Desert climates are sectioned into High and Low zones as a function of their respective elevation. High Desert defines the western Mojave from Victorville north to Ridgecrest. Intermediate to High Desert includes most of the desert west of the Mojave River and areas east and south to Twentynine Palms. The High Desert is a more or less flat region at 2,500–3,000 ft (760–910 m), interspersed with isolated mountains. In areas such as Lancaster and Twentynine Palms, precipitation drops 30% within 10–20 mi (16–32 km) of the mountain crests. Well inland from any maritime influence, the clear dry atmosphere of desert summers allows blisteringly intense solar radiation to penetrate with next to no heat lost to reflectivity or evaporative cooling. Summers experience more than 100 days above 90°F (32.2°C) with highs from 111°F to 117°F (43.8°C–47.2°C). By late fall the heat abates, and during winter average temperatures fall below 50°F (10°C).

The Low Desert includes the eastern Mojave and the remaining Southern California deserts below 1,000 ft (ca. 300 m). Elevations drop to sea level or below in the Imperial Valley and the Salton Trough. Average summer highs are 106°F–108°F (41.1°C–42.2°C) and summer rains are normal (Figure 2.1). As an example, Blythe receives on average more rain in August than it does for any of the winter months. The winters in the Low Deserts are short and generally mild. The average winter minimum temperature is 37°F (2.7°C) with lows reaching 13°F–19°F (–10.5°C to –7.2°C).

3 Exploration, Collections and the Museum Tradition

Columbus's voyage to the West Indies in 1492 ignited the Great Age of European Exploration to the New World and beyond. At its heart, the ocean voyages were commissioned for the purpose of enriching the crowns of Europe, and among the hoped-for treasures would be new plants as a source of food and medicine or those that could be used in manufacturing and horticulture. Plants are comparatively easy to collect, preserve, or keep alive. Botany by default provided the entry into the study of nature in a systematic way. On the other hand, vertebrate animals are difficult to obtain and time-consuming to preserve, which early on required precious stores of consumable alcohol. On that score it is a stretch to imagine any sailor forgoing his rum ration so that some screwball could pickle lizards in it. Consequently, animal collections tended to focus on those that were easy to prepare, like insects and shells.

With each successive voyage from Europe, a ship's crew included at least one person familiar with, if not trained in, botany and zoology. By the latter half of the eighteenth century, expeditions were being launched for the expressed purpose of scientific discovery.

In the Americas, scientific exploration proceeded on two fronts. One of those was in British American colonies and its eventual westward expansion in the early nineteenth century. The other was along the Pacific Coast where Russian, British, French and Spanish interests vied for influence and commercial advantage. Most of the early forays along the West Coast were inspired by the exploits of Captain James Cook (1728–1779), the famed English navigator whose three voyages into the Pacific were, at the time, unparalleled for their discoveries. On his third voyage, Cook sailed along the Pacific Northwest in part to locate the supposed Northwest Passage, a waterway that would cross North America and link the Atlantic Ocean with the Pacific. By the mid-1800s, California, along with what are now the states of Oregon and Washington, were the possession of the United States. Toward the end of that century, natural history was maturing from a discipline preoccupied with collections and identification to one more focused on how plants and animals survive in the natural world.

3.1 EARLY OCEAN VOYAGES TO THE WEST COAST

Exploration of the Pacific Coast began in 1542 when Juan Rodríguez Cabrillo sailed the galleon *San Salvador* up the west side of Baja California north into Alta California as far as Monterey Bay. Spain did not immediately follow up on Cabrillo's discoveries; it was in no particular hurry to expand its empire farther north than Mexico. Only the galleon ships returning from the Philippines made subsequent and sporadic visits to Alta California. In 1602, one veteran of the Philippine campaign, Sebastían Vizcaíno, was hired by Spain to map and explore the California coast. Rather defiantly Vizcaíno disregarded instructions to use the place names that had been designated by his predecessors. Instead he renamed almost everything, including the location of California's first settlement named by Cabrillo as *Bahía de San Miguel*. Vizcaíno called it *San Diego de Alcalá*.

Spain did not begin settling Alta California until almost 170 years later when it awakened to the fact that it needed to establish its presence to discourage other nations from doing so. The Russians were already exploiting the lucrative fur trade in Alaska and were moving south; they eventually built a colony at Fort Ross along California's north coast. Great Britain was establishing itself in Canada, and the French were becoming curious themselves.

The French frigates *L'Astrolabe* and *La Boussole* anchored in Monterey Bay in 1786 under the command of Jean François de Galaup, Comte de la Pérouse. Louis XVI commissioned this expedition, having been inspired by the voyages of Captain Cook. The French vessels were on a scouting mission but also were outfitted with scientific instruments, naturalists, and illustrators. During their short 10-day visit, botanical specimens were collected along with several birds (e.g., the Oak Titmouse, California Thrasher, Nuttall's Woodpecker, and Western Scrub Jay). None of these was formally described. For California, the legacy of this French expedition is the first painting of the California Quail by Jean Robert Prévost. It was named the state bird in 1931.

Not long after the Pérouse expedition, Britain sent the *Discovery* and its sister ship the *Chatham* to the Pacific under the command of George Vancouver. They reached the West Coast in the spring of 1792. Vancouver had served as a midshipman on Cook's second and third voyages. A part of the admiralty's orders included Vancouver accepting Spain's relinquishment of sovereignty to the territory around Nootka Sound and Vancouver Island, a claim that dated back to 1775. In October the two ships sailed south along the coast of Alta California.

Accompanying the mission was the Scottish botanist Archibald Menzies, who inarguably had the greatest influence of any of the early naturalists that visited California. He was the first to study the length of the coast during the *Discovery*'s nearly 2½ years in the Pacific (which included a trip to the Sandwich Islands and another visit to the Pacific Northwest). Menzies returned to England with a bonanza of plant specimens that included ferns, grasses, annuals, shrubs, and trees. It was the largest botanical collection ever brought back from California and was studied by botanists well into the twentieth century. Menzies died in his home outside London in 1842; he was 88 years old.

Spain launched its own around-the-world expedition as well. It reached the West Coast in September of 1791 under the command of the gifted captain Alessandro Malaspina. Unlike the older French vessels of the Pérouse expedition, the new Spanish corvettes *Descubierta* and *Atrevida* were fully outfitted for scientific and polar exploration. They had the latest scientific instruments in addition to a well-appointed library. Malaspina, like Pérouse, had only a short stay in Monterey, and because of various forms of attrition, he had but one naturalist left on board, Thaddeus Haenke. Haenke, an Austrian, received a doctorate degree from the University of Prague and was well trained in botany, physics, and medicine. He was thus the first PhD to visit California. Haenke lost little time in getting ashore and making the most of is opportunities. He collected all manner of plants and even seaweed. Among his collections were specimens of the Coast Live Oak and Valley Oak, which later were formally named and described by the botanist Luis Née; they were the first western tree specimens brought to Europe. Haenke was also the first to observe and collect specimens of the California Redwood, which he referred to as "Red Cypress" and noted that it was a valuable source of timber. In 1926 a 125-ft-tall Redwood growing in the Alhambra of Spain was verified as one of his specimens. Haenke cataloged some 250 species of plants from his short time in Monterey, applying an admixture of scientific names, like *rhusradicans* for Poison Oak, *millefolium* for Yarrow, and *sambucus-racemosa* for Red Elderberry. The expedition also recorded numerous birds and such mammals as mountain lion, bears, rats, and rabbits.

By the early 1800s, sailing voyages to the California coast were routine. The itineraries focused mainly on northern California. Noteworthy is the Russian expedition of 1816 that resulted in the collection of the California Poppy in the San Francisco Bay area by Adelbert von Chamisso and Johann Friedrich Eschscholtz. The poppy was designated the state flower in 1903.

3.2 SETTLING CALIFORNIA—PORTOLÁ AND ANZA

As the seafaring expeditions of the Pacific Coast were underway, Spain moved ahead with its settlement of Alta California. José de Galvez, the inspector general of New Spain, was put in charge. Galvez had the conquistadores and their armies at his disposal, but he also had the intrepid Christianizing priests, or *padres*, who came from various orders of the Catholic Church—the Jesuits, Dominicans, Franciscans, and Carmelites. The padres all had a steadfast resolve to establish outposts in remote places. Previously, Father Eusebio Kino had set a precedent through his missionary work in northwest Mexico by demonstrating the success of passive conversion of the heathens over the rough-and-tumble ways of a soldiers' garrison. Galvez set up his base in San Blas along the Pacific coast of mainland Mexico, and from there he launched his campaign. He selected as his commander Don Gaspar de Portolá, and the padre-presidente Fray Junípero Serra. In spring 1769, four parties set out for San Diego Bay—two by ship and two by an overland route. One of the overland parties included both Portolá and

Padre Serra who arrived in San Diego at the end of May. Portolá had hoped to sail on to Monterey, but one of the supply ships had failed to arrive; so he sent the other back to San Blas for more provisions. Portolá bravely opted to take a large contingency overland in hopes of finding Monterey by a coastal and inland route, a distance of some 450 mi.

Portolá and his bedraggled men returned to San Diego after 6½ months. The historic expedition never located Monterey, nor did it establish a presidio anywhere along the way. Their efforts were hardly a bust, however, because they learned much about the terrain, identified a dozen places for missions, and discovered San Francisco Bay by land. Into the wilderness the Portolá expedition established the trail that would become the main route between San Diego and San Francisco, *El Camino Real* (the Royal Road or King's Highway), and along it, the settlements and missions that were to follow as well as the route traveled by naturalists as they explored California's interior.

No other Spanish venture was as responsible for so many California place names. At a camp along a river on the day of the Virgin Mary's Feast, one of the Portolá expedition padres, Juan Crespí, christened it in her honor as *El Río de Nuestra Señora la Reina de Los Angeles* (the river of Our Lady, the Queen of the Angles). In 1781, the pueblo near their camp would become California's largest city, modified from Crespí's name to El Pueblo de Nuestra Señora la Reina de Los Angeles, which was eventually shortened to Los Angeles, and nowadays simply L.A. Portolá's men also discovered the large pits of liquid tar, or *brea*, and recorded the unsettling experience of shaking ground (*temblores*). At one stop they awed over an encounter with their first grizzly (*brutos ferocímos*), near what is today San Luis Obispo's Los Osos Valley.

Still, many elites of New Spain remained disdainful of Alta California, regarding it as crude and remote. Some of Portolá's soldiers, however, saw opportunity and many of them stayed on to influence California's early history in numerous ways. Two of the most prominent were Lieutenant Pedro Fages and Captain Fernando Rivera y Moncada who, in succession, became *comandantes* and acting governors of Alta California.

After Portolá, the most important Spanish expeditions were in the mid-1770s led by Juan Batista de Anza, a military officer stationed near present-day Tucson. Anza's superiors longed for an overland route to the new settlements in California from the east, and they implored Anza to lead an expedition. With around 30 men Anza set out across Padre Kino's *Camino de Diablo* (devil's highway), a 200-mi trek across desolate desert. With the help of Yuma Indians, they crossed the Gila and Colorado Rivers then trudged across the waterless dunes of the Colorado Desert, eventually reaching Borrego (lamb) Valley in 1774. They were now in sight of the San Jacinto Mountains to the north and eventually found their way over them and down to the mission in San Gabriel, much to the astonishment of the resident padres. Having shown that an overland route from Mexico was doable, Anza returned the following year with a much larger contingent of around 240 soldiers and colonists (*pobladores*) and hundreds of head of cattle. The prosperity of the new settlements was now at hand.

Accompanying Anza on his second crossing was Padre Garcés, who set out to explore on his own by heading north for several hundred miles following the Colorado River. He turned west into the Mojave Desert (the first white man to cross it) and followed the Mojave River into the San Bernardino Mountains. From there he descended to the San Gabriel Mission. Garcés named San Antonio Creek while crossing the San Bernardino Mountains, a name that was later applied to the range's highest peak known familiarly today as Old Baldy in reference to the wind-swept nearly denuded mountain top. Fifty years later U.S. trappers and explorers followed segments of Garcés itinerary, which became known as the Santa Fe, or Old Spanish, Trail.

Spain at last had a viable presence in Alta California with Monterey as its capital. Still, the homeland was far away, and with its growing affairs in South and Central America, not to mention Mexico, Spain had little motivation to expend further effort and money in California other than at a leisurely pace. By the end of Spanish rule in 1821, four presidios and 21 Franciscan missions were in place from San Diego to San Francisco. Additionally, Spanish courts had authorized around 20 concessions of large tracts of land to retiring soldiers who wished to stay on. Cattle, horses, and sheep became their economic foundation. Under Mexican rule, various governors extended the Spanish land grants in parcels up to 48,000 acres; some of which were deeded to white settlers—*Californianos* who had taken Mexican citizenship. The land grants continued after secularization of the missions and were occasionally expanded through marriage or by the acquisition of other parcels. Around 700 land grants had been issued by the end of Mexican rule in 1848. That brief quarter-century of Mexico's presence celebrates the golden age of the California Ranchos, overseen by the elegant *señoras* and gentlemen *dons* that we romanticize to this day.

3.3 NATURALISTS OF THE EARLY 1800s

Alta California was politically ill at ease in the early 1820s as Mexico took over the territory from Spain. This did not, however, inhibit the continued exploration of the land by both foreign and domestic naturalists. In the early going, most of them arrived on merchant ships that usually stopped first at either Ft. Vancouver or Monterrey. Six of those explorer-collectors earned a permanent place in the state's history.

Paolo Emilio Botta was a physician of Italian descent who was schooled in France. Botta is probably best known for his archaeological discoveries in the Middle East, although he was also a respectable naturalist and was aboard the French vessel *Héros* exploring the California coast in 1827. The ship and its crew spent nearly a year along the coast experiencing Fort Ross, Bodega Bay, Santa Cruz, Monterey, San Pedro, and San Diego. Botta's numerous zoological contributions covered reptiles, mammals, and insects. The Rubber Boa (*Charina bottae*) bears his name, as does Botta's Pocket Gopher (*Thomomys botta*). In San Diego, he collected the type specimen of the Greater Road Runner, the most iconic bird of the American Southwest.

David Douglas was working for the London Horticultural Society when he arrived at Ft. Vancouver on the Columbia River in 1825. He was an experienced plant collector and had already visited the East Coast where he made the acquaintance of political and scholarly luminaries, like Thomas Nuttall of the Philadelphia Academy of Natural Sciences. While exploring the Northwest in 1827 Douglas discovered the Sugar Pine (*Pinus lambertiana*—the largest Western Pine) and the conifer that bears his name, the common Douglas-Fir (*Pseudosuga menziesii*). Douglas reached Monterey Bay in 1830 aboard the *Dryad*, a brig of the Hudson Bay Company. Most of his time in California, about 1 year, was spent collecting plants along the Camino Real between Monterey and Santa Barbara. At the end of his California sojourn, Douglas had amassed around 500 species of plants, at least 60 of which were new to science. Perhaps his most unusual, and unlikely, find was discovered years later in the roots of a Douglas-Fir that he had collected in the Santa Lucia Mountains—trace flakes of California gold.

In the fall of 1831 Douglas had a chance encounter with Dr. Thomas Coulter, a fellow British botanist who had arrived at Monterey a month earlier from Mexico where he had been studying the flora. The two spent the next several months together hunting, fishing, and collecting. Coulter had a keen interest in California's southeastern deserts and a particular fascination for lizards and snakes. His trek through California led him south from Monterey into the Santa Lucia Mountains, whence to Santa Barbara, San Gabriel, and the settlement of La Pala located in the San Luis River Valley below Mount Palomar. Following in part the route established by Anza some 50 years previously, he traveled southeast along San Felipe Creek in what is today Anza Borrego Desert State Park. Coulter became the first naturalist to cross the Colorado Desert when he reached the Colorado River near Yuma. Among his collections was the tall and showy Matilija Poppy (*Romneya trichocalyx*). Coulter's name lives on in numerous plant taxa including the Coulter or Big-cone Pine (*Pinus coulteri*).

At some point in 1832 yet another naturalist showed up in Monterey, the German zoologist Ferdinand Deppe from the Berlin Museum. Deppe had been in western North America for about 6 years, mainly in Mexico, when he arrived in Alta California from Baja by way of a pack train. His primary interest was collecting mammals and birds for the Berlin Museum. Deppe's most spectacular specimen was a California condor that was eventually put on display in the Berlin Museum.

Thomas Nuttall was a botanist and ornithologist at Harvard University when he published his two-volume *Manual of the Ornithology of the United States and Canada* in 1832. In 1834, sick of the academic life, he joined the second Nathaniel Wyeth Expedition to the Columbia River, accompanied by his friend and fellow bird collector John Kirk Townsend. These two men became the first experienced naturalists to reach the West Coast by an overland route while traversing the breadth of the country from east to west. The numerous birds they collected in the Pacific Northwest were shipped to their colleague John James Audubon at the Academy of Natural Sciences in Philadelphia. Audubon used them as models for his renowned *Birds of America*.

Following their stint in the Pacific Northwest, Nuttall and Townsend sailed to the Sandwich Islands (Hawaii) where Nuttall studied seashells and marine crustaceans. Townsend collected specimens of native birds, several of which are now extinct. The pair returned to the Pacific Northwest in the spring of 1835. Townsend stayed on while Nuttall returned to Hawaii. In the spring of 1836 Nuttall made plans to return to Boston with a stop in Alta California, reaching Monterey in March. From there he sailed on board the merchant ship *Pilgrim* to Santa Barbara. Among his finds there was the Scrub Oak he named *Quercus dumosa* and the Birch-leaf Mountain-mahogany (*Cercocarpus betuloides*). Nuttall also collected or reported on such birds as the Yellow-billed Magpie (*Pica nuttalli*), Black Phoebe, Anna's Hummingbird, the first specimens of Tricolored Blackbirds, which Audubon would name, and a new race of White-crowned Sparrow.

From Santa Barbara, the *Pilgrim* sailed on to San Diego with a brief stop in San Pedro. In San Diego, Nuttall's botanical achievements included what at the time were eight new genera of plants, such as *Pholisma* (Desert Christians Tree), *Streptanthus* (jewel flower), *Isomeris* (bladderpod), *Apiastrum* (wild celery), *Nemacladus* (threadplant) and three in the sunflower family (Asteraceae): *Pentachaeta* (golden-ray flower), *Uropappus* (silver puffs), and *Rafinesquia* (California chicory). During his comparatively brief time in Alta California Nuttall collected a remarkable 44 new species of plants, most of them from the southwestern sector.

A short time before departing San Diego for Boston, Nuttall was casually combing the beach for shells when he encountered a former student from his teaching days at Harvard; a young sailor named Richard Henry Dana, who had been serving as a common seaman aboard merchant vessels trading for hides along the California coast. Nuttall, it turns out, is whom Dana refers to as "Professor N______" in his chronicles about his time at sea in *Two Years Before the Mast*. The pair returned to Boston together on the same ship (the recently arrived *Alert*), although apparently, they saw little of each other on the homeward voyage. The ship's crew didn't quite know what to make of the unusual man and his collections, and so they took to calling him "Old Curious." Upon returning to Boston, Nuttall continued his botanical studies and became a strident voice for avian conservation.

In his later years Nuttall was a mentor to the budding naturalist William Gambel of Philadelphia. Gambel proved an able student and became an apprentice bird collector for his older teacher. In 1841 Gambel was only 18 when he had an opportunity to travel with a cadre of traders on their way to California following the Santa Fe Trail. He arrived on the coast of Southern California in November of that year and remained in California until the end of 1843. As a trained naturalist, Gambel achieved some notable firsts: he was the first to spend an extensive amount of time in the state, the first to cross the Southwest deserts from Santa Fe to Los Angeles, and the first to visit Santa Catalina Island. His enthusiasm for natural history was seemingly limitless, even at one point during his travels approaching the U.S. Navy about protection from local Indians. Instead he was hired as secretary to Commodore Ap-Catesby Jones.

In 1842 Gambel sent Nuttall descriptions from Puebla Los Angeles of 11 southwestern birds, 4 of which were new to science: Wrentit, California Thrasher, Oak Titmouse, and Nuttall's Woodpecker that he named for his mentor. Indeed, so prolific were Gambel's bird studies that he rightly can be regarded as California's first ornithologist. In 1845 he returned to Philadelphia and completed his medical training. Yet he longed for a return to California; so when the Gold Rush began, he joined up with a group of settlers for the cross-country trip. Sadly, he died of typhoid after crossing the Sierra Nevada late in 1849. He was only 28 years old. Gambel's name persists in the taxonomy of numerous Southwestern plants and animals, as for example Gambel's Quail (*Callipepla gambelii*), the Mountain Chickadee (*Poecile gambeli*), the Leopard lizard genus *Gambelia*, and Gambel's Oak (*Quercus gambelii*).

3.4 COLONIAL AMERICA

Universities and museums did not yet exist in the American colonies and the earliest naturalists were mainly those few souls who came from abroad, collected specimens, and returned them to the great museums in London, Berlin, and Paris. Sir Hans Sloan (1660–1753) was one of them. He was a personal physician to the British crown who studied in America from 1687–1689. His collections and library became an early component of the British Museum. America's own identity in the natural sciences began when Mark Catesby (1683–1749) arrived in 1712. Catesby was a self-taught naturalist who lived in Virginia where he wrote the first authoritative treatment of "British North America" before the American Revolution. The Bullfrog, *Lithobates* (*Rana*) *catesbeiana*, is named after him. His *Natural History of Carolina, Florida and the Bahamas* took 20 years to compile. Although not especially profound, it was carefully done and he hand colored all the plates; the work had a strong influence on Audubon a century later. By the mid-eighteenth century, natural history was becoming an increasingly active discipline in and by Americans. The American Philosophical Society was founded in Philadelphia in 1742, and similar scientific societies followed in New York and Boston.

Thomas Jefferson was in the second year of his presidency (1801–1809) when he set about securing access to western lands and waterways for the United States. Diplomats were dispatched to France to negotiate the purchase of New Orleans, an acquisition that would be essential for U.S. trade along the Mississippi River. About that same time, he commissioned Meriwether Lewis to lead an expedition across the country to the Pacific. Some of Jefferson's contemporaries held hope for the existence of an inland waterway—that elusive Northwest Passage—which would link the Atlantic with the Pacific. Jefferson, on the other hand, presciently suspected a continental divide and he wanted to find a way over it before the British did. Thus, Lewis and William Clark set off up the Missouri River in 1804. By then, the United States had completed the Louisiana Purchase ("New France"), which included most of the land that Lewis and Clark traveled through. During their 2-year expedition through the new territory Lewis and Clark learned the hard way that there was no water passage linking the Columbia and Missouri Rivers, nor was there any easy route over the Rocky Mountains.

The Louisiana Purchase was a not a tidy transaction. It ignored Indian land rights and failed to delineate clear boundaries. For example, the Red River, now the border between Oklahoma and Texas, was the reputed boundary between U.S. and Spanish territory. It was here in 1806 that Zebulon Pike, on a survey mission for the general governor of the Louisiana Territory, maintained he was camped. His claim was likely a ruse. Pike, in fact, was camped well to the south on a tributary of the Rio Grande. More aggressive congressional interests favored the Rio Grande as the border with Spanish territory because they also had designs on Spanish-held Santa Fe. Pike was arrested by Spanish troops, taken to Santa Fe, and on his release was able to supply information on its economic potential and military weaknesses, further stimulating U.S. interests in the West.

Jefferson had doubled the size of the country for a scant U.S.$15 million yet was roundly criticized for acquiring such vast tracks of land when the country already had so much. The complaints didn't last long. The War of 1812 had weakened Indian resistance in the Mississippi Valley and settlers flooded into the likes of Missouri and Illinois; by 1820, the proportion of Americans living west of the Appalachians had doubled. Sixty years after the American Revolution, the fauna and flora of the original colonies was well known, but the same could not be said for the territory west of the Appalachians. That land was the last frontier of unexplored territory with a new flora and fauna to be collected and described.

3.5 EXPLORING THE WEST

It began with the mostly independent fur trappers and explorers that followed in the wake of Lewis and Clark. This hardy and irreverent bunch learned the lay of the land, identified the most promising travel routes, and introduced Indians to some of the white settlers' ways. Their fearless, if not colorful, exploits are thoroughly embedded in Western lore, men like Jim Bridgers and the legendary Jedediah Smith. Smith was a self-made mountain man born in New York who began his intrepid campaigns around 1823. His second expedition for a Missouri fur company began with a fierce Indian attack. Soon thereafter his scalp was ripped, and an ear torn off by a grizzly bear; a companion sewed it back on. That particular expedition ended with the discovery of Wyoming's South Pass, which later became a key link in the California–Oregon Trail. Comanche Indians killed Smith in 1831 when he was 32 years old, but by then he had opened large parts of the West to the fur trade. From 1826 to 1829, he crossed the Mojave Desert, traveled the length of California well north into the Oregon territory, and was the first explorer to cross the Sierra Nevada.

Meanwhile the U.S. government was busy establishing the Army Corps of Topographic Engineers, a rugged group of surveyors and cartographers who made possible the great western migration that was to follow. The expeditions were both military (to pacify Indians) and scientific, and they were the blueprint for later government surveys. Major Stephen H. Long was one of the Corps' most prolific explorers, covering some 26,000 mi over five expeditions. Unfairly or not, Long is probably best remembered for what he got wrong. In 1820 he had been sent to survey the Great Plains and southern Rockies. His report characterized the region as a wasteland unfit for habitation. Settlers were already crowding the plains' eastern edges, but Long's report delayed further settlement for almost another 40 years. By the end of the century, the Great Plains had become the nation's breadbasket.

The first U.S. survey of the West Coast was the famous United States South Seas Exploring Expedition under the command of Charles Wilkes that commenced in 1838. It was one of the largest voyages of discovery in the history of Western exploration, including previous ones by the British, French, and Spanish. For the young nation, decades before it fully surveyed and mapped its own interior, the venture was an audacious undertaking. The United States was not interested in claiming distant land; it already had plenty of its own. Rather, its intent was to establish itself as a player coequal with the Europeans and to strengthen commerce by developing maps and charts for U.S. whalers and trade with China. The agenda included surveying islands of the western Pacific as well as Australia, New Zealand, and the Antarctic coast, which at the time was still *terra incognita*. Another priority was the Pacific Northwest. Since the Lewis and Clark expedition, the British and their Hudson Bay Company dominated what was then called the Oregon Territory, a region that the United States one day hoped to obtain. To that end, the Wilkes Expedition surveyed 100 mi of the Columbia River, 800 mi of the Oregon coast, and an overland route between Oregon and San Francisco.

The Exploring Expedition consisted of six vessels with 346 crewmembers and a party of zoologists, botanists, geologists, surveyors, and illustrators. The expedition returned in 1842 with a staggering number of collections, artifacts, charts, and data. Plants alone numbered around 50,000 specimens representing 4,000 species. Titian Peale (son of the famous painter Charles Wilson Peale of Philadelphia) had collected 2,150 bird specimens, 134 mammals, and 500 species of fish. Ethnographic objects numbered more than 4,000, and there were charts of 280 Pacific Islands done with heretofore-unseen precision. There was only one problem: The federal government had neither the facilities nor the means to accommodate this vast trove. After a few false starts and some political deal making, the bounty from the Exploring Expedition was shipped to Washington D.C. It became the nucleus of the first national museum. Fortuitously, the United States had received a bequest from the wealthy Englishman James Smithson to establish an institute "for the increase and diffusion of knowledge among men." The Smithsonian Institution was established in 1846.

The Smithsonian's first secretary, Joseph Henry, viewed the Smithsonian Institution as a research organization, not a repository for specimens and artifacts.

But by 1850, Henry accepted the fact that the growth of collections was inevitable. Moreover, he recognized the need for someone to oversee them and handle the museum's daily affairs. He hired the young Spencer Fullerton Baird for the job. In 1858 when the Smithsonian Institution finally accessioned the collections from the Wilkes Expedition, they accounted for only about one-fifth of the museum's holdings; such was Baird's success at obtaining new material from exploration of the West. To be sure, the collections from the Wilkes Expedition added enormously to the museum's prestige. Today, the Smithsonian Institution not only covers the National Museum of Natural History but also 18 other entities plus the National Zoo. Other outcomes of the Exploring Expedition include the U.S. Botanical Garden, the National Herbarium, the National Observatory, and the U.S. Hydrographic Office.

Most all of the collections and accompanying data from the Exploring Expedition were eventually studied and published. Baird himself, along with his assistant Charles Girard, described most of the reptiles and amphibians, for instance. Publication was made possible because Wilkes persistently badgered Congress for appropriations to publish the reports. Each one added to the stature of U.S. science. For his part, Wilkes wrote up his own narrative of the expedition in five volumes, a project that was overseen by a congressional committee. In the end, only 100 copies were made available of the elaborately illustrated, bound, and gilded government edition. Wilkes finagled a way to retain the copyright, however, and he proceeded to publish his own commercial edition. Although his critics found it padded with irrelevant information and digressions, it nonetheless sold fairly well. For many readers the most captivating accounts were the descriptions of California and Oregon.

Almost simultaneously, another publication appeared of a government-sponsored expedition to the West; this one led by John C. Frémont. Frémont's account of his overland journey to the Columbia River and on into California brought him considerable fame. The book's success was due in large measure to his collaborator wife and her flare for overly romantic prose.

Frémont circled the West for 14 months between the years 1843 and 1844. Like Long, he was an officer in the Army Corps of Topographic Engineers. Frémont's party logged some 3,500 mi during this (his second of five) expedition. The maps and logs fascinated the public, and his account of the Great Salt Lake Valley inspired Brigham Young to make it the Mormon's new Zion. The book was also the source of Joseph Ware's 1849 *Emigrants Guide to New Mexico, Oregon and California*, a book that became the standard reference for the forty-niners that flooded into California during the gold rush years.

Frémont's admirers labeled him bold and daring; his critics deemed him impetuous and reckless. He became a war hero, one of California's first senators, and the first Republican nominee for president. President Rutherford B. Hayes later appointed him territorial governor of Arizona (1878–1881).

Henceforth, a steady flow of U.S. government surveys ensued, launched in part by Mexico's refusal to acknowledge the annexation of Texas by the United States. War with Mexico was imminent.

At the outbreak of the war in 1846, President James K. Polk sent General Stephen Kearny to secure the Southwest. Kearny's Army of the West marched mostly unopposed through Santa Fe to San Diego where they finally met resistance from Mexican forces at the Battle of San Pasqual. Accompanying Kearny was Lt. William Emory who mapped the arid land along the way and made valuable botanical collections later studied by Professor John Torrey of Columbia College. Emory pronounced the land unsuitable for agriculture and in doing so deflected fears that the new land would become slave states if admitted to the Union.

During his single term as president that began in 1845, Polk completed the broad perimeter of U.S. geography. Texas was admitted to the union and the country's boundaries now extended to the Pacific, from Southern California to Washington.

3.6 MEXICAN BOUNDARY SURVEY

At the end of the war with Mexico in 1848, Mexican and U.S. diplomats were hashing out the details of their treaty only to discover that the map they were relying on was inaccurate. For one thing, the boundary near El Paso was about 50 mi north of what the United States intended, which was ample easement for wagons and railroads that the new border was supposed to accommodate. For another there was no accurate measure of just where the Gila River joined the Colorado River near Fort Yuma, nor was there reliable information on the exact starting point of the border in San Diego. A federal commission was thus established to survey, erect markers, and make a comprehensive assessment of the land through which the border would pass. Emory was put in charge of field operations.

From 1849 to 1855, six successive commissioners were appointed (the last being Emory himself), other personnel came and went, and funding dried up later to be restored. In the field the two survey crews, one each in El Paso the other in San Diego, experienced their own delays, equipment failure, flash floods, insane heat, a shortage of rations, irritable bowels, miscommunications, and just-for-the-hell-of-it forays into Mexico by key personnel. Even so, Emory's remarkable skills won out, and the first survey was completed in August 1852. Yet it soon became clear that the United States still did not have sufficient land for a railroad, and the Mexican border was readjusted through the Gadsden Purchase of 1853. The new boundary, completed in August 1855, moved the border to just north of Juárez, Mexico, and about 20 mi south of the junction of the Gila and Colorado Rivers.

During the on-again, off-again surveys over 6 years, a wealth of plants, animals, and minerals were collected and studied in the field. The names of many of the men who participated persist today in the botanical, zoological, and geological literature. They include John Russell Bartlett, George Henry Thomas, George Thurber, Amiel Weeks Whipple, Edmund Hardcastle, Charles Christopher Parry, Charles Wright, and Arthur C. V. Schott.

The final report of the Mexican Boundary Survey was released in 1856 in three volumes: a narrative and overview of geology, one on botany, and another on zoology. Baird wrote up the birds, mammals, reptiles, and amphibians (Figure 3.1); Charles Girard detailed the fishes; John Torrey, Asa Gray, and George Engelmann handled botany; Charles Parry reported on geology and ecology; James Hall described geology and paleontology. Emory drafted the maps and charts. For decades the *Report of the United States Boundary Survey* remained the most important treatise on the natural history of the American Southwest.

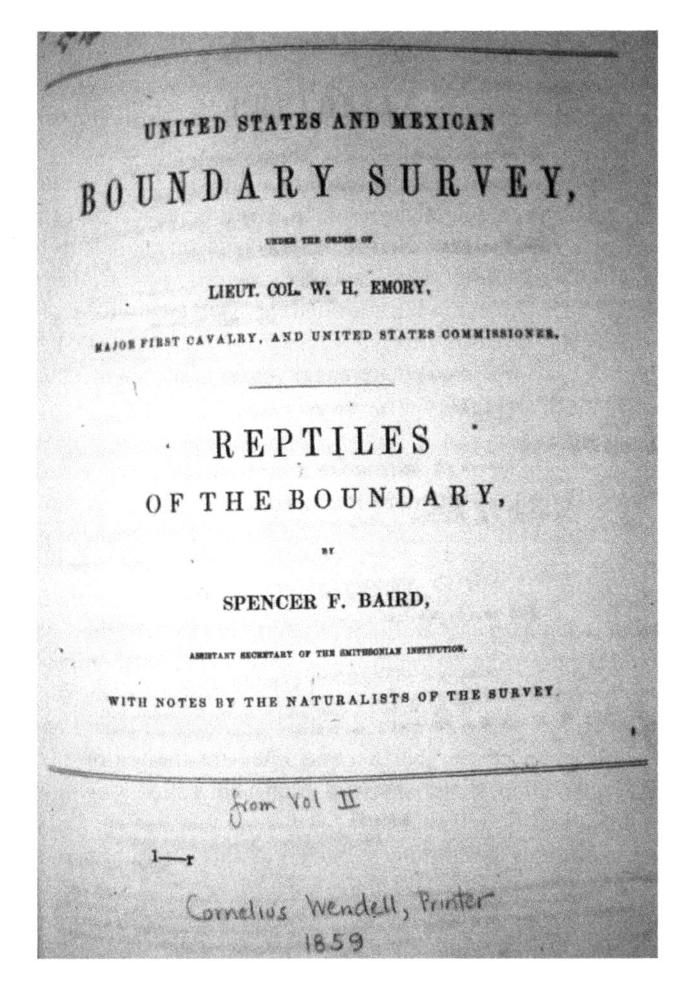

UNITED STATES AND MEXICAN

BOUNDARY SURVEY,

UNDER THE ORDER OF

LIEUT. COL. W. H. EMORY,

MAJOR FIRST CAVALRY, AND UNITED STATES COMMISSIONER.

REPTILES

OF THE BOUNDARY,

BY

SPENCER F. BAIRD,

ASSISTANT SECRETARY OF THE SMITHSONIAN INSTITUTION.

WITH NOTES BY THE NATURALISTS OF THE SURVEY.

from Vol II

1—r

Cornelius Wendell, Printer

1859

FIGURE 3.1 The *Report of the United States Boundary Survey* published between 1856 and 1859 was a major milestone in the natural history of the American Southwest. It appeared in three volumes, one of which reported on the expedition's zoological collections like this account of the reptiles by Spencer F. Baird of the Smithsonian Institution.

3.7 RAILROAD SURVEYS

Thoughts of a transcontinental railroad were already in the public's mind when California achieved statehood in 1850. In fact, it was a national obsession. Nearly a dozen potential routes were proposed, and of those, Congress authorized four to be surveyed in 1853. One was across the Northwest and the other three across California. The northern California route was to survey the 38th parallel beginning in Colorado. To the south, the two other surveys were to be conducted along the 35th parallel beginning in New Mexico and along the 32nd parallel from Texas to San Diego.

The southernmost survey that began in California assembled in Benicia in eastern San Francisco Bay under the command of Lt. Robert S. Williamson and Lt. John G. Parke of the Army Corps of Topographic Engineers. Their mission was to locate a suitable pass over the southern Sierra Nevada as well as a favorable route across the Colorado River and on into San Diego. As time permitted, they were also charged with studying geology, meteorology, ethnography, and natural history. The survey geologist was the renowned William P. Blake, and the naturalist and physician was Adolphus Heermann. Heermann, accomplished on both fronts, was friends with Gambel, Baird, and was an acquaintance of the eminent British geologist Charles Lyell. He was already familiar with California, having traveled there in 1851 surveying birds around the Sacramento River. He had also visited the Coronado Islands off northern Baja California where he had collected nests and eggs of a red-billed gull later named by John Cassin of Philadelphia as *Larus heermanni* (Heermann's Gull).

Williamson's party set out in July 1853 southeast across the San Joaquin Valley reaching the Sierra Foothills near present-day Bakersfield early in August. They surveyed the southern Sierra for railroad passes and then spent several weeks around the Tehachapi Mountains exploring *Cañada de las Uvas* (Grapevine Canyon), Tejon Pass, and the area now called Antelope Valley. While Williamson established a depot in Tejon and made detailed surveys of the region, Parke led a crew over Tejon pass into the Los Angeles basin. There he discovered a pass (San Gorgonio) east into the desert (upper Coachella Valley), which would eventually be used by the Southern Pacific Railroad. Meanwhile Heermann, who remained with Williamson's party, collected the first-ever Pacific Pond Turtle, the first specimen for the region of the stream bird, the American Dipper, and amassed additional collections of other birds, small mammals, lizards, and snakes.

Williamson and Parke reunited in early October and commenced a study of the San Gabriel Mountains, Cajon Pass, and the upper Mojave Desert along the Mojave River near present-day Victorville. They explored east to Soda Dry Lake where Williamson mistakenly believed he was still more than a hundred miles from the Colorado River. Accordingly, he abandoned his search for a railroad route across this part of the desert. A few months later the party surveying the 35th parallel, led by Lt. Amiel Whipple, crossed the

Colorado River and reached Soda Dry Lake from the east, which eventually became the route of the Santa Fe Railroad.

Williamson's party concluded its mission and disbanded in San Diego late in December 1853, having covered more than 1,300 miles that included a transect across the Colorado Desert from Fort Yuma to San Diego. Their careful assessment of Tejon Pass led to the eventual establishment of a military depot there.

The Whipple Expedition of the 35th parallel headed west from Albuquerque in autumn 1853, not long after Williamson and Parke departed Benicia for the southern Sierra. Among Whipple's staff were the naturalists Dr. John M. Bigelow (a veteran of the Boundary Survey), Caleb B. Kennerly (a former undergraduate student of Baird), Heinrich B. Möllhausen (an artist and zoologist), and the geologist Jules Marcou. The party reached the Colorado River at a wide point across from present-day Needles, California, late in February 1854.

Crossing the river was a daunting challenge. It took 4 days to accomplish. Prefabricated pontoons were used to float the supply wagons and carts across to the other side, several of which capsized, tossing overboard their scientific instruments and reference books. Some was recovered; most of it not. Once on the western shore, Whipple's party headed to the Providence Mountains and then on to Soda Dry Lake where Williamson and Parke had abandoned their eastward transect months before.

Just west of Soda Dry Lake the Mojave River makes a brief above-ground appearance in Afton Canyon. With water caches replenished, Whipple's party followed the Mojave River around Barstow and up to the narrows near Victorville. They passed over the summit of the Cajon Pass, the narrow gap between the San Gabriel and San Bernardino Mountains delineated by the San Andreas Fault and its striking geologic formations. They worked their way down the west side to Rancho Cucamonga and finally on into Los Angeles to conclude the expedition.

The railroad survey of the 38th parallel set out from eastern Kansas in mid-June 1853. That survey, led by John W. Gunnison, successfully crossed the continental divide in southern Colorado and was foraging ahead into Utah when they met with disaster. In October, Paiute Indians massacred Gunnison and seven other men. Among the dead were the party's naturalists Frederick Creutzfeldt and Richard Kern. The survey continued on with Lt. Edward G. Beckwith, Gunnison's second in command, leading the party. He appointed Kern's young assistant, James Snyder, as botanist, a role for which he showed little enthusiasm following the loss of his mentor. They departed Salt Lake late in winter 1854 with a new directive to follow the 41st parallel across northern California. From Nevada's Pyramid Lake, Beckwith crossed into California north of Susanville and followed the Susan River over the Sierra down into the Sacramento Valley; their destination was Fort Reading on Cow Creek, a lush riparian area feeding the Sacramento River. Beckwith correctly stated in his report that the mountainous regions of northern California were simply too steep and snowy for a railroad route, and none of the main passes were ever used for a rail line.

Like the Mexican Boundary Survey, the reports of the railroad surveys appeared in government publications (Figure 3.2) and other outlets. Results of the Williamson and Parke survey were the most extensive and detailed. Natural history collections were well cataloged, and Blake's geological accounts included a colored geologic map of California. (Blake would go on to become director of the school of mines at the University of Arizona.) Torrey wrote up the plant collections, Heermann compiled the bird data, and Baird covered the mammals. Edward Hallowell described the amphibians and reptiles. Data collected by

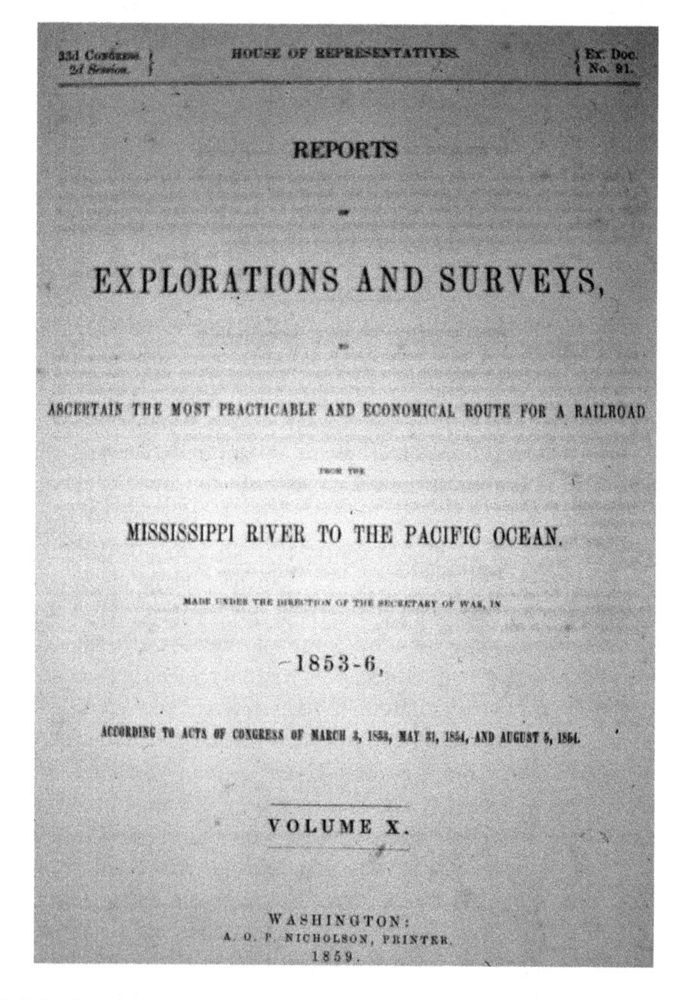

33d Congress, 2d Session. HOUSE OF REPRESENTATIVES. Ex. Doc. No. 91.

REPORTS

EXPLORATIONS AND SURVEYS,

ASCERTAIN THE MOST PRACTICABLE AND ECONOMICAL ROUTE FOR A RAILROAD

FROM THE

MISSISSIPPI RIVER TO THE PACIFIC OCEAN.

MADE UNDER THE DIRECTION OF THE SECRETARY OF WAR, IN

1853-6,

ACCORDING TO ACTS OF CONGRESS OF MARCH 3, 1853, MAY 31, 1854, AND AUGUST 5, 1854.

VOLUME X.

WASHINGTON:
A. O. P. NICHOLSON, PRINTER.
1859.

FIGURE 3.2 By the time California achieved statehood in 1850, the United States Congress was already addressing the need for a transcontinental railroad. Of the numerous possible routes that were proposed four were selected, two of which included Southern California. Besides identifying suitable routes over mountain passes, the railroad survey expeditions were also charged with documenting the routes' geological, botanical, and zoological resources.

the Whipple naturalists included new or rare species of plants and animals and unique profiles of the geology and flora of the Cajon Pass.

3.8 JOHN XÁNTUS AND FORT TEJON

In addition to the boundary and railroad surveys, other government-sponsored expeditions penetrated many areas of the West during the 1850s. Captain Lorenzo Sitgreaves, for instance, was charged with exploring the Zuni and Colorado Rivers. His party dispatched from Albuquerque eventually reached San Diego by a route that the Santa Fe Railroad in part later followed. The party's naturalist, Samuel Woodhouse, made significant collections along the way while suffering the misfortune of taking an arrow in his leg, only later to be bitten by a rattlesnake. His recompense was having a toad named after him, *Anaxyrus* (*Bufo*) *woodhousii* (Woodhouse's Toad).

Captain Howard Stansbury surveyed in detail the basin of the Great Salt Lake. Among the zoological collections made was a new lizard named for the Ute Indians that Baird and Girard named *Uta stansburiana* (the widespread Side-blotched Lizard).

Certainly one of the most outrageous figures from that era was John Xántus, a Hungarian immigrant expelled from his own country who had come to the United States in 1851. Possessed of a glorified self-image and singular hubris, Xántus tested the patience of most everyone he ever met. In the United States, he struggled to earn a living and was eventually forced to enlist as a private in the U.S. Army in 1855. Xántus was so humiliated that he took the assumed name of *Louis de Vésey.*

Xántus's first duty station was Fort Riley, Kansas, a post he detested. But to his good fortune, he became an assistant to Dr. William Hammond, an army physician who would go on to become Surgeon General of the United States. Hammond was a generous and intelligent man, an amateur ornithologist, and friend of Baird at the Smithsonian to whom Hammond would ship specimens from time to time. In Xántus, Hammond found a willing and capable student who had a knack for, and took an interest in, preparing natural history specimens. Hammond eventually introduced Xántus to Baird.

Baird used his connections in Washington D.C. to have Xántus transferred to Fort Tejon, a new outpost in the Tehachapi Mountains carefully surveyed by the Williamson railroad party in 1853. Tejon Pass was the main north–south route between the San Joaquin Valley and the Mojave Desert and to the Los Angeles basin (today occupied by I-5). Don Pedro Fages first brought attention to *Cañada de las Uvas* (Grapevine Canyon) during a reconnaissance in 1772. This is where the fort was built. Fages, as you recall, was a lieutenant under Don Gaspar de Portolá. The area was in the possession of the Catholic church until 1834 when Lt. Antonio del Valle was granted the land, which he renamed *Tejon* (badger). The land was still in Mexican hands when Lt. Edward Beale secured part of it for the Tejon Indian Reservation in 1853. The Tejon Pass was rightly described as an Eden of magnificent oaks, grasslands, numerous springs, and teeming with wildlife. It is easy to appreciate Baird's enthusiasm for having a collector on site.

Upon his arrival, Xántus wasted little time in souring officers and fellow enlisted men with his insufferable arrogance, sense of entitlement, and insistence on special privilege. He routinely shirked his duties at the post dispensary by taking time off to hunt for specimens. His reputation as an eccentric malingerer was only enhanced when the grizzly bear cub he kept for a pet attacked and ate the commanding officer's dog. Unbowed, Xántus continued his fieldwork and became an accomplished collector. He artfully stuffed bird skins, mounted insects, pressed plants, and carefully packaged them for shipment back to Washington D.C. Xántus was posted at Fort Tejon from May 1857 to January 1859. During that time, he maintained a running correspondence with Baird, and by the time, he left the fort (and the army) he had sent 1,794 bird skins, 211 nests, 740 eggs, 145 mammals, 229 containers of fish and reptiles, 107 bottles of insects, 140 skulls, and 14 bales of plants. Today, Fort Tejon is a state historic park.

3.9 INSTITUTIONS

When California achieved statehood in 1850, there was already an impressive league of geologists, botanists, zoologists, and meteorologists who called California home. It was inevitable that societies would form, and institutions would be built to further their interests in natural history and serve as repositories for collections. The first of those was the California Academy of Natural Sciences founded in 1853 by Dr. John Trask, Dr. Henry Gibbons, Dr. Albert Kellogg, Lewis Sloat, and Dr. Andrew Randall, who would become the academy's first president. It was the first natural history institution west of St. Louis. In 1868, the institution was renamed the California Academy of Sciences, and its first official museum opened in 1874 in what is now China Town. It moved to a larger facility on Market Street in 1891 where it remained until the earthquake of 1906 destroyed the facility and practically everything in it. The "Cal Acad" was rebuilt in Golden Gate Park in 1916 where it remains to this day.

In its early years, the academy had as members and associates many of the states' A-list historical figures in the natural sciences. Joseph Le Conte became the University of California's first professor of geology, botany, and natural history. Edward L. Greene was named the first director of the University of California's botany department. Josiah Whitney was the State Geologist and later the academy's president. James G. Cooper (for whom the Cooper Ornithological Society is named) was an indefatigable champion of California's natural history and regarded as the "last of (Spencer) Baird's Boys." Cooper collected the first specimens of Lucy's Warbler (*Vermivora luciae*) along the Colorado River and named it after Baird's daughter. However, the familiar Cooper's Hawk is named for his father, William Cooper, founder of New York's Lyceum of Natural History. James Lick, a prominent (and wealthy) academy member underwrote the observatory on Mount Hamilton (Santa Clara County) that bears his name. John Van Denburgh consolidated much of what was known about California's vertebrates, especially amphibians and reptiles. And in a display of enlightenment far ahead of the time, the academy encouraged women to participate. Dr. Mary Katharine "Kate" Curran, a sophisticated botanist, modernized the herbarium in the 1880s.

In the 1960s, Stanford University divested itself of its considerable holdings in botany, entomology, and vertebrates and donated them to the academy, which by then had vast collections of its own and a worldwide reputation as a research and education facility.

The Museum of Vertebrate Zoology at the University of California–Berkeley is another distinguished institution in the Bay Area. The museum was the brainchild of philanthropist Annie Alexander who founded it in 1908. Alexander insisted that zoologist Joseph Grinnell be its director, and he held that position until his death in 1939. During his tenure Grinnell became the authority on vertebrate biology of the West Coast. Grinnell was a relentless field person and prolific writer. His instance on rigorous field data brought long overdue standardization and clarity to an essential component of collecting that had been haphazard, sometimes sloppy and often incomplete in the nineteenth century. His reforms included separate catalogs for specimens, and accurate and precise locality information along with all pertinent environmental and ecological observations at the time the specimen was obtained. Grinnell conducted extensive surveys of California including the Colorado Desert and Colorado River in 1908 and 1910, Mt. Whitney in 1911, and Southern California's San Jacinto Mountains in 1913. The museum today is a prestigious research facility and renowned center for graduate education in vertebrate biology.

In Southern California, amateur naturalists founded the San Diego Society of Natural History in 1874 as a portal for the exchange of ideas, scientific literature, and personal observations. The society began with casual conversations between Daniel Cleveland, a lawyer and amateur botanist, and Oliver Sanford, a surveyor and beetle collector. Joining them in drafting the first articles of incorporation were Charles Coleman, George W. Barnes, and E. W. Hendrick. The first women members were Rosa Smith [Eigenmann] who was elected in 1879, and a few years later Kate Sessions who was a prominent voice in the establishment of Balboa Park. In the early decades the society met in several locations around downtown San Diego until relocating permanently to their new museum in Balboa Park in 1933.

Several members brought early recognition and stature to the society through their own scientific endeavors: Anthony Vogdes, Charles Orcutt, Henry Hemphill, Laurence Huey, and Frank and Kate Stevens. The society took an early and active role in conservation, as for example in establishing the Torrey Pines Reserve (now state park) and was a major advocate in the creation of Anza Borrego Desert State Park.

The San Diego Natural History Museum, renovated and expanded in 1999, has six major research departments each with collections numbering in the tens to hundreds of thousands of specimens. The geographic emphasis is the western United States, eastern Pacific, northern Mexico, and Baja California.

The Santa Barbara Natural History Society is another venerable Southern California institution, founded in the 1880s largely by the tireless effort of local botanist Caroline Bingham. The society built the Santa Barbara Museum of Natural History in 1916 along idyllic Mission Creek. William Dawson was an ornithologist from Ohio and one of the museum's early directors. Dawson brought with him his large collection of birds' eggs and nests. I make a point of this because my own interests in natural history began as a small boy in Santa Barbara. I visited the museum often, and some of my lasting memories were of standing on my tiptoes peering into the soft, backlit cases of eggs and nests. Today, museum staff are active in numerous environmental concerns throughout Southern California.

The Natural History Museum of Los Angeles County, located in Exposition Park, opened its doors to the public in 1913 as the Museum of History, Science, and Art. It split from the Los Angeles County Museum of Art in 1961, renamed itself the Los Angeles County Museum of Natural History, and finally settled on its present name (NHMLAC). The museum has grown steadily and impressively over the years, acquiring abandoned regional collections and adding to their holdings by the fieldwork of its staff. Today it has some 35 million specimens and is the largest natural history museum in the western United States.

The San Bernardino County Museum, located in Redlands, is the fourth of Southern California's institutions that arose from the dedication of local naturalists wanting to preserve and celebrate the area's biological and geological heritage. It began as the San Bernardino Museum Association in 1952, opened to the public in 1957, and became a county department in 1961. In addition to exhibits and public outreach programs, the museum houses around 200,000 specimens in entomology, herpetology, ornithology, mammalogy, and botany. There are also important collections in paleontology and geology.

3.10 COLLECTIONS

When visiting a natural history museum, you will see only a tiny fraction of the vast number of specimens that the museum has on hand. Those on display are selected not only for their aesthetic appeal, but also for what they can teach us about nature. They inform us about the world we live in and the great diversity of life forms past and present. To the researcher, collections document variation; they inform us about organismal growth, development, and lifestyle; and they are a calendar of events in both space and time.

Like a library, collections must be well organized and accessible. Each specimen needs to be accurately identified, numbered, cataloged, and then properly stored. Special techniques are required to ensure their preservation. Plants, for example, are carefully pressed, mounted, and stored in specially designed cabinets. Some vertebrates such as lizards, snakes, and frogs are preserved whole in jars of ethyl alcohol after initially being posed in proper anatomical position (e.g., hands and feet displayed) and fixed in formaldehyde-based solutions. As with birds and mammals, some are made into skeletons, a delicate process that requires several lengthy steps. Birds and mammals are often preserved as whole mounts or flat skins, which is a labor-intensive procedure demanding hours of skilled attention. Fossils usually must be extracted from their sedimentary matrix even before they can be identified.

Through the preparation, pertinent information about the individual specimen must be recorded, such as the age and sex of an animal, the date, time, place, and circumstances of collection in the field, and in the case of fossils, the geologic age and structure of the sediments in which they were found. This brings us back to accurate notes taken in the field. Essentially a specimen is only as useful as the data that accompany it. A perfectly preserved specimen may be worth less (and usually is) than one in poorer condition but with complete data. Collections do not sit unattended on shelves and in cabinets; they are always under curatorial oversight for the benefit of future generations.

4 The Three Pillars of Natural History

The basis of modern science traces back to the natural philosophers of the eighteenth and nineteenth centuries. Compelled by the utter beauty of nature, they sought to understand its processes through objective analysis without appealing to mysticism or the divine. Carolus Linnaeus, Alfred Russel Wallace, and Charles Darwin were among the luminaries, but they were hardly alone.

Even today many biologists get started in their youth, enthralled with some particular animals or plants. It is something of a paradox, then, that in many contemporary science education environments (universities especially) natural history is often viewed dismissively as a quaint pastime of a former era—the European Age of Exploration when naturalists returned from the field with crates of plant and animal specimens that were named, cataloged, and then supposedly disappeared into the dark corners of museums. Neither the European naturalists nor their U.S. counterparts were earning a living at it; their incomes derived from other sources. As we saw in the previous chapter, many of the U.S. naturalists were physicians who provided medical care for survey expeditions.

In the twentieth century, natural history began losing its prestige after World War II. Science became confused if not synonymous with technology. In the biological sciences, advances in instrumentation led to increasing numbers of cellular and genetic investigations. The research money followed, and the notion that science must be *experimental* became a mantra. Many (but not all) universities favor molecular research and would just as soon jettison such field courses as herpetology, ornithology, and mammalogy. To carve out a niche and avoid being dismissed as simply a botanist or zoologist, many field people have rebranded themselves as "evolutionary biologists" or some other moniker with more cachet. Unfortunately, doing so simply reinforces the belief that natural history was the bygone pursuit of expired dilettantes.

The irony in all this is that natural history is as germane as it ever has been. It has taken on more responsibilities and developed principles to guide what it does. I have great respect for molecular biology and other laboratory disciplines. In recent decades their discoveries have been extraordinary and have yielded fascinating insights on how plants and animals regulate their internal environments while coping with the vicissitudes of the external one they live in. But to hold the position that experimentation is the only means of testing hypotheses is absurd. Biology can also test by *observation*, much like astronomy does.

We are compelled to describe and interpret the diversity of life by observing it firsthand, directly in the field where the plants and animals live, interact, reproduce, and eventually become extinct. That endeavor is the historical foundation of natural history—a source of uniquely precious information that has no expiration date. Natural history is more than an assemblage of specimens and facts. Properly arranged, it raises intriguing questions and provides pathways for exploring them. The biodiversity crisis facing the planet today has recruited legions of conservation biologists who require the fundamental skills of natural history to slow the relentless loss of species and habitats.

Natural history embodies three components, or pillars. Although they can be (and often are) studied independently of one another, a singular forcefulness is achieved when they behave synergistically. First, we ask what is it (a plant or animal) that we are observing (systematics). Second, we want to know why it occurs where it does (distribution), and third we ask how it survives. The first two are interrelated such that solving one leads to an understanding of the other, and vice versa. They often involve exploring into deep time in search of evolutionary events that led a species to be what it is and exist where it does. The third Pillar (survival) requires a grasp of the first two, yet it also serves as a counterweight by keeping a rein on wayward speculation.

4.1 SYSTEMATICS: TAXONOMY, PHYLOGENETICS, AND CLASSIFICATION

On the surface, identification of a plant or animal ("what is it?") would seem simple enough. We might say, for example, "It's a Great Blue Heron!" However, our question involves more than a name even if we recognize this bird as *Ardea herodias*. We also ask if it is a juvenile or an adult, a male or female. (In this case the sexes of Great Blue Herons (GBHs) are indistinguishable in the field.) Most importantly, identification implies group membership. The GBH is a long-legged wading bird with a long narrow bill, deeply cambered wings that inhabits marshes, tidal flats, streams, and lakeshores. The herons, egrets, and their allies are an evolutionary unit, a natural group that shares a more recent common ancestor not shared with other avian lineages. Until recently they were regarded as a separate taxonomic order, the Ciconiiformes. They are now included with pelicans and cormorants in the Pelicaniformes.

Recognize that the name *Pelicaniformes* is a taxon (taxa, plural form), which is any group of organisms worthy of being formally named, that is, taxonomy. The group can be a species, genus, family, or any other higher category. Worthiness derives from a demonstration that it includes all individuals that share a suite of uniquely derived characteristics and excludes those do not share them. Characters can be morphological or molecular but they must be heritable. In other words, a taxon is made up of a common ancestor and all its descendants. Arriving at that decision is often an arduous process. Large samples of prospective group members must be ransacked for characteristics that point to a unique, shared evolutionary history. When we are satisfied that our group(s) meets the criteria, we say that it is *monophyletic*, meaning "single tribe." A monophyletic group is also called a *clade*.

Next, we want to answer another fundamental question: to what other group(s) is the taxon most closely related? A *phylogeny* is the evolutionary history of a taxon or group of taxa. All taxa past and present are related at some level, and our task is to discover that "tree of life." It is assembled a few branches at a time by experts knowledgeable about particular groups of organisms. Developing a phylogeny is a search for ancestor–descendant relationships. It is a historical estimate depicted in the form of a branching diagram, or tree. Phylogenetic inference is a weighty, dense exercise impassioned with philosophical underpinnings and arcane technical procedures. Fortunately, one need not know all the methodology behind phylogenetic inference to "read" a tree and understand its power in organizing biological diversity.

The phylogenetic trees of some vertebrate lineages are included in the chapters that follow. Trees are designed to show patterns of descent not patterns of similarity. True, closely related species tend to resemble one another, but this is not always the case if the rates of evolution are not the same. Nor do trees imply specific ages, only relative ages based on recency of shared ancestry. The correct way to read a tree is as a series of hierarchically nested groups. Each branch point, or node, indicates a common ancestor. The node closest to a cluster of branched taxa indicates a more recent ancestor than a node further down the branches of the tree.

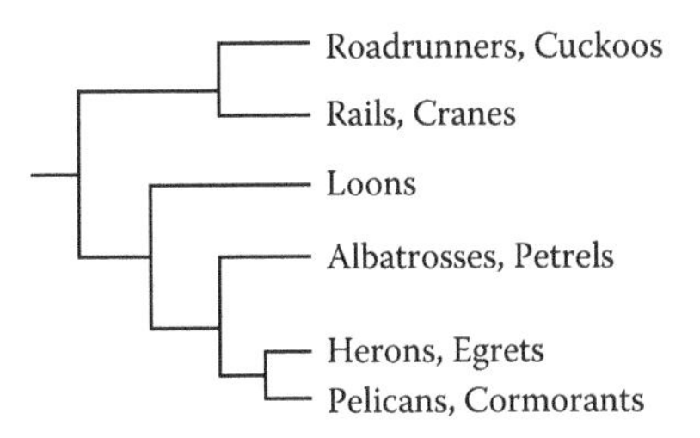

For example, the phylogenetic tree above (taken from Figure 10.3) shows that herons and egrets share a recent common ancestor with pelicans and cormorants (together comprising the Pelicaniformes). The Pelicaniformes themselves are most closely related to albatrosses and petrels. Put another way, the albatrosses and petrels are *not* more closely related to loons; rather, the closest relatives of loons are the albatrosses and petrels *plus* the Pelicaniformes (herons and egrets + pelicans and cormorants). More distant still are the roadrunners and cuckoos + rails and cranes. Notice also that it matters not if the taxa are oriented vertically as shown here or horizontally (rotate the image 90° counterclockwise); the information conveyed is the same. Trees are not absolute; precision will always be lacking; and trees will be revised as new information becomes available. Our understanding of the directions in which life evolved will always be inexact.

Classification is the organization of taxa in to inclusive, hierarchical groups. Ideally, a biological classification should be based on phylogeny, that is monophyletic lineages sharing common ancestry. In the Linnean system (first proposed by the Swedish botanist Carolus Linnaeus in the eighteenth century), species are grouped into genera, genera into families, families into orders, and so forth. From the standpoint of evolution, the Linnean rank category beyond the species level (e.g., order) is irrelevant, despite its prevalent use in biology textbooks. For one thing, what constitutes an order, say, may have a different meaning to an entomologist than it does to a mammalogist. What matters is not the rank category but the proper nesting of monophyletic, hierarchical lineages. For example, as vertebrates, mammals can be classified in an indented scheme as follows without applying rank names:

Vertebrata
  Gnathostomes (jawed vertebrates)
    Sarcopterygia (the fish ancestry of tetrapods)
      Tetrapoda
        Amniota
          Synapsida
            Mammalia

In this example, Mammalia is the most *exclusive* taxon and therefore subordinate to the group(s) above it (e.g., Synapsida). Vertebrata is the most *inclusive* taxon. Put another way the criteria for group membership is more restrictive for mammals than it is for Vertebrata. Linnean categories are still widely used, and, in this book, I largely follow that convention for ease of reference.

4.2 DISTRIBUTION

The natural historian is innately curious about the geographic patterns of plant and animal distribution in time and space. All species evolve somewhere, and *biogeography* attempts to ascertain where that was and what if any spatial changes have happened since then. Some of the basic questions asked by biogeography are as follows:

1. How did a species come to occupy its present range?
2. Who are the species closest relatives and where are they found?
3. Why are some pairs of closely related species found in adjacent areas, whereas other such pairs are widely separated?
4. How does species diversity change when ascending a mountain, say, or moving inland from the coast?
5. How have past events in earth history, such as glacial cycles, affected species diversity and distribution?

Potential answers to these questions are formulated from many sources including systematics, geology and tectonics, paleontology, climatology, and ecology. The process begins with determining a species present range.

A few vertebrate species are cosmopolitan, occurring nearly worldwide, but most occupy a limited part of the earth's surface where they are endemic to a particular region or area. Endemism is more or less synonymous with "native" or "indigenous" except that it has biogeographic implications at a variety of spatial scales. A species may have originated in one place and never dispersed, or perhaps now survives in a small part of its former range. Good examples of the latter are the four species of desert pupfish (*Cyprinodon*, *Siphateles*) inhabiting a few isolated localities in the Mojave and Colorado Deserts. They are relicts of a wider distribution in the wetter period of the Pleistocene when surface water was more plentiful. These pupfish illustrate how ecological circumstances permit the continued survival of endemics, yet knowledge of historical events is required to explain how that came about.

Like taxonomic categories, endemism is hierarchical. A family has a broader range than the individual genera that comprise it, and each genus is more broadly distributed than its individual species unless the genus is monotypic (composed of a single species). Seen this way, the American Southwest has a high degree of endemicity among, for example, phrynosomatid lizards and heteromyid rodents.

Most vertebrate field guides include maps depicting the present distribution of a species. The maps are assembled from museum records and direct observations in the field. Outline maps circumscribe the entire known range, or that part covered by the field guide if it is a regional account. For migrating species like many birds, outline maps show where the species occurs in winter, summer, and year-round. Although providing a useful, at-a-glance distribution, outline maps generally cannot accurately define the geographical limits of a species. Dot maps show actual locality records, but inevitably they reflect only a small sample of the population. Contour maps attempt to combine relative abundance in a three-dimensional graphic.

Range maps are simplifications and have limited accuracy, if for no other reason than it is impossible to map every individual. They are also scale-dependent: the larger the area covered the less resolution there will be. In all cases, field guide users have an advantage if they have working knowledge of the geography that the map covers.

Distributions oscillate in time and space and individuals are neither homogeneously nor randomly distributed. In most cases the majority of individuals occupy a relatively small part of the species' total range and tend to be clumped in some areas but fewer in number at the edge of the range. An examination of spatial patterns can often reveal which resources are affecting a species' distribution and its relative abundance. The familiar Western Scrub Jay, for example, is an opportunist common in urban settings (bird baths and feeders) throughout Southern California. Nonetheless, it is most abundant in oak woodlands and adjacent areas of chaparral where a primary resource, acorns, is located.

Now, all animals have some capacity to move from their place of birth. This usually occurs at a particular stage or time in their life cycle, for example as larvae or juveniles or later as adults. This process of dispersion is about the spatial distribution of individuals within a population. As an ecological process, dispersion reduces overcrowding and thus minimizes competition among siblings and adults. Most of this type of movement does not result in significant change in the distribution of the entire species. Biogeography on the other hand is interested in the historical phenomena that lead to range extensions or contractions. Many species never do extend their range significantly; they evolve and become extinct in the same place. On the other hand, some present-day distributions show disjunct patterns, where two or more populations are geographically isolated from one another. Species like some of the rodents and birds that occupy mountaintops in our Transverse and Peninsular Ranges show such disjunctions. The "sky islands" of Arizona and northern Mexico host numerous examples as well. These high elevation species are isolated by inhospitable terrain of the deserts below.

Disjunct distributions suggest either that (1) the species became extinct in the intervening areas where it does not now occur (like the desert pupfish mentioned previously), leaving the outlier population as a relic, or (2) the outlier represents a dispersal event under historically more favorable conditions.

A vicariant event is the historical erection of a physical barrier like a mountain range, the formation of a peninsula like Baja, the appearance of a large body of water, or a desert that divides a once contiguous distribution into two or more populations. The plants and animals involved with the event are passive participants. In the 1970s, the emergence of plate tectonics as a geophysical force drove biogeography toward vicariance explanations. One reason is that testable information is recoverable, unlike most dispersal hypotheses. Nonetheless, long-distance dispersal does happen, as for example by Pacific island birds, but it is generally rare and difficult to study. Successful long-distance dispersal across a formidable barrier, akin to winning the lottery, requires the ability to travel to a new area and to withstand potentially unfavorable or lethal conditions in route. If it is successful, then viable populations need to be established upon arrival. Diffusion is the slower expansion by populations over generations into areas with suitable resources. Familiar examples include Africanized honeybees that spread northward from Brazil; the spread of fire ants in the southern United States; the diffusion of European Starlings (*Sturnus vulgaris*) and House Sparrows (*Passer domesticus*) across the United States following their introduction to the East Coast in the nineteenth century.

Collectively, individual species of plants and animals comprise communities that live under the same environmental conditions. They are defined by structure and function, albeit somewhat arbitrarily. Large aggregates of communities form entities called *biomes*, *life zones*, or *biotic provinces*. Consider the two major biomes of Southern California. Deserts occur in low to intermediate elevations in hot dry belts around the world at round 30°–40° latitude. Rainfall is scant, seasonal, often unpredictable, and is always less than the evaporation potential. Scelerophyllous Woodland like our coastal sage scrub and chaparral consists of plants with hard, tough evergreen leaves. It is found in mild temperate climates with moderate winter rain as in Southern California and also the Mediterranean, South Africa, Chile, and Western Australia. Biomes are distributed along latitudinal gradients from the equator to the poles. In addition to Deserts and Sclerophyllous Woodland, they include Tropical Rainforest, Topical Deciduous Forests, Thorn Woodland, Tropical Savanna, Tropical Deciduous Forest, Temperate Deciduous Forest, Temperate Rainforest, Temperate Grasslands, Boreal Forest (also known as Taiga), and Tundra.

Biomes are shaped by history, geography (e.g., a continent, peninsula, or island) and the relative homogeneity or heterogeneity of the physical setting (e.g., mountainous terrain versus vast plains). Climate, however, is the most important parameter. Climate is determined by temperature extremes and precipitation, which in turn are a function of latitude, altitude, continental versus maritime geography, and seasonality. Latitudinal gradients in climate determine the world's biomes from the equator to the poles.

Widely separated but similarly appearing biomes do not necessarily share the same phylogenetic histories, however. Each one is composed of different species that, through convergence, respond to similar environmental influences and so come to resemble one another in physical form. Likewise, continental regions sharing the same latitude can differ markedly in the distribution of rain depending on geography and the patterns of global wind circulation. One region may receive precipitation throughout the year and another for only for a few months, resulting in quite different vegetation profiles. Zonation of biomes can also be observed attitudinally. From the base to the summit of high elevation mountain ranges biomes will replace one another in a fashion that is similar to changes in latitude.

Besides geography and climate, biomes are also shaped by major upheavals that occur over geologic time, such as mountain building and plate movements of the earth's crust. For instance, the evolution of the Baja California peninsula and the formation of Southern California's Transverse Ranges began around 5.5 million years ago and had an obvious and profound effect on our region's character (see Chapter 1 on Land Forms). The ebb and flow of glacial stages of the Pleistocene that began 1.2 million years ago is the most recent of large historical perturbations. During glacial maxima, the latitudinal zonation of Northern Hemisphere biomes was compressed, particularly in the mid-continent, and Southern California itself experienced cooler summers and a more mesic (wetter) environment overall. Too, with so much water locked in ice, sea levels dropped 100 meters or more. This allowed continental islands (like our Channel Islands) to approach or reconnect with the mainland or coalesce with themselves to create larger entities.

Overall, the physical environment is the single-most important factor in determining natural distribution, but it is not the only one. Nor is it discrete from the regional features that, on a smaller scale, determine where in that biome a species can exist under conditions that are favorable to it or at least benign. In other words, distribution is a function of the continuum of large (i.e., biomes) to smaller scale phenomena. Smaller, regional influences include temperature variation, moisture and humidity, substrate and cover, natural disturbances, and biotic interactions of the species themselves.

Temperature variations across the seasons can mean frost, freezing, or excessive heat. Temperature extremes usually are growth-inhibiting, and the plant communities and the animals that depend on them must be able to respond with seasonal dormancy or some other survival mechanism. If they lack such adaptations, they cannot inhabit the area. Temperature fluctuations and extremes are generally more important to ectotherms like lizards and snakes that have narrower thermal limits than birds and mammals. Moreover, temperature in itself may not be a limiting factor, but its effects on other resources could be (e.g., food availability for small endothermic vertebrates). Endothermy, maintaining a steady body temperature through metabolism, is energetically expensive. Small endotherms like most birds and many rodents must either migrate seasonally to other areas (birds) or greatly reduce their activity during temperature extremes. Interestingly, among mammals, more large-bodied species are found at higher latitudes then small-bodied ones. Carl Bergmann first made this general observation in the mid-nineteenth century, and it became known as *Bergmann's Rule*. Bergmann suggested that large mammals have a lower (body) surface-to-(body) volume ratio and thus are better able to retain metabolic heat. Nowadays, surface-to-volume ratio per se is not generally regarded as the causal mechanism of increasing body size among mammals toward the poles. However, that is not say that it is unimportant, only that other physiological issues come into play.

Moisture and humidity gradients clearly affect the distribution of plant species as well as the density of individuals. In Southern California and the Southwest, generally the arid climate forces concessions for living in a dry, xeric environment. The xerophytic plants that dominate our region have numerous adaptations for conserving water. These adaptations, reflected in the various plant communities, evolved according to moisture and humidity gradients from the coast, to the mountains, to the deserts. Vertebrates, especially small ones, favor some of these communities over others.

Substrates and cover affect a spectrum of needs. Soil type, for instance, is often critical for plants. For small vertebrates, cover equates to shelter, which can mean soil that is conducive for constructing a burrow or a suitable abundance of rocks, logs, bushes, or trees that serve as hideouts and places to forage and feed. Substrate and cover can at times be the single-most important factor affecting distribution. A good example is the strong correlation of the Granite Night Lizard with outcrops and exposures of granitic rocks in Southern California and northern Baja.

Unlike historical upheavals over geologic time, recurring natural disturbances have less influence on distribution per se than they do on population densities. Natural disturbances include irregular but inevitable events like Atlantic hurricanes, and in our region, fire. Some plant species like certain pines (cones and seeds) are dependent on periodic burning, and many chaparral plants have strategies for recovery after fire. For small vertebrates, here is a case where a retreat and suitable cover could be a lifesaver.

Apart from the physical environment, distribution also can be influenced by any number of biotic factors. These include nutrition, meaning food quality and its abundance, as well as specialized diets and the foraging tactics used to locate food. Another is the density of competing species and individuals of the same species that seek out similar resources such as nesting and denning sites. Nesting requirements can, like some substrate essentials, restrict a distribution if specific types of burrows, tree cavities, or old growth forest are necessary.

4.3 SURVIVAL

This third pillar of natural history considers basic ecology, or how animals make a living. From an evolutionary standpoint, they need to accomplish only one thing: stay alive long enough to reproduce. In doing so, they face an array of challenges, many of which can also affect a species' distribution, and where within its range the individuals have the best chances for success. Survival tactics can be studied in their own right, and often are, independent of their potential role in distribution. Food and sex are at the top of that list of essentials, followed by predator avoidance.

4.3.1 Feeding and Foraging

The diets of most vertebrates are known in a general way. Carnivores are meat eaters, herbivores eat plants, and omnivores like humans consume other animals as well as plant matter. Within these domains we find numerous subdivisions like carnivores that specialize on, say, insects (insectivores), fish (piscivores), or vertebrate carcasses (carrion feeders). Among herbivores are the fruit eaters (frugivores), seedeaters (granivores), grass feeders (graminivores), and nectar specialists (nectivores) to name some possibilities. Still, the full dietary spectrum of any one species is elusive, and many solid natural history papers deal with that subject alone. Data can come from field observations (including catch-and-release after pumping the stomach), museum specimens, and the analysis of scats (droppings).

Vertebrate diets correlate with anatomical specializations such as jaw and tooth structure and gut morphology. Large mammalian herbivores like sheep and deer have evident adaptations for cutting, chewing, and processing the tough cellulose of plant stems, leaves, and dense seed coats. Grinding molars pulverize the food. The long, complex alimentary canal (gut) converts the chopped plant material into a nutritious meal. Seed-eating birds such as sparrows and finches have evident conical bills. Those of hawks and other raptors use their curved bills to tear flesh. Nectar-feeding birds have long thin bills and agile tongues. Mammalian carnivores have specialized jaws and teeth that function like serrated scissors for slicing meat.

Remember that communities are assemblages of species that interact with one another. The most important dynamic is the trophic structure formed by the interacting participants. Trophic structure, popularly known as a food chain, is the transfer of food energy from plants, the primary producers, to the consumers (primary, secondary, tertiary, etc.). For example, a grasshopper (primary consumer) feeding on a leaf is eaten by a lizard or a rodent (secondary consumers), which is eaten by a snake (tertiary consumer) that in turn is preyed on by a hawk (quaternary consumer).

Food chains usually have only about four or five links. One reason is that only around 10% of the energy store in one trophic level is transferred to organic matter in the next level, thus setting a limit on the number of apex predators. Another limitation is that long chains are inherently unstable because natural population fluctuations at lower trophic levels are exacerbated at the higher ones, potentially eliminating top carnivores from the chain.

Food chains do not operate as isolated units. They are linked together as food webs based on who eats whom. Many insectivores will feed on almost any arthropod, but others are specialists, preying on a particular type; horn lizards, for example, have a fondness for ants. How do horn lizards deal with formic acid buildup from a diet of ants or manage the venomous sting of harvester ants? These and other questions extend the topic of nutrition into areas beyond dietary preferences. Some, like the issue faced by horn lizards, are physiological; others are behavioral. Among carnivorous vertebrates, we want to know how prey is located. What senses are involved (visual, acoustic, and olfactory)? How is prey subdued: by ambush or pursuit, solitary or in a group?

4.3.2 Reproduction

Reproduction replaces the individuals lost through natural mortality. In that sense only two offspring per parental generation need to survive to maintain a static population size. Yet reproducing is more than a matter of numbers. It also maintains genetic and phenotypic diversity in a population, and as a behavioral phenomenon it reinforces species integrity. Potentially, any member of a species can reproduce with any other member. However, panmixis is seldom the rule. Individuals tend to breed with their neighbors. To minimize inbreeding, local immigration and emigration provide a degree of randomness in mate choice. Reproductive biology is a complex subject that covers the following topics:

Frequency: Among vertebrates, mating generally takes place once a year but multiple matings are possible in some species. In long-lived species, mating may be every other year. Reproduction is energetically expensive and therefore sets limits on frequency. The costs are high in females; not so for males. Sperm is cheap, and there is generally less male involvement all around in the reproductive effort. Breeding is seasonal in most vertebrates so as to coincide with resource opportunities.

Courtship: This is the first behavioral response following hormonal arousal. It begins with the sexes locating one another and developing a pair bond if the female is receptive. The courtship behavior itself can be a minimal series of cues to elaborate rituals involving voice and posture. In many vertebrates the sexes are distinct from each other in size (dimorphic) or color and pattern (dichromatic). In numerous bird species, for instance, brightly colored males advertise that they are reproductively primed. Usually the females determinc if mating will proceed based on the perceived quality of the male's courtship.

Mating schemes: Each reproductive effort is guided by the distribution of nesting and denning sites and the overall sexual strategy that includes the availability of mates and whether the scheme is monogamy or polygamy (one sex has several partners). Monogamy may be for part of a season, all season, or a lifetime. For example, if all females in the population are receptive simultaneously, a single male would have difficulty dominating because he can only mate with one female at a time. If, on the other hand, females become receptive over a period of weeks then the individual male has the opportunity to mate more than once during the breeding season. Polygyny (one male mates with multiple females) is fairly common among birds and mammals, such as in resource-limited habitats where a male dominates through display or defends a harem. Among birds there are instances of a female having multiple male partners (polyandry). This usually occurs where there is a need for multiple, rapid clutches; the males incubate the eggs.

Mating: Is fertilization external (most amphibians) or internal? If internal, how is sperm transfer accomplished?

Sex determination: In most mammals, males have a pair of sex chromosomes denoted X and Y. Females have homologous X and X chromosomes. During the production of gametes (eggs, sperm) the cell division process known as *meiosis* results in cells (gametes) with half the number of chromosome sets as the original cell. Mammalian sperm cells have either the X or Y chromosome. Therefore, males determine the sex of the offspring. A gene on the Y chromosome initiates male gonadal development. If the gene is absent, female gonadal development takes place. Once the sex is established, the sex hormones are produced, either testosterone (male) or estrogen (female). These hormones are instrumental in developing secondary sex characteristics.

In birds, the sex chromosomes of males are designated ZZ, in females ZW. Females therefore determine the sex. The presence of the W chromosome causes the embryonic gonad to produce estrogen and the left ovary and oviduct. Generally only the left ovary and oviduct develop in birds.

In some reptiles, sex chromosomes (ZZ and ZW as in birds) are known. However, in many species, sex is determined by temperature during the middle third of the incubation period. Known as temperature-dependent sex determination (TSD), three pathways have been identified. In Type A, males are produced at higher temperatures, females at lower temperatures (most crocodilians and some lizards and snakes). In the Type B pathway, females result from higher temperatures and males at lower temperatures (many turtles), and in Type C females are produced at high and low temperatures and males at intermediate temperatures (three crocodiles, one lizard, a few turtles). The adaptive significance of environmental sex determination (incubation temperature) is complex and still being investigated.

Within amphibians there is no uniform type of sex determination. Heterogametic sex chromosomes are known in a few cases (either XY/XX or ZZ/ZW) but for most amphibians, the mechanisms of sex determination are unclear.

Nest site selection: In many birds, amphibians, and some mammals, nest site selection may precede actual mating. Key factors include resource availability, competition, and predator avoidance.

Fecundity: This refers to the number of offspring produced, that is, the number of eggs in a clutch or newborns in a litter. Some amniotes (reptiles, birds, and mammals) are determinate in that there is a fixed number. Fecundity often correlates with embryogenesis—the period of time from fertilization until the offspring is born or hatches. A longer period of embryogenesis typically sees a decline in fecundity.

Parental care: The extent and duration of parental care varies from almost none to both parents rearing and seeing after the young for many months. The degree of parental care is often correlated with fecundity and the length of embryogenesis. It is highest in birds and mammals.

Survivorship: This is the measure of reproductive success. How many offspring survived to become parents themselves? In general, higher survivorship of individual offspring accompanies an increase in parental investment. Yet this can be a risky and expensive strategy. If the ultimate objective is to maintain a stable population size, the alternative is to produce large numbers of offspring with less investment so that the odds favor that at least some will become reproducing adults.

4.3.3 Predator Prey Relationships

Referring back to trophic structure, it should be obvious that most predators except for the top carnivores may become prey themselves. Therefore, predator-avoidance mechanisms evolve with the tactics of the predator; adaptation and natural selection are unlikely to act on one independent of the other. Predation falls disproportionately on the young along with the weak and infirmed, yet all individuals possess the antipredator defenses characteristic of the species. Some of the defenses are specific to a particular species, but more commonly they apply to a group of species because phylogenetically closely related prey tend to share similar predators. Other more basic defenses are common among vertebrates generally.

Among the trends in antipredator defense are that if something can be used as a defense mechanism it will be, and the most effective defenses are those that stop the predator sequence early. We also see a shift from passive to active mechanisms in later stages of the predation sequence, and those mechanisms that work most efficiently in later stages are noxious, injurious, or distasteful.

Likewise, different sensory modes may be used at different stages (e.g., vision and hearing early, chemo, and tactile late).

Predation follows a sequence of often-stereotypical events by both the predator and the prey. Detection is the first of these. The predator relies on its primary sense (vision, olfaction, or auditory) to recognize the prey as an object distinct from its background, whereas the prey species is responsible for avoiding detection in the first place. The prey does this by some form of concealment such as a cryptic color and pattern (camouflage), by limiting its activity, or simply by being rare and uncommon.

Next the predator makes an identification: is the object edible or not? Here, as in avoiding detection, crypsis may be the prey's best defense. An alternative strategy is to be unpalatable or noxious and to advertise oneself as something to avoid. Skunks with their bold black-and-white pattern are a familiar example. Another clever ploy is to mimic the color and pattern of another species that the predator species has learn to shun. Mimicry is fairly common among arthropods, less so in vertebrates. Some of the better-verified examples are certain harmless Neotropical snakes that mimic the color and pattern of a noxious or lethal snake species in their region.

If the predator commits to an attack, it is either by a sudden lunge from an ambush position or by a slow stalk and a rapid close. At this juncture the prey has little choice but to flee. This might be an all-out sprint or evasion by quick, erratic movements.

If captured, the prey then must rely on matters of last resort. These can be shear strength or some mechanical, structural, or other behavioral defense. Many salamanders exude noxious mucous secretions over their skin. Turtles withdraw into their shells. Many lizards can shuck their tail, which keeps wiggling to distract the predator while its owner makes good her escape. Other vertebrates fight back with their jaws, claws, a bad taste, or by emitting a foul odor.

A predator's success rate is associated with its place in the trophic structure. Those species at the top typically have a lower capture rate than those species occupying a lower link in the food chain.

Section II

Settings

5 Vegetation Past and Present

5.1 CENOZOIC VEGETATION AND GLACIATION

Southern California's plant communities and the animals that inhabit them evolved with mountain building, a cooling and drying trend over the past 30 million years, the onset of glacial and interglacial cycles 2 million years ago, and the arrival of western Europeans. Floristic history is inferred from fossils that consist mostly of the durable parts of plants such as pollen, spores, seeds, cones, hard fruits, and leaves. The informative soft tissues of flowers are, unfortunately, rarely preserved. As is the case with animals, the fossil record for plants is not without potential bias. It needs to be interpreted carefully.

A single fossil-bearing sediment sample is a static view of one particular place in time. To trace historical changes in diversity, comparison is required with other samples that are separated temporally and spatially. Prudence and experience are needed here. Consider that because most preservation occurs in sediments deposited along lakes and flood plains, the fossil record may disproportionately sample species that favor wet habitats. These deposits may not necessarily even include the most common species in the area at the time. Secondly, plants that produce cones or seeds resistant to decay are more likely to be preserved over those that do not. Finally, misinterpretations can arise if there is an over-reliance on the composition of modern plant communities as a basis for extrapolating climates and assemblages of the past.

Volcanic ash deposits are another source of plant fossils that have the potential to sample a much broader area than lake or flood plain sediments. Yet volcanic eruptions are not continuous through time, and their contribution to an extended history is limited. This does not mean that we are hopelessly frustrated in the effort to uncover the past. Some areas of the West have produced a record rich in fossil plants that in total reveal images of major vegetation regimes that have come and gone and whose vestiges are detectable in Southern California's remarkable plant life today.

5.1.1 Wet Forest

In the early Cenozoic 65 million years ago, most of western North America was steeped in a mild wet climate. Seasonal changes were practically negligible. Compared with today, the winters were warmer and the summers wetter, especially in the interior. The diversity of plant species exceeded anything known elsewhere on the continent. First appearing were lineages of some of California's modern forest icons, such as fir, oak, pine, hemlock, giant sequoia, redwood, spruce, alder, and rhododendron. Other taxa present at the time exist today only in wet regions of Asia (e.g., *Gingko*) or in eastern North America (e.g., beech and sweetgum). The vegetation was not homogeneous but, like today, distributed in particular associations that responded to regional climate, topography, and soil. This rich flora, which persisted for several millions of years, was the source of at least 50% of California's modern genera and species.

Beginning in the late Eocene 35 million years ago, a cooling-and-drying trend advanced across the continent, and the western rainforests of the early Cenozoic contracted northwesterly toward more favorable refugia. By the early Pliocene 5 million years ago, a reduction of summer rain was becoming evident, and the remnants of these forests endured only in moister upland sites and stretchers of the humid coast. The survivors adapted to a climate of winter rain and summer drought. Further differentiation of species and phenotypes followed with the uplift of the Sierra Nevada, the Coast Ranges, and in Southern California, the Transverse and Peninsular Ranges.

5.1.2 Tropical Savannah

At its southern extent, the dense coastal rainforest of the early Cenozoic merged with a tropical savannah. This southern flora consisted of, for example, cycads (*Cycas*, *Dioon*), fig (*Ficus*), fan palms (*Sabal*), various *Euphorbia*, and other broadleaf tropical evergreen angiosperms (flowering plants) such as camphor trees and avocado (*Cinnamomum*, *Persea*). The nearest counterparts of that flora today are found in southern Mexico and in the middle of the United States, the West Indies, and parts of southern Asia. Rainfall of 80 inches was normal. The same cooling and drying that affected the northern forests at the end of the Eocene forced the retreat of this tropical element ever southward. Most of the California constituents disappeared by the end of the Miocene 5 million years ago. A few (like some palms) persisted to the end of the Pliocene only to vanish when summer rains were effectively eliminated. No direct descendants of this humid, subtropical flora occur in California today, although members of some current genera are thought to be secondary derivatives.

5.1.3 The Madro-Tertiary Geoflora

A third vegetation element of the early Tertiary (Paleogene) developed in southwestern North America. Its expansion and evolutionary fate had the greatest impact on Southern California as a floristic unit today. Because most of the surviving members of that element persist in the Sierra Madre of Mexico, it was given the name "Madro-Tertiary Geoflora" by Daniel Axelrod. (A Geoflora, or paleoflora, is a vegetational unit that maintains its identity in time and space.)

This flora was composed of drought-deciduous, sclerophyllous (small, leathery leaves) plants that were a dry-tropical to warm-temperate derivative of more mesic tropical species. It may have also included taxa that evolved from components of the northern forest in the middle and later Tertiary. The Madro-Tertiary Geoflora adapted to dry climates and, beginning in the middle Eocene, spread over the Southwest from the vicinity of the present-day southern Rocky Mountains. Several major vegetation types and subunits were represented, for example, thornscrub, oak-conifer woodland, chaparral, desert grasslands, sagebrush, and piñon-juniper woodland. Some familiar constituents were Madrone (*Arbutus*), Cypress (*Cupressus*), Locust (*Robinia*), Cottonwood (*Populus*), and oak (*Quercus*). Also present were Walnut (*Juglans*), Digger and Piñon pine (*Pinus*), Lyontree (*Lyonothamnus*), Bay (*Umbellularia*), Mountain Mahogany (*Cercocarpus*), Bitterbush (*Purshia*), Cherry/Plum (*Prunus*), and Cream Bush (*Holodiscus*).

This was an admixed flora that differentiated regionally according to climate and altitude. With the reduction of summer rains in the late Pliocene, most Sierra Madrean species and their community configurations were eliminated from the far West. However, some persist in the Southwest today from southeast Arizona to west Texas and northern Mexico, regions with ample summer rain. In California, those adapted to rainless summers and lower temperatures held fast, for example Piñon-Juniper on cold, semi-arid desert slopes. A pine woodland (Foothill Pine, *P. sabiniana*, also called Digger Pine) matured mainly in low elevations of northern and central California and, as a more recent adventive, to the Southern California mountains.

The lowland flora of Southern California differentiated in the late Pliocene/early Pleistocene with, for example, the relatives of California Walnut, Lyontree (now restricted to the Channel Islands), Laurel Sumac (*Melasoma laurina*), Sugar Bush (*Rhus ovata*), Palmer's Oak (*Quercus palmeri*), and numerous chaparral taxa.

Chaparral of the Madro-Tertiary Geoflora consisted of a "California element" mixed in with plants of a "Southwest element." Together they interfaced in places with subtropical thornscrub. The Southwest chaparral element is now restricted to southern Arizona and adjacent parts of Mexico. The California element included Manzanita (*Arctostaphylos* spp.), California-Lilac (*Ceanothus* spp.), Mountain Mahogany (*Cercocarpus* spp.), Tree Poppy (*Dendromecon* spp.), Silk Tassel Bush (*Garrya* spp.), Toyon (*Heteromeles arbutifolia*), Cherry (*Prunus* spp.), Oak (*Quercus* spp.), Buckthorn (*Frangula* and *Rhamnus*), and Sumac (*Rhus*). The presence of chaparral relatives in the Pliocene of eastern California, Nevada, and Oregon demonstrates that the present-day area occupied by California chaparral developed late, after the cessation of summer rains. Along with the expansion of dry climates, the near simultaneous rising of the Sierra Nevada and the Transverse and Peninsular Ranges brought even drier conditions leeward, and the desert flora came into existence.

Chaparral ranged widely in the late Tertiary (Neogene) of the Southwest, but the subtropical thornscrub with which it was associated was largely confined to warmer regions of Southern California and areas to the east and south. It included species whose nearest relatives now comprise the subtropical scrub vegetation of northern and central Mexico: for example, Acacia (*Senegalia, Vachellia* spp.), Elephant-tree (*Bursera* spp.), Nakedwood (*Colubrina* spp.), Hopbush (*Dodonea*), Fig (*Ficus* spp.), Feather Bush (*Lysiloma* spp.), Mesquite (*Prosopis* spp.), and Indigoberry (*Randia* spp.). With the expansion of dry climates, much of the thorn forest was eliminated from Southern California except for a few survivors in the southeast deserts like Mesquite, Acacia, Elephant-tree, and Condalia (*Condalia globosa*).

5.1.4 Glaciation

Following sporadic cooling episodes beginning 5 million years ago, major glacial events began in earnest in the late Pliocene about 1.8 million years ago. In North America the Ice Ages consisted of four major glacial (cold)–interglacial (warm) cycles, and as many as 48 shorter episodes that were imposed on them. Major glaciations lasted about 40,000 years on average although some were much longer. The last major glaciation, called the *Wisconsinan*, began 120,000 years ago and lasted 100,000 years before beginning its retreat 18,000 years ago. We are now in an interglacial period, the Holocene, which began 12,000 years ago. The Wisconsinan experienced at least six of the shorter warm-cold cycles during its lifetime. The last of these (the Tioga cycle) occurred from 26,000 to 20,000 years ago.

Glaciation had a profound impact on the distribution (latitude and altitude) of California's plant communities. Some species presently restricted to higher latitudes extended farther south during glacial events. In the moist coastal zones, Douglas-Fir, Monterey Pine, and Redwood moved as far south as the Los Angeles Basin. Inland, away from the moister coastal zone, more drought-tolerant conifers like Ponderosa and Sugar pines expanded their range. In the mountains, conifer and oak woodlands migrated down slope. Piñon and Juniper displaced chaparral and live oak in the eastern Transverse Ranges as well as the Peninsular Ranges. On the east-facing slopes, sagebrush scrub advanced into higher elevations despite the cooler temperature, indicating that summers were dry. Many of Southern California's rare, montane species, perhaps 300 in number, are Sierra Nevada relics that moved southward during glacial periods, became isolated, and adapted. Others moved upward from lower elevations as warmer, drier conditions followed glacial retreat. With plant migration occurring north and south as well as up and down slopes, opportunities were created for hybridization and the appearance of novel adaptations specific to their terrain, soil, and microclimates.

Temperature cycles were not the only expressions of glaciation as they affected plant communities. Pluvial (wet) periods resulted in an array of large freshwater lakes in the Great Basin, Mojave Desert, and Salton Trough; regions that are bone dry today. For instance, Owens Lake south of Lone Pine was 250 ft higher and spilled into the basins of China Lake and Searles Lake, both of which are now dry lakes of the northern Mojave. Searles Lake then overflowed into the Panamint Basin to form a lake 900 ft deep. Lake Panamint overflowed into Death Valley producing a lake 190 miles long and 600 ft deep. There were also basin lakes in the San Joaquin Valley and in the Salton Trough of extreme southeast California, such as Lake Cahuilla that covered an area approximately 3,000 mi^2 to a depth of over 300 ft. All of this expanse of freshwater helped to mitigate temperature and promote the association of moisture-loving plant communities.

5.1.5 Sea Levels

With so much water locked up in ice during glacial episodes, sea levels dropped about 360 ft (110 m) at glacial maximum of the Wisconsinan 26,000–20,000 years ago. As a result, large areas of the continental shelf off the California coast were exposed. This, in turn, created opportunities for the expansion of coastal plant communities. The northern Channel Islands combined into a single, larger island called "Santarosae Island" with its easternmost point a scant couple of miles from the mainland. Several coastal plant taxa that are now rare or restricted, for example Torrey Pine in the northern San Diego County and Santa Rosa Island and Gowan Cypress isolated in patches on the Monterey Peninsula, are both relics of broader distributions during this most recent period of lowered sea level.

The most recent chapter in Southern California's botanical history began 450 years ago with the arrival of Spanish colonists and other Western Europeans who followed them. The introduction of livestock, non-native grasses, and other plants like several species of mustard (*Brassica*) was quick and severe. The ensuing agricultural transformation and urbanization has continued unabated ever since. Huge tracts of the original native vegetation no longer exist.

5.2 MODERN PLANT ASSOCIATIONS

California is home to about 5,800 species of vascular plants, a number that represents 32% of all plant species in the United States. Of these, 2,150 species and subspecies are endemic; they are found nowhere else. The sheer size of the state alone, 156,000 mi^2, accounts for some but not all of this richness. Also contributing are the 10° of latitude, adjacency to the Pacific Ocean, topography, climate, and soil varieties. These all result in a mélange of macro and micro conditions for plants and the communities they define. How can we organize this complexity?

The *Jepson Manual* on *Higher Plants of California* employs a hierarchical system of geographic units from provinces to regions, subregions, and districts. The classification is logically based on naturally defining features of the landscape, which in turn broadly encompass vegetation types. Of the three recognized provinces, the most extensive is the *California Floristic Province* (abbreviated

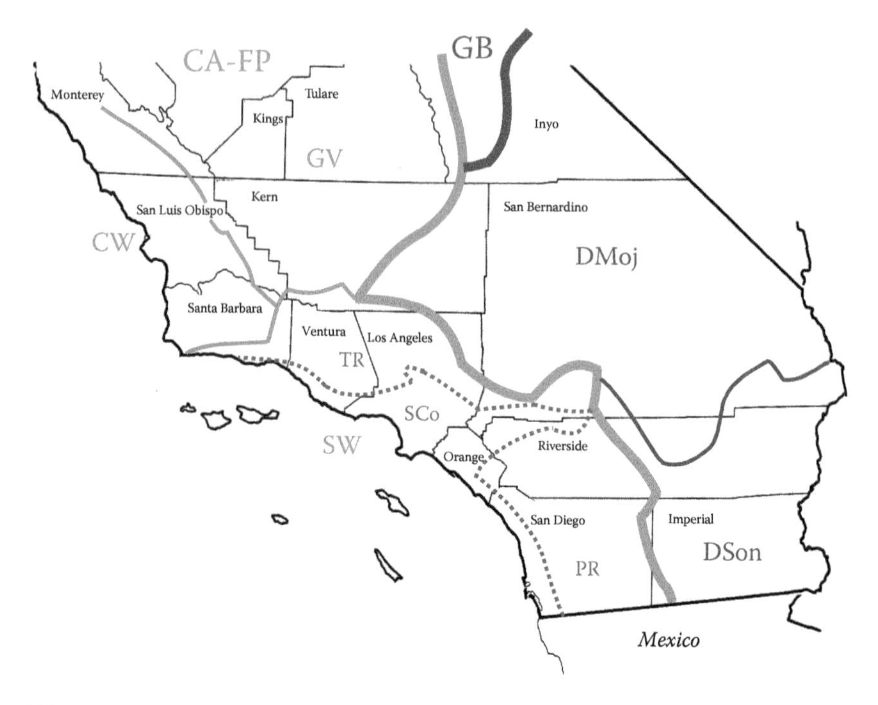

FIGURE 5.1 California's three floristic provinces replicate major geographic features. Most of California consists of the California Floristic Province (CA-FP, green) with six regions, three of which are identified on this map: Great Central Valley (GV), Central Western California (CW), and Southwestern California (SW). Regions are divided into subregions (outlined in blue). Southwestern California, for example, is made up of four subregions: the South Coast (SCo), Transverse Ranges (TR), Peninsular Ranges (PR), and Channel Islands (not labeled). The Desert province (brown) consists of two regions, the Mojave (DMoj) and Sonoran (DSon). The third province, the Great Basin (GB), is a minor component in California compared with the CA-FP and deserts. (Data from Hickman, J. Ed., *The Jepson Manual—Higher Plants of California*, University of California Press, Berkeley, CA.)

CA-FP) that includes six regions. The *Great Basin Province* (GB) and the *Desert Province* (D) each include two regions (Figure 5.1).

5.2.1 California Floristic Province

The California Floristic Provence extends from southwest Oregon around the Klamath Mountains to northern Baja California, Mexico. It subsumes all the state west of the Great Basin and the deserts. Its six regions are:

Northwest California (three subregions)
Cascade Ranges (two subregions)
Sierra Nevada (three subregions)

Great Central Valley (two subregions)
Central Western California (three subregions)
Southwestern California (four subregions)

Southwestern California region: This region borders the Pacific Ocean (beyond the Channel Islands) on the west and northern Baja to the south (Figure 5.1). It therefore embraces most of Southern California west of the deserts. Its northern limit is the crest of the Santa Ynez Mountains where they transition with the southern end of the Central Western Region (i.e., the central coast) and the grassland/woodland interface of the Great Central Valley Region and the Tehachapi Mountains (itself a Sierra Nevada subregion but with much floristic mixing from the Transverse Ranges, lower Sierra, Central Valley, and desert). The Transverse and Peninsular Ranges define the northeastern and eastern border of the Southwestern California Region with the Desert.

The four subregions are:

South Coast subregion: This subregion extends from around Point Conception to northern Baja California. It extends inland to San Gorgonio Pass at Banning where it transitions to desert. To the south, its eastern limits are the foothills of the Peninsular Ranges. The South Coast subregion is composed of narrow, discontinuous bands of coastal strand and salt marsh, and broader areas of coastal sage scrub and lower chaparral. Most of the native vegetation in this subregion has been lost to urbanization and agriculture.

Channel Islands subregion: Although floristically similar to the South Coast subregion, there is a sufficient diversity of endemics to justify status as a separate entity. It is divisible into two districts. One encompasses the northern Channel Islands of San Miguel, Santa Rosa, Santa Cruz, and Anacapa, which geologically are western peaks of the Santa Monica Mountains. The other district is geologically and floristically more isolated and includes the southern islands of Santa Barbara, San Nicholas Santa Catalina, and San Clemente.

Transverse Range subregion: The east-to-west alignment of the Transverse Ranges is at odds with most of California's north–south geological orientation. The San Gorgonio Pass separates the eastern end (i.e., the San Bernardino Mountains) from the northern end of the Peninsular Ranges (San Jacinto Mountains). The San Gorgonio Pass is also the interface between the South Coast subregion and the desert province. The lower slopes of the Transverse Ranges are largely chaparral. Higher up are the oak forests and montane forests of White fir and pine (Jeffrey, Sugar, and Lodgepole). The highest peaks (e.g., Mt. San Antonio, or "Old Baldy") are treeless. The three districts of this subregion are physiographically discreet and progressively higher and dryer from west to east.

A western district comprises Mt. Pinos and the ranges of Santa Ynez, Topatopa, Santa Suzana, Santa Monica, Liebre, and Sierra Pelona.

The next district east is the San Gabriel Mountains, which is separated by Cajon Canyon from the San Bernardino Mountains, which is the third and easternmost district of the Transverse Range subregion. The Little San Bernardino Mountains are included in the desert Province.

Peninsular Range subregion: All terrain south of the Transverse Ranges that is not part of the South Coast subregion comprises this unit. It therefore includes Mount Palomar and the Santa Ana, Santa Rosa, Cuyamaca, Laguna, and Jacumba Mountains. The San Jacinto Mountains are a separate district within this subregion because they support unique alpine plant species at the highest elevations. As with the Transverse Range subregion, the foothills of the Peninsular Ranges are largely chaparral. Oak and pine woodland occur at higher elevations.

5.2.2 Great Basin Province

The Great Basin defines much of the intermountain west between the Rocky Mountains and the Sierra Nevada and the Cascades. Hence its excursion into California is limited to the eastern and northern two-thirds of the state. Most of Modoc and Lassen Counties in the extreme northeast is a high plateau of juniper woodland and sagebrush steppe with enclaves of Ponderosa and Jeffrey Pine forest. South of Lake Tahoe, the Great Basin province varies considerably in elevation from around 3,300 ft (ca. 1,000 m) at Owens Lake to over 14,000 ft (ca. 4,250 m) in the White Mountains east of Bishop. Sagebrush associations, coniferous forest, and riparian communities are characteristic. At its southern extent in California, the province transitions to the creosote bush communities of the Mojave Desert.

5.2.3 Desert Province

The Desert province, where vegetation is largely dominated by creosote scrub, occupies Southeast California. To the north the Desert province is bordered by the southern end of the Great Basin and to the west lays the CA-FP. The two subregions are the Mojave Desert and the Colorado–Sonoran Desert. The Mojave accounts for two-thirds of the province and generally exhibits greater extremes of temperature and elevation. Wide, low valleys separate several impressive mountain ranges. Many are high enough to support Piñon–Juniper woodlands. The most prominent are the Grapevine, Last Chance, Panamint, Coso, Argus, Kingston, Clark, Ivanpah, New York, Providence, Old Woman, and Little San Bernardino Mountains. The Panamint, Kingston, Clark, and New Yorks also support white fir. Some of the eastern mountain ranges of the Mojave sustain plants not otherwise known in California.

The Colorado–Sonoran Desert lies southeast of the Mojave, although the transition is not abrupt. The Colorado–Sonoran is lower and flatter, and although characterized by creosote bush scrub like the Mojave, it does have some floristic peculiarities of its own.

6 Plant–Vertebrate Communities

Recall that the floristic provinces along with their regions and subregions are delimited by broad scale geographical criteria. Plant formations or vegetation types offer another approach to organizing plant diversity. Regions are defined by the physiognomy or structure of the species themselves. Thus, we can speak of coastal sage scrub, oak woodland, grassland, chaparral, mixed hardwood forest, or coniferous forest scrub. Plant formations are somewhat similar to ecosystems except they do not deliberately entertain such abiotic factors as soil and moisture, or other biotic factors like the animal species that use them. If they did it would be more appropriate to refer to them as *biotic communities*. Of course, topography, soil type, moisture availability, and temperature extremes largely dictate which plants (and animals) will occur where, along with the structural characteristics that they exhibit. At another level, vegetation types can be treated as a *community*, an *alliance*, or an *association* based on the dominant plant species present, either in stature or abundance. The point to keep in mind is that plant formations are a hierarchical continuum from large scale to small, regardless of any category or rank that is applied along that continuum. Some botanists might argue that certain categories only split hairs, whereas others see value in formally recognizing subtle variation.

California's complex microclimates and diverse topography mean that plant communities are more often than not difficult to define and characterize precisely, at least their distributional limits. Most are not geographically uniform but discontinuous and dissected. Some will occur as isolated islands within another. North-facing slopes may be woodland, whereas south-facing slopes will be composed of drier-adapted species. Discontinuity is especially pronounced at community margins. Transitional areas are numerous, and the recognition of certain plant communities can be subject to varied interpretation. That said, a plant community will consist of species more or less restricted to it but may include some from an adjacent community. A number of California's plant species exhibit a high degree of polymorphism, meaning they have evolved adaptive "ecotypes" (physical characteristics) depending on where they grow. Still others, like manzanita (*Arctostaphylos*) are represented by dozens of discrete species found in a variety of situations.

The following sections provide general descriptions of Southern California's biotic communities (plants and associated vertebrates) in a west-to-east sequence, that is from the coast, to the foothills, to the mountains, and in to the deserts (Figure 6.1). You will note that numerous species of vertebrates range across several types of communities, especially in the coastal, foothill, and mountain habitats.

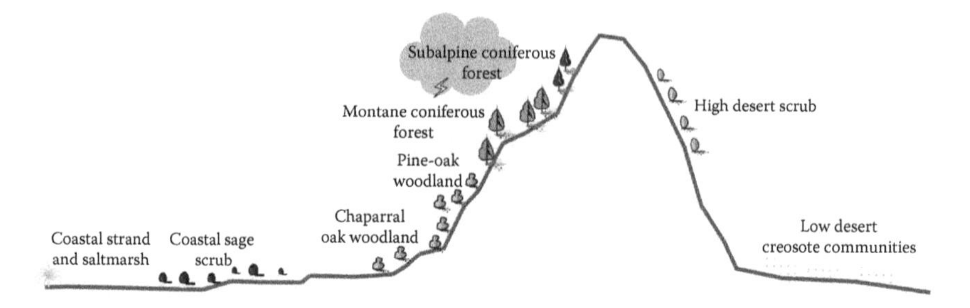

FIGURE 6.1 This schematic profile illustrates the relative distribution of the biotic communities as a function of altitudinal zonation. Because winter storms approach the coast from off the Pacific, the west-facing slopes of the mountains capture most of the moisture and create a rain shadow that effectively deprives the deserts from receiving significant accumulation.

6.1 COASTAL STRAND

Sandy beaches and dunes comprise the interface between sea and land. Neither in time nor place are they static. Sand accretes from bluffs and watersheds as the beaches themselves are reshaped and eroded by wind, rain, and tides. The plant communities associated with beaches and dunes, called *strand vegetation*, consist of low-growing prostrate, often succulent, forms subject to salt spray and fog. Many are perennials, but functionally respond as annuals because of the ongoing dynamics of sand accretion and erosion. In all cases they are important in dune formation and stabilization. Some species occur over the length of the state and into Baja California, whereas others are restricted to particular stretches of coast line. A few, such as bush lupine (*Lupinus* spp.), may occur inland where they comingle with coastal sage scrub. Characteristic species of coastal strand include sand verbena (*Abronia* spp.), Beach Salt Bush (*Atriplex leucophylla*), Beach Burr (*Ambrosia chamissonia*), Rye Grass (*Elymus mollis*), beach primrose (*Camissonia* spp.), beach morning-glory (*Calystegia* spp.), and Sea Rocket (*Cakil maritima*).

Recreational development, roads, berms, sea walls, and other forms of human activity have hit strand communities hard. Beach grooming is another deleterious practice. Heavy equipment used to remove wrack (e.g., kelp) and debris interferes with the natural process of beach-building and erosion. Kelp removal has the additional effect of dislodging many invertebrates that are a food resource for such shorebirds as sandpipers and plovers. Invasive, non-native plants also have overwhelmed strand communities in some areas. Two of these are Hottentot Fig and Pampas Grass. Hottentot Fig, often called "ice plant," is a South African species (*Carpobrotus* = *Mesembryanthemum edule*) widely used as a ground cover in Southern California. It is a common sight along coastal areas including beaches, dunes, and estuaries. Pampas grass (*Cortaderia selloana*), native to Argentina, germinates easily from its wind-borne seeds and has become thoroughly established around some dune deposits.

6.2 COASTAL SALT MARSHES AND ESTUARIES

Estuaries consist of subtidal habitats and adjacent wetlands that are partially enclosed by land. Access to open ocean is often obstructed by sand bars, and the influx of seawater is diluted by freshwater runoff from the land, creating the brackish conditions characteristic of these habitats. Unique combinations of substrate, nutrients, tides, depth, and salinity require corresponding adaptations by the plants, invertebrates, fish, and birds that depend on estuaries for their existence.

Estuaries are not homogenous but vary according to physical structure, configuration, and water sources. River mouth estuaries in Southern California are limited to the few large rivers with unimpeded perennial flows to the coast like the Santa Clara, San Diego, and Tijuana Rivers. They are permanently brackish.

Seasonal creeks that flow from canyons draining interior watersheds form canyon mouth estuaries. The creeks form small estuaries subject to varying frequency of tidal flushing.

Sandbars enclose lagoon estuaries most of the year. As a result, salinities are not as high, especially if freshwater input is enhanced by urban and agricultural runoff. The fringes are immersed in brackish water rather than being exposed vegetated flats. Mugu Lagoon in Ventura County is one example. Others are Buena Vista, Batiquitas, San Elijo (Figure 6.2) and Agua Hedionda Lagoons in San Diego County.

FIGURE 6.2 Estuaries like San Elijo lagoon located between Solana Beach and Encinitas in San Diego County are enclosed by sandbars much of year. San Elijo lagoon is the terminus of the Escondido creek water shed. It contains a variety of plant communities such as coastal strand, salt marsh, coastal sage scrub, and mixed chaparral. Like all estuaries, nutrients, depth, substrate, tidal influence, and salinity all combine in unique ways to support the plants, invertebrates, fish, and birds that depend on it.

Coastal dune-creek estuaries form as wetlands near dunes. They have seasonal sandbars, connect with the ocean, and are brackish most of the year.

Bay estuaries are large subtidal areas (bays) and salt marsh habitats with a strong marine influence because of a wide mouth consistently open to the sea. Intertidal mudflats are extensive, as for example at Bolsa Chica, Upper Newport Bay, Anaheim Bay, and Mission Bay in San Diego.

Structural basin estuaries were created by tectonic forces (down faulting and folding), which resulted in landforms below sea level and estuaries of moderate size. Primary examples are in Santa Barbara County at the Goleta Slough and Carpenteria Salt Marsh.

An estuary's configuration and size significantly influence the kinds of plants and animals it will support. Because salinity gradients vary in time and space, plants are distributed based on their tolerance for salinity intrusion into the substrate. Many species are succulents and grasses. The bayward portion of an estuary is dominated by the more salt-tolerant species such as Cord Grass (*Spartina foliosa*) and pickleweed (*Salicornia* spp.). Pickleweed also dominates the middle (less saline) areas along with salt grass (*Distichlis*), Alkalin Heath (*Frankenia salina*), and Sea Lavender (*Limonium californicum*). Plants become less salt tolerant in areas receiving irregular flooding by freshwater. Cattails (*Typha* spp.) and rushes (*Juncus* spp., *Scirpus* spp.) only occur at the inland extreme where salinity is consistently low.

When healthy, estuaries and salt marshes are intricately productive habitats. They abound in mollusks, crustaceans, and other invertebrates as well as fish, shorebirds, ducks, and wading birds like the Great Blue Heron shown in Figure 6.3.

FIGURE 6.3 The Great Blue Heron (*Ardea herodias*) is a familiar sight around Southern California's estuaries, tidal flats, lakes, reservoirs, and even damp meadows. Great Blue Herons feed on a variety of invertebrate and vertebrate prey. By day they are usually seen singly, but at night they will often roost communally in tall trees.

Humans also easily impact them. Historically, commercial interests and governments tended to view salt marshes as unattractive swamps, little more than breeding sites for mosquitoes that should be converted to more productive uses. Southern California's estuaries have been drained, dredged, channeled, compacted, and squeezed by agriculture and urbanization and further degraded by invasive exotic species while being trampled and littered through unregulated pubic access. The result is that around 90% of coastal wetland habitat has been mangled or destroyed. Recent efforts in the past couple of decades to restore and rejuvenate estuaries and salt marshes were spurred by conservation concerns or because of a mitigation requirement for wetland loss somewhere else. The effectiveness of these efforts is mixed. Assessment is difficult, expensive, and time consuming in part because estuaries vary in their ecological functions. Hence, restoration may simply convert one type of wetland to another with no significant gain. Still, any and all ventures should be encouraged and supported that preserve and maintain our remaining coastal wetlands.

6.3 COASTAL SAGE SCRUB

6.3.1 Coastal Sage Scrub and Chaparral

Mediterranean climates feature summers that are hot and dry, and winters that account for more than 80% of the annual rain. The plant communities (warm-temperate scrublands) growing adjacent to and inland from the coast consist of many families and species. The structurally dominant forms have dense, compact crowns, small, hard-thick (sclerophyllous) leaves, and deep wide-spreading root systems. They are all adapted to fire and reproduce prolifically from fire-scarred seeds or enlarged roots. At lower elevations, with less rainfall, the more xeric community is called *coastal sage scrub*. Higher up in the foothills the scrublands are called *chaparral*.

Coastal sage scrub once extended uninterrupted from south of San Francisco well into Baja California, and it more or less remained that way through much of the nineteenth century. Southern California was a collection of pueblos and filthy outposts through most of that period while San Francisco was gaining status as one of the grand cities of the world. For the southland, this all changed when a reliable water supply was imported from the Owen's Valley in 1913. Henceforth eradication was the fate of coastal sage scrub. People needed the land, and the dull, gray-green scraggly growth with its incendiary tendencies, then as now, do not inspire the kind of botanical reverence bestowed on say the coast redwoods farther north. Nowadays, coastal sage scrub communities are among the most endangered habitats in the United States. Probably less than 15% remains intact and undeveloped, mostly around military establishments like Miramar and Camp Pendleton. It still grows on some mesas and inland valleys and along dry, gravelly canyons and slopes to the lower limit of the chaparral.

FIGURE 6.4 Coastal sage scrub once extended uninterrupted from south of San Francisco well into Baja, California. It is now among the most endangered habitats in the United States. Compared with chaparral mature, coastal sage scrub forms relatively open stands of low-growing aromatic plants, although they share a number of the same plant species.

Mature stands of coastal sage scrub (Figure 6.4) form relatively open communities compared with chaparral because of poorer site conditions and lower rainfall. Coastal sage scrub, or simply coastal scrub or "soft chaparral," develops inland from the beach and consists of low-growing (up to about 6 ft [1.8 m]) aromatic shrubs that evolved with the drying trends beginning in the middle Miocene 14 million years ago and culminated in the late Pliocene when meager rainfall became restricted to winter months. Were it not for the ameliorating effects of the maritime climate (i.e., coastal fog and moderate temperatures), this community would be more like desert scrubland.

Plants conserve water in several ways. In arid regions like ours, the leaves of plants are small, thick, and waxy and may be covered in fine filaments of hair that reflect sunlight, which in turn helps minimize evaporative water loss. Many will simply become dormant or shed their leaves in the summer (drought deciduous). Hence, for much of the year, coastal sage scrub looks dead to the casual observer.

Common species include Black Sage (*Salvia mellifera*), White Sage (*S. alpiana*), California Sagebrush (*Artemesia californica*), California Buckwheat (*Eriogonum fasiculatum*), Lemonade Berry (*Rhus integrifolia*), Fuschia-flowered Gooseberry (*Ribes speciosum*), Scrub Oak (*Quercus dumosa*), Encelia, (*Encelia californica*), monkey flower (*Mimulus* spp.), Goldenweed (*Haplopappus* spp.), Our Lord's Candle (*Yucca whipplei*), poison oak (*Toxicodendron diversilobum*), and such cactus as Coast Cholla (*Cylindropuntia prolifera*), and Coast Prickly Pear (*Opuntia littoralis*).

Occasionally, isolated individuals of species more typical of chaparral can be found, like, Holly leaf Cherry (*Prunus ilicifolia*) and Sugar Bush (*Rhus ovata*). Toward the southern end of its range near El Rosario, Baja California, increasing numbers of desert species intrude like Desert Apricot (*Prunus fremonti*) and Jojoba (*Simmondsia chinensis*) along with various species of cactus.

6.3.2 Coastal Sage Scrub Vertebrates

Demise of coastal cage scrub was inevitable. Agriculture, urbanization, invasive non-native grasses, pollution, and disruption of natural fire regimes each contributed. In Southern California there are now more than 100 species of plants and animals under legislative protection or are so proposed. Whereas some coastal sage scrub vertebrates survive in adjacent areas of chaparral or lower montane habitats, for instance the Coast Horned Lizard (*Phrynosoma corornatum*), those more dependent on coastal scrub have not fared well. The California Gnatcatcher (*Polioptila californica*—possibly a disjunct subspecies of the Black-tailed Gnatcatcher *P. melanura*) and the San Diego subspecies of Cactus Wren (*Campylorhynchus brunneicapillus sandiegensis*) are conservation flag bearers. Both of these small birds are localized and barely hanging on as isolated breeding populations in protected or inaccessible areas—the gnatcatcher in Riverside, Orange, and San Diego Counties. The same goes for Belding's Orange-throated Whiptail Lizard (*Aspidoscelis hyperythra beldingi*), a small, slender lizard of coastal sage scrub with a limited and spotty distribution in San Diego and Riverside Counties. Stephens' Kangaroo Rat (*Dipodomys stephensi*) is endangered because of its narrow, fragmented range in San Diego, Riverside, and San Bernardino Counties. The Coast Horned Lizard mentioned previously inhabits chaparral, oak woodland, piñon-juniper woodland, and coniferous forests to around 6,000 ft (ca. 1,820 m). However, in coastal sage scrub, it has become rare to nonexistent because of habitat loss, predation by feral cats, and the displacement of several species of native ants, a preferred food item, by the introduced Argentine ant (*Linepithema humile*), which the horned lizard will not eat.

6.4 CHAPARRAL

The next time you fly in to Ontario or San Diego, take in the terrain below. Better yet, hike up a foothill trail to the ridgeline. From either perspective, the panorama is an uninterrupted gray-green hue of chaparral accented by occasional steep canyons and granite outcrops. This is nature's response to a Mediterranean climate. Chaparral cloaks the slopes of the Transverse and Peninsular Ranges from about 6,500 ft (1,975 m) westward to the coastal lowlands that are not otherwise grown to coastal sage scrub and oak woodland (Figure 6.5). It is *the* signature biome of our region apart from the deserts.

FIGURE 6.5 California has nearly 7 million acres (ca. 280,000 km^2) of chaparral. In Southern California, it is the dominant plant community west of the deserts. Arid adapted and growing on boulder-strewn slopes and hillsides, chaparral consists of woody shrubs from 3 to 10 ft (1–3 m) in height crowded into nearly impenetrable stands.

People unfamiliar with chaparral may be dismissive of its scrubby appearance and unruly attitude. Some are derisively amused that it accounts for the majority of four national "forests" (Cleveland, San Bernardino, Angles, and Los Padres), the thinking being that only large tracts of timber reserves can designate a forest. This is simply an uninformed point of view. As a forest, chaparral differs only in stature. Although diminutive, it is as complex as any forest, as expansive in the territory it covers, and ever as representative of nature's constitution. Its constituents comprise the "Elfin Forest"—Francis Fultz's fitting appellation for chaparral in his delightful account written in 1923.

Chaparral extends north along the Coast Ranges and the Sierra foothills to southwest Oregon; nearly 7 million acres occur in California alone. To the south, it courses into northern Baja California as far south as the Sierra San Pedro Martir. Still, chaparral remains identified with Southern California. The fact that so much of it comprises the area's national forests speaks to its dominion and persistence in the region's history. Recipients of the Spanish land grants that began in 1822 justifiably prized the lowland valleys of Southern California. Chaparral surrounded the foothill periphery of their ranchos, but it made for poor pasture and essentially was ignored. Decades later the Forest Reserve Act of 1891 gave the president of the United States authority to "set apart and reserve in any state or territory having public land bearing forests … as public reserves." So it was that when the federal government got around to Southern California in 1907–1908, it found the chaparral almost entirely intact and in its possession.

The term “chaparral” is sometimes applied to any scrubland plant community. But botanically, California (or coastal) chaparral is a specific biome apart from other scrub vegetation with which it associates or overlaps, for instance coastal sage scrub. Another type of chaparral also describes an important plant community in Arizona and northern Mexico, with outlier populations in New Mexico and west Texas. Although structurally similar, interior chaparral differs in species composition as well as in its bimodal precipitation regime of sporadic winter rains and summer monsoons; the driest periods are in the spring from April to June.

California chaparral, as with other maritime chaparral communities around the world, thrives in a Mediterranean climate of cool, wet winters and hot, dry summers. In the Mediterranean it is called *maquis*, in South Africa *cape scrub*, in Australia *mallee scrub*, and in Chile *matorral*. The word derives from the Spanish *chaparro* meaning “scrub oak.” The early Spaniards in California coined chaparral, “a place of scrub oak,” because of its similarity to habitats they knew from southern Spain. “Chaps,” the leather leggings worn by horsemen in heavy brush, comes from the same root. Southern California chaparral consists of shrubs typically from 3 to 10 ft in height, crowded in dense, nearly impenetrable stands. It grows on a variety of thin, low-nutrient soils best developed on weathered, coarse-grain granite and gneiss—what locals refer to as decomposed granite, or “DG.” Soils, shallow or deep, are well drained and afford deep percolation. On steep slopes, soil horizons are indiscernible because of high erosion rates. Chaparral is therefore critically important in erosion control, and that fact alone has invited intensive study of its composition and growing habits. Its importance as a watershed cannot be overstated. This is easily appreciated when heavy rains trigger mudslides on burned over slopes, sometimes dislodging boulders the size of cars that careen into neighborhoods below.

In Southern California, around 85% of the rainfall occurs from November to March. From May to September, chaparral plants must cope with increasingly higher temperatures and seasonal drought. Plants begin their dormancy or dieback at that time, and by August and September they become highly flammable. With the onset of Santa Ana winds in the fall, chaparral is at its most vulnerable to devastating wildfires. Despite the rigors they face, chaparral plants are accustomed to seasonal drought and occasional fire. Shrubs maintain small, hard “sclerophyllous” leaves that resist water loss and thus wilting. Most are evergreen in the sense that leaves of one year or not shed until next year’s leaves are set. So, some leaves persist even during the driest days of the year. This feature distinguishes Southern California chaparral from the periodic leafless montane scrub of central and northern California. Many chaparral plants grow from a large woody base, or *burl*, which is a wide and thickened underground trunk. Root systems are deep, and individual plants grow close to one another such that intertwining branches provide mutual shade. Healthy plants often bear numerous dead stems and branches as a form of self-pruning. This of course contributes to the structural density of the plant community and the heavy fuel load that feeds wildfires.

Abundant annuals and short-lived perennials put on a spectacular flower show as the first plants to appear following initial rains after a burn. Meanwhile, the dominant chaparral associates sprout from their bases and burls or from fire-scarred seeds. In a few years, the original chaparral reclaims the land. Accordingly, all the plants in a particular area are generally the same age.

More than 1,000 plant species comprise California's chaparral. Only a few are dominant and common and belong to a handful of families: the rose family Rosaceae (e.g., Chamise, Toyon, Holly-leaf Cherry, Mountain Mahogany, Red shank), heath family Ericaceae (e.g., manzanita), buckthorn family Rhamnaceae (e.g., ceanothus), oak family Fagaceae (e.g., Scrub Oak), buckwheat family Polygonaceae (e.g., Wild Buckwheat), and sumac family Anacardiaceae (e.g., Sugar Bush, Poison Oak).

Chamise (*Adenostoma fasiculatum*) (Figure 6.6): The common name of this plant derives from the Spanish *chamiza* meaning "brushwood." Chamise, sometimes called *greasewood*, is the most widespread of the dominant plants in Southern California chaparral. Vast stands, either pure or mixed, range over Los Angeles, San Bernardino, Riverside, Orange, and San Diego Counties. It is common in northern Baja California and in chaparral communities in the central and northern parts of the state. A single Chamise plants grows from a burl into multiple stems 4–6 ft (1.2–1.8 m) tall, each bearing tiny, narrow, dark-green leaves arranged

FIGURE 6.6 Chamise (*Adenostoma fasiculatum*) is the dominant plant in Southern California chaparral, much like creosote is in the deserts. Sometimes referred to as *greasewood*, a single Chamise plant grows into a tangle of multiple stems that form dense, woody arrays. Chamise is the primary source of fuel in chaparral wildfires.

in bundles (fascicles). The growing season is from late winter into June. Small off-white flowers appear in mid to late spring into early summer.

Chamise is the primary fuel of chaparral fires not only because of its pervasiveness but also because the dense interwoven branches carry dead and drying stems. Herbaceous understory is negligible except for occasional postfire components that may persist for several years. Chamise will achieve about 90% cover in 25 years, and individual plants may live hundreds of years. It is slower to recover from fire than other chaparral species largely because of the poor soils on which it grows. Periodic, but not frequent, fires maintain the vigor of Chamise stands. When not growing in pure stands Chamise occurs with ceanothus, manzanita, California Buckwheat, White and Black sage, Scrub Oak, Sugar Bush, and Laurel Sumac.

Toyon (Christmas Berry or California Holly—*Heteromeles arbutifolia*): Recognized by its clusters of red berries appearing in the winter months (Figure 6.7), Toyon is a common chaparral shrub of cooler slopes and canyons from northern California to Baja. Mature plants are about 15 ft (4.5 m) tall. Plants are evergreen, with dark-green oblong serrated leaves. The young leaves are pungent from a cyanide-based chemical that functions as a deterrent to herbivores. However effective it is, they are still eaten by Mule Deer. Toyon berries are consumed by birds, which then disperse the seeds. As a result, Toyon rarely grows in pure stands.

FIGURE 6.7 Toyon (*Heteromeles arbutifolia*), sometimes called Christmas berry, grows in canyons and on cooler chaparral slopes. The red berries are consumed by birds, which disperse the seeds. Young leaves contain a cyanide compound as a deterrent to herbivores, but Mule Deer will eat the leaves nonetheless.

Usually it mixes with other chaparral shrubs like scrub oak, manzanita, and ceanothus. Toyon is a common horticultural plant in gardens and was introduced for that purpose to Great Britain and elsewhere in Europe some 200 years ago.

Holly-Leaf Cherry (*Prunus ilicifolia*): Found largely in shadier, moister slopes, Holly-leaf Cherry occurs in chaparral along the coast ranges north of the Bay Area south to Baja California. This is another easily recognizable plant, not only by its leaves but also its in-season fruit. The leaves, as the name "Holly-leaf" implies, are 1–2 in. (ca. 25–50 mm) long, deep shiny green and have wavy, spiny margins (Figure 6.8). The fruits mature from white flowers into purple cherries, although they are smaller than the commercially grown variety of the same genus. Birds gorge themselves on them. Plants generally achieve 4–6 ft (1.2–1.8 m) but can grow to double that size. The Channel Island subspecies, called the Catalina Cherry (*P. i. lyonii*) is notable for its height (over 15 ft) and larger leaves with smooth margins.

Mountain Mahogany (*Cercocarpus betuloides*): Mountain Mahogany ranges from southern Oregon to Baja California. It is common on rocky slopes in Southern California chaparral, including on the Channel Islands and in the transition zones with the deserts. Plants can grow to 20 ft (6 m) and are generally found with manzanita, ceanothus, and Chamise. The whitish stems are conspicuous, as are the rounded, fan-shaped leaves. Mountain Mahogany is most easily recognized, however,

FIGURE 6.8 Holly-leaf Cherry (*Prunus ilicifolia*) is a resident of chaparral slopes from the San Francisco Bay area south to northern Baja, California. White flowers mature into small purple fruits that are fed on by many birds.

by the showy white flowers, especially when the fruits develop in summer. Fruits consist of a silvery, feather-like plume whose base is attached to an enclosed seed. Plants resprout from the base or germinate from seeds dispersed by the plumes. A second species, the small-flowered *C. minutiflorus* is confined to San Diego County and northern Baja where it grows closer to the coast.

Red Shank (*Adenostoma sparsifolium*): A handsome plant (Figure 6.9), Red Shank superficially looks like Manzanita because its loose, shaggy red bark peels away in strips, which has led to its alternative name, "Ribbonwood." Red Shank is restricted to Southern California from San Luis Obispo south through San Diego County into Baja. Curiously, it does not occur naturally in the Transverse Ranges (yet is successfully planted). It is otherwise represented in our chaparral and the desert transition zones. Red Shank is in the same genus as Chamise but prefers richer, moister soils. Similarity is evident in the shape of the leaf and their arrangement on the stems. Like Chamise it is also multistemmed, but unlike Chamise it can grow to 20 ft (6 m) and become almost tree-like.

Manzanita (*Arctostaphylous* spp.): About 60 species of the Heath family genus *Arctostaphylous* are recognized. All but three species grow in California and many of them are regional or local endemics with restricted distributions. Approximately 30 species occur in Southern California. Manzanita are somewhat inconspicuous nearer the coast.

FIGURE 6.9 Red Shank or Ribbonwood (*Adenostoma sparsifolium*) is restricted to Southern California chaparral communities from San Luis Obispo County south into Baja California. It is most common in the Peninsula Ranges. Red Shank resembles Manzanita with its smooth, reddish contorted stems, but it is a relative of Chamise (*A. fasiculatum*).

At higher elevations where moisture is greater and soils are deeper, they will grow into large arborescent structures up to 15 ft (4.5 m) tall. Their contorted, smooth, purplish-brown trunks and branches are distinctive, and the thin, paper-like bark peels away in curling sheets. Sections of the root system routinely die along with the associated parts of the trunk and stems; only the purplish brown sections are living. (*Ceanothus* exhibits a similar growth pattern.) Some species of *Arctostaphylous* sprout from burls after a fire, whereas others germinate from seeds. Leaf shape and color, bracts (leaf-like structures subtending a flower), and the size of the fruits and seeds are some of the morphological features useful for identifying individual species. Two of the more widespread forms in Southern California are Big Berry Manzanita (*A. glauca*, Figure 6.10), a nonsprouting species, and Eastwood Manzanita (*A. glaundulosa*), a sprouting species. Mexican Manzanita (*A. pungens*) thrives at higher elevations in the mountains and in transitional zones with the desert. Closer to the coast we find Mission Manzanita, formerly a species of *Arctostaphylous* now assigned to the genus *Xylococcus* (*X. bicolor*). The bark is pale brown and not as dark.

FIGURE 6.10 Around 30 species of manzanita (*Arctostaphylos* spp.) grow in Southern California. Their twisted trunks with smooth reddish bark make them easily recognizable. Many are of small stature and have limited distributions. A few like Big-berry Manzanita (*A. glauca*) are common in the chaparral and reach 15 ft (4.5 m). The thin, exfoliating bark of Manzanita is distinctive. Manzanita takes its name from these (inset) chickpea-sized fruits ("little apples") produced in profusion in mid-spring. The sticky, resinous exterior adheres to almost anything, like your shoes, and is a superb mechanism for dispersing the seeds.

In the winter months from December to March small urn-shaped pink to white flowers emerge in bunches dangling upside down from the branch tips. Pollinated flowers produce sticky dark red berries in abundance, which may remain on the tree for months. Appropriately, manzanita is Spanish for "little apple." The generic name *Arctostaphylous* literally means "bear berry," a reference to the fact that the fruits are eaten by Black Bears and formerly by the extinct California Grizzly. Coyotes and Grey Fox also include the fruits in their diet. Native Americans and the early Spaniards rendered them into tea, flower, and even wine.

Ceanothus (*Ceanothus* spp.): These somewhat spiny shrubs of the Buckthorn family go by various common names such as "lilac" ("wild lilac," "white lilac," "blue lilac") because of their superficial resemblance to the ornamentals (*Syringa*) of the Olive family. More typically they are simply referred to by their generic name preceded by a descriptive modifier, for example Hoary-leaf Ceanothus (*C. crassifolius*), a common constituent of Southern California chaparral. Approximately 40 species occur statewide, which represents around 80% of all *Ceanothus* in North America. In California, half of them have limited distributions, and many of these are state-listed as species of special concern. All areas of chaparral will have at least one species. Ceanothus prefers wetter sites than Chamise and are most prevalent on the coastal slopes of the Peninsular Ranges to about 6,000 ft. Others can be found along desert slopes and the interior mountains of the Southwest, for example Desert Ceanothus (*C. perplexans*).

Some species grow as small, dense shrubs. Others develop tall, open crowns around 15 ft (4.5) tall. The small fragrant flowers, blue to white, put on spectacular blooms from January to June (Figure 6.11). In a grand display of swagger, all individuals bloom at the same time in a particular area. Like manzanita, about half the species of Ceanothus germinate from seeds, such as Hoary-leaf Ceanothus. Others, like the widespread Chaparral Whitethorn (*C. leucoderma*) have burls and sprout from the base. Ceanothus occasionally grows in pure stands; more often it is a codominant in the chaparral along with Chamise, Toyon, scrub oak, and Sugar bush.

California scrub oak (*Quercus berberidifolia*): California hosts more than 20 species of oaks (*Quercus* spp.), one of the commonest hardwoods in the state. Oaks, of course, are known for their acorns. The difference between scrub oak and oak trees is simply one of size and structure. Oak trees, for example the Coast Live Oak, *Q. agrifolia*, are larger, up to 75 ft (23 m) tall and have a single main trunk. Scrub oaks usually top out at around 15 ft (4.5 m) growing from multiple stems or trunks that sprout from a common base. Leaves are small, leathery, and spiny. Stems of smaller scrub oaks produce limbs that are gnarled and twisted. Those of larger individuals become more linear and develop dense overhead canopies.

Numerous species of *Quercus* are referred to as scrub oak of one kind or another. The most significant in our chaparral is the California

FIGURE 6.11 Ceanothus (*Ceanothus* spp.) is a co-dominant plant of the chaparral that blooms in late winter in Southern California. Many, like these, produce flowers in shades of blue. Others are white. Approximately 40 species occur in California, which represents around 80% of all Ceanothus species in the United States.

Scrub Oak, *Q. berberidifolia*, particularly in areas approaching 20 in. (ca. 50 cm) of rain. Mature trees create heavy leaf litter with little or no understory. The Coast (or Nuttall) Scrub Oak, *Q. dumosa*, replaces it at lower elevations and in coastal sage scrub. Scrub oaks are susceptible to red, golf-ball–sized growths on their stems, called "galls" (Figure 6.12). Galls are not fruits but a systemic growth response by the plant in reaction to egg deposition by "gall wasps" (Cynipidae). The eggs and later the larvae induce the stems to produce the gall, a spongy water and sugar-rich medium that protects and nourishes the larva. Eventually the gall turns brown and woody but remains on the stem after the wasp as emerged.

California Scrub Oak associates with the other dominant chaparral species like Chamise, manzanita, and Ceanothus. They may also host mistletoe (*Phoradendron* spp.), a plant parasite.

California (Flat-top) Buckwheat (Eriogonum fasciculatum): This is a widespread plant of Southwest scrublands in Nevada, Utah, Arizona, and northern Mexico. In California it occurs in the central and southern part of the state from the coast to the deserts. California Buckwheat is scattered in drier areas of chaparral and coastal sage scrub as a multistemmed, sprawling shrub from 3 to 4 ft tall (ca. 1.2 m). The leaves are minute like those of Chamise. The flowers are small, pinkish white and grow in flat, terminal clusters in the spring and early summer (Figure 6.13). Also, like Chamise, they turn burnt orange to brown in the fall. California Buckwheat is an important resource for

FIGURE 6.12 This growth that resembles an apple is actually a gall produced by the stems of the California Scrub Oak (*Quercus berberidifolia*). Eggs deposited by gall wasps (Cynipidae) induce the plant to develop this spongy, sugar-rich capsule to protect and nourish the wasp larvae. The gall dries up and turns brown after the adult wasp has emerged.

FIGURE 6.13 A common plant of chaparral, California (Flat-top) Buckwheat (*Eriogonum fasciculatum*) is abundant in arid scrubland throughout the Southwest. The small white flowers grow in terminal clusters in the spring and early summer, turning a burnt orange in the late summer and fall. Many small vertebrates depend on Buckwheat for cover and its edible seeds.

small vertebrates in providing both cover and edible seeds. Seeds also were harvested and ground into flour by Native Americans and early European settlers.

Sumacs and relatives: Sumacs are members of the Anacardiaceae. This is a large, diverse family that includes poison oak and poison ivy as well as such tropical trees as cashews. The clear to milky sap of some species causes skin irritations, rashes, blisters, and other forms of dermatitis. Chaparral sumacs occur only in Southern California. Laurel Sumac (*Malosma laurina*) is common on lower elevation slopes and in coastal sage scrub of Southern California and northern Baja. They can be large at times, up to 20 ft (6 m) in height. Distinctive are their large (4–10 in.; 10–25 mm), green leaves folded up and inward along the midrib. Small, white flowers bloom in the summer, producing fruits that, like other sumacs, are attractive to birds and mammals. Laurel Sumac is intolerant of hard freezes; plants die back to the ground if exposed to successive nights of freezing temperatures. An unusual feature for a chaparral plant is that new stems sprout from sections along the roots instead of from a base or burl. Southern California's early citrus growers recognized Laurel Sumac's frost sensitivity and accordingly considered it an indicator of where they could safely plant their orchards.

Like Laurel Sumac, Lemonade Berry (*Rhus integrifolia*) is a shrub of low-elevation slopes and coastal sage scrub from Santa Barbara and the Channel Islands south into northern Baja. It is smaller, denser shrub than Laurel Sumac, rarely more than 4 ft (1.2 m) tall, though occasionally some plants will reach 15 ft (4.5 m) on north-facing slopes and in protected areas. The small, thick leaves are from 1 to 2 in. (26–52 mm). Pink to white flowers appear in the winter and into spring, which in turn yield red sticky fruits. Mixed with water, the fruits impart a taste of lemon. Native Americans valued it for medicinal properties.

Sugar bush (*Rhus ovata*, Figure 6.14): is freeze-tolarant and replaces Laurel Sumac and Lemonade Berry in mid- to high-elevations away from the coast. Plants are multistemmed and may grow to 15 ft (4.5 m) as a rounded dome. The shiny bright green leaves, 1–3 in. (26–76 mm), are only moderately folded along the midrib. Pink flower clusters are produced in the winter, and the fruits are similar to Lemonade Berry in their sugary, crystalline coating.

Poison oak (*Toxicodendron diversilobum*): Abounds in California in shady habitats, especially near creeks, ravines, and other damp places. A member of the Sumac family (Anacardiaceae) poison oak is notoriously toxic. Although not restricted to chaparral, it is a common component encountered along trails, in disturbed areas, and in canyon bottoms. Anyone who spends time hiking and camping is wise to be familiar with it. I always introduced poison oak to my students at the beginning of the semester. All of them knew of it, but few could actually identify the plant (Figure 6.15).

FIGURE 6.14 Sugar Bush (*Rhus ovata*) is abundant from near the coast to the western slopes of the mountains, especially in chaparral where it largely replaces its close relatives Laurel Sumac (*Malosma laurina*) and Lemonade Berry (*R. integrifoli*). The small sweet fruits are the source of its common name.

FIGURE 6.15 Poison oak (*Toxicodendron diversilobum*) is common throughout the West Coast, typically growing in damp shady places often near creeks and drainages. The shiny trifoliate leaves are bright green in the spring and turn red in the fall. The plant shown here is growing beside a nest of a Big-eared Woodrat (*Neotoma macrotus*).

The deciduous compound leaves are in three parts with each leaflet from 1 to 4 in. (25–100 mm). In the spring, the new growth is bright, glossy green. Leaves turn red in autumn. Plants grow as shrubs or as vines that twist around and over supporting structures like oaks, fences, and abandoned buildings. Old growth poison oak may have stems more than a foot in diameter. Flowers are yellow to white. Being immune to the plant's toxic effects, birds and small mammals forage on the small white fruits. An oily secretion known as *urushiol* causes the blistering, itchy rash that results from contact. People usually develop symptoms from inadvertently brushing up against the leaves, but any part of the plant can induce the rash. Equally, clothing, backpacks, and a dog's fur may pick up secretions that can be easily transferred to a human. Even burning poison oak is hazardous because the smoke can cause severe respiratory inflammation.

Other chaparral plants: The foregoing introduces the major large shrubs and trees that are definitive and common in Southern California chaparral. Other important and interesting species include, for example, chaparral currents, several types of sage, yuccas, succulents, and vines. They all add resources and character to the community.

Among the Gooseberry family (Grossulariaceae) the large genus *Ribes* is known for its red to purple berries called *currants*. Winter Currant or White-flower (*Ribes indecorum*) is the best known in Southern California chaparral woodlands.

The aromatic California or Coastal Sagebrush (*Eriogonum californica*), a member of the Sunflower family Asteraceae, is abundant in coastal sage scrub and in chaparral up to about 3500 ft (1060 m). Another sage, but from the Mint family Lamiacea, is the broadly distributed (from coast to deserts) White Sage (*Salvia apiana*) distinguished by its pungent pale green whorled leaves. Black Sage (*S. mellifera*) and Purple Sage (*S. leucophylla*) are especially likely to be found in open and disturbed areas.

Agaves (Agavaceae) are represented by Mojave Yucca (*Yucca shidigera*) also called Spanish Bayonet, and Our Lord's Candle (*Hesperoyucca whipplei*; Chaparral Yucca, Figure 6.16). These yuccas are notable for their basal rosettes of stiff, sword like leaves with sharp pointed tips. When in bloom, their magnificent tall flower stalks stand out in the chaparral. Our Lord's Candle shares an intriguing symbiotic relationship with its lone pollinator, the Yucca Moth (*Tegeticula maculate*). Each depends on the other for survival. The female Yucca Moth has a specific anatomical apparatus to remove the pollen, which she carefully deposits in another yucca flower. The larvae that emerge from her eggs that she deposits in the carpel feed on the flower's developing seeds.

Two interesting succulents of the Stonecrop family (Crassulaceae) belong to the genus *Dudleya*, a diverse clade of about 40 species

FIGURE 6.16 The tall flower stalk of Our Lord's Candle (*Hesperoyucca whipplei*) emerges conspicuously in Southern California chaparral. Occasionally nibbled on by deer, its lone pollinator is the Yucca Moth (*Tegeticula maculata*).

mostly confined to the Southwest. Dudleyas are low-growing rosettes of succulent leaves that produce tall, showy stalks of tubular flowers. White Live-forever or Ladyfingers (*D. edulis*), and *Chalk Live-forever* (*D. pulverulenta*) are found on granite slopes and clay soils from the coast to the mountains.

Bush Monkeyflower (*Mimulus aurantiacus*) is a common colorful shrub when in bloom. Most species of *Mimulus* are small and herbaceous, but *M. aurantiacus* is woody at the base and sprawling. Cup-shaped flowers are a light creamy orange. *Wild Pea* (*Lathyrus vestitus*) is a familiar vine that climbs over almost any adjacent shrub suitable for a scaffold. It dies back in the winter, but in the spring, the magenta flowers are striking. *Wild Cucumber* (*Marah macrcarpus*) is another vine of coastal scrub and chaparral. This member of the Gourd Family (Cucurbitacea) has a perennial tuber, but the aboveground branches and tendrils that emerge in late winter die back after setting fruit, an inedible and bazaar-looking bright green prickly ball.

6.4.1 Chaparral Vertebrates

As extensive as chaparral is in Southern California the vertebrates that inhabit it are as equally varied and abundant. Almost any regional species can occur in the chaparral except for those restricted to the deserts or higher elevation pine forests. Nor are any of them limited to chaparral.

Amphibians are not especially diverse in Southern California and, with few exceptions, their activity is restricted to the late fall and winter months when rains are falling and the ground is wet. The small Pacific Treefrog (also called Pacific Chorus Frog) is common, although more often heard calling than seen. On the other hand, the large Western Toad is easily found moving around on damp nights. The Western Spadefoot remains below ground most of the year and occurs in chaparral only where soils are friable enough for burrowing. The diminutive Slender Salamanders (*Batrachoseps* spp.) are difficult to detect because, in addition to their small size, their reclusive habits keep them sheltered beneath rocks and logs in damp places.

Some lizards can be observed anytime of the year; sunny and warmer winter days are as likely as those in the spring and summer. The most frequently encountered are the Side-blotched Lizard and the Western Fence Lizard (Figure 6.17). In general, seasonal activity is the norm for most lizards (and snakes). Spring and summer are the active periods for the Coast Horned Lizard, Granite Spiny Lizard, Granite Night Lizard (both associated with large boulders), Western Whiptail, Western Skink, and Southern Alligator Lizard. Among the more common chaparral snakes are the Striped Whipsnake, Gopher Snake, California Kingsnake, Spotted Night Snake, Southern Pacific Rattlesnake (Figure 6.18), Southwestern Speckled Rattlesnake, and Red-diamond Rattlesnake.

FIGURE 6.17 Familiarly known as "blue-bellies," the Western Fence Lizard (*Sceloporus occidentalis*) is a common sight around Southern California sitting atop fence posts, boulders, and rock walls. This individual has a rejuvenated tail (look closely near the tip). Many species of lizards have the ability to detach their tail as a predator defense mechanism and occasionally as a result of combat with other individuals (see text).

FIGURE 6.18 Perhaps only a month old, this neonate Southern Pacific Rattlesnake (*Crotalus helleri*) is easily overlooked among a bed of dried leaves. Notice the button at the end of the tail—a keratinized nob that anchors the rattle as it grows. Once this baby snake starts growing it will add a rattle segment each time it sheds its skin. The Southern Pacific is the most common rattlesnake in Southern California west of the deserts.

Various species of bats like Myotis can be seen in summer evenings. Typical nonvolant mammals include the Ornate Shrew, Broad-footed Mole, Coyote, Gray Fox, Raccoon, Striped Skunk, Long-tailed Weasel, Bobcat, Mountain Lion, Mule (Black-tailed) Deer, California Ground Squirrel, Merriam's Chipmunk, Botta's Pocket Gopher, Agile Kangaroo Rat, Dulzura Kangaroo Rat, California Pocket Mouse, San Diego Pocket Mouse, Big-eared Wood Rat, California Deer Mouse, and Audubon's Cottontail.

The diversity of bird species in the chaparral increases sharply with the addition of seasonal migrants along with the winter and summer residents. Plant composition of a particular chaparral community will dictate their diversity and abundance. For example, robust stands of Manzanita, Scrub Oak, Ceanothus, Sugar Bush, and a nearby seasonal creek will all favor greater numbers. Common year-round species likely to be seen or heard are California Quail, Mourning Dove, Red-tailed Hawk, Red-shouldered Hawk, Coopers's Hawk, Roadrunner (Figure 6.19), Great Horned Owl, Anna's Hummingbird, Northern (Red-shafted) Flicker, Acorn Woodpecker, Nuttall's Woodpecker, Scrub Jay, American Crow, Common Raven, Wrentit, Bushtit (Figure 6.20), Bewick's Wren, California Thrasher, California Towhee, Spotted Towhee, Lesser Goldfinch and House Finch. Common wintering birds are the Hermit Thrush, White-crowned Sparrow, and Dark-eyed (Oregon) Junco.

FIGURE 6.19 The Greater Roadrunner (*Geococcyx californianus*) ranges across the southern United States from California to Arkansas and Louisiana. It occupies a good deal of Mexico as well. This well-known bird is as at home in open chaparral as it is in rocky deserts and grasslands.

FIGURE 6.20 The Bushtit (*Psaltriparus minimus*) is common in mature chaparral and brushy woodlands of Southern California where it is often seen moving through an area in straggling flocks. This one is perched in a Scrub Oak (*Quercus berberidifolia*). These tiny birds with long tail feathers build elegant pendulous nests.

6.5 CALIFORNIA EVERGREEN WOODLAND

Mixed hardwoods growing in oak-pine associations with Digger Pine and Blue Oak are a hallmark of the Central Valley foothills. The southernmost of these associations only reach northern Los Angeles County. In Southern California, mixed evergreen forests are restricted to cooler, moister interior valleys and foothills west of the mountain ridges to around 3,000–4,800 ft (910–1,460 m) as in the San Gabriel, San Bernardino, Palomar and Cuyamaca Mountains. Common species include deciduous Black Oak (*Quercus kelloggi*), Canyon Oak (*Q. chrysolepis*), Incense Cedar (*Calocedrus decurrens*), Pacific Madrone (*Arbutus menziesii*), California Bay (*Umbellularia californica*), Coulter Pine (*Pinus couleri*), and Big-cone Douglass Fir (*Pseudotsuga macrocarpa*), a Southern California endemic.

At lower elevations up to around 3,000 ft (910 m) Coast Live Oak (*Quercus agrifolia*) is the predominate evergreen. To a lesser extent Engelmann Oak (*Q. engelmannii*) is also featured. Seedlings of Coast Live Oak are intolerant of grazing, which along with urbanization has reduced much of this iconic oak community throughout Southern California. The best-developed stands are now in San Diego County. It may dominate on north slopes, in deep soils, and alluvium of canyon bottoms. At lower elevations it usually occurs as savannah-like *encinals* (from the Spanish for live oak) with few other woody plants interspersed. With increased elevation, especially southward, it becomes confined to riparian areas of dense chaparral where it associates with, Engelmann Oak, Toyon, manzanita Sugar Sumac, Ceanothus, and other chaparral shrubs.

Englemann Oak is a large, semi-evergreen "white oak" growing in deep clay soils. Never as extensive as the live oak series, most of the Engelmann woodland has been lost to agriculture and development. Associations of Engelmann Oak persist on the Santa Rosa plateau, Mesa de Burro, and Mesa de Colorado in the Santa Ana Mountains. South into San Diego County, it is most robust in the areas around Ramona and Santa Ysabel. From there it ranges sporadically in other interior valleys south to Tecate in northern Baja. Most chaparral plants are absent from the Engelmann Oak series because of soil conditions and level ground. The understory is otherwise composed of grasses and forbes.

6.5.1 Oak-Evergreen Woodland Vertebrates

No vertebrates are restricted to Oak/Evergreen woodland. All of the species that can be found there also occur in adjacent habitats of chaparral or montane forest. Among amphibians, for example, the Arboreal Salamander does have an affinity for oaks and sycamore, but it also occurs in coastal sage scrub and chaparral; likewise for the Garden Slender Salamander. The Monterey Ensatina also inhabits oak woodland but is generally more common in pine, and pine-oak forests at higher elevations. The Western Toad is a resident from the coast to the deserts and is one of the most common amphibians in Southern California. The same is true of the Pacific Treefrog, which is usually (though not necessarily) associated with ponds, creeks and man-made water sources. The more restricted Western Spadefoot is found in grasslands, chaparral, and oak woodland where friable soils are available for burrowing. This alluring, toad-like frog is listed as a Federal Species of Special concern, a California species of Special Concern, and a Bureau of Land Management Sensitive Species. Common reptiles of Oak/Evergreen Woodland include species that are also at home in a variety of other communities. Among these are the Western Fence Lizard, Side-blotched Lizard, the burrowing and secretive California Legless Lizard, Southern Alligator Lizard, Gopher Snake, California Kingsnake, Spotted Night Snake, and the region's most commonly encountered rattlesnake, the Southern Pacific Rattlesnake.

As with amphibians and reptiles, no bird species are confined to Oak/Evergreen Woodland. Some of the more frequently encountered residents are the Western Screech Owl, Acorn Woodpecker, Nuttall's Woodpecker, Northern Flicker, Western Wood-Pewee, Plain Titmouse, Bushtit, House Wren, Western Bluebird, and Hutton's Vireo.

Wild turkeys (*Meleagris gallopavo*) are common in oak woodlands, but they are not native to California. Turkeys were present here in the Pleistocene, but those birds were a different species that became extinct 10,000–12,000 years ago. Modern, North American turkeys are now established in oak woodlands in many parts of the state including Southern California, particularly in San Diego County. Attempts to establish turkeys as game birds started early. Wild birds were introduced to Santa Cruz Island in 1877. In 1908, stock acquired from Mexico was released in the San Bernardino Mountains. Neither of these transplants succeeded. Between 1928 and 1951, more than 3,000 farm-raised birds were released in 23 counties across the state, and again establishment failed. Finally, in 1993, California Department of Fish and Game released 234 wild birds on to private land in the foothills of San Diego County. Seven years later they had spread to southern Riverside County and to within 10 mi of the Mexican border. Although they remain concentrated in oak woodland with scattered underbrush, wild turkeys have shown up in the chaparral, the coastal lowlands, and even the high desert. Conservation groups and native plant societies opposed the 1993 introductions because (1) introducing non-native species of any kind is simply bad practice, and (2) wild turkeys have the potential to negatively affect sensitive plants and rare amphibians and reptiles. On the other side, they are attractive to sport hunters.

Mammals common in Oak/Evergreen Woodland are the California Ground Squirrel, Western Pocket Gopher and, in undisturbed areas, Mule deer.

6.6 MONTANE CONIFEROUS FOREST

Vast tracks of conifers and winter deciduous trees stretch across the northern United States and southern Canada. Substantial regional variation in structural and taxonomic composition correlates with differences in climate and geologic history. Mixed deciduous forests dominate in the east, whereas the singularly impressive redwoods define California's north coast. In between are the Rocky Mountains, the Great Basin woodland of piñon and juniper, and the Sierran-Cascadean coniferous forests.

Their woody fruit or "cones" (except for the berry-like fruits of Junipers) make conifers easily recognizable. Leaves are long and slender (needles), or small, overlapping, and scale-like (cypress and junipers). Most are evergreen. In the West they include larches (conifers of cold, high latitudes, absent from California), spruce (*Picea* spp.), firs, (*Abies*), Douglass-fir (*Pseudotsuga* spp.), hemlock (*Tsuga*), California Torreya (*Torreya californica*, a Sierra Nevada relic of a previous geologic time), yew (*Taxus* spp.), Redwood (*Sequoia sempervirens*), Giant Sequoia (*Sequoiadendron giganteum*), "cedars" (e.g., Western Redcedar [*Thuja*] and Incense-Cedar [*Calocedrus*]), cypress (*Cupressus* spp.), juniper (*Juniperus* spp.), true cedars (*Cedrus* spp., introduced in gardens and parks from Asia, north Africa, and the Middle East), and pines including piñon (*Pinus* spp.).

Overall, Southern California's montane forests are not as diverse in species as those of the Sierra Nevada. The standard conifers in our mountains are pines, piñons, firs, "cedars," Douglas-firs, junipers, and cypress. Well-developed stands grow above the chaparral and evergreen woodlands beginning at around 4,500 ft (1,370 m) in the San Gabriel, San Bernardino (Figure 6.21), San Jacinto, Santa Rosa, Palomar, and Cuyamaca Mountains, and into the Sierra San Pedro Martir of northern Baja California. Areas of chaparral and desert isolate our coniferous forests from one another, but they share most of the same species. Species composition is dictated by elevation, soils, moisture, and slope orientation much like at higher latitudes. They grow and germinate best on bare, inorganic (mineral) soils with annual precipitation of 25 in. (700 mm) or more, much of that falling as snow. The growing season is abbreviated by the lack of late summer rain.

FIGURE 6.21 Above 4,500 ft (1,365 m) Jeffrey and Ponderosa Pines usually dominate Southern California's montane forests of the Transverse and Peninsular Ranges. Other confers include firs, junipers, and cypress.

Jeffrey Pine (*Pinus jeffreyi*) and Ponderosa Pine (*P. ponderosa*), members of the "yellow" pines, are typically the dominant conifers in our mountains. Yellow and white are the principle groups of pines. Yellow pines, sometimes called "hard pines," have denser, more resinous wood; the needles are in clusters (usually) of two to three, and the egg-shaped cones have thick, thorny scales. White, or "soft pines" (e.g., Sugar Pine [*P. lambertiana*] and Limber Pine [*P. flexilis*]) have soft clear wood and thin blue-green needles in clusters (usually) of five; cones are elongate with thin scales lacking spines and thorns. In Southern California, Jeffrey and Ponderosa Pines may exist together in mixed stands, but more often they grow as separate series.

Ponderosa forests and Jeffery pine forests are similar in appearance. Both have an open, parkland character, and share numerous understory species of shrubs. Learning to differentiate them is a matter of becoming familiar with their cones, needles, and bark. Jeffrey pine has larger cones (6–8 and up to 10 in; ca. 160–250 mm) with moderate-sized prickles that are turned inward. The bark is darker and has the fragrance of pineapple or vanilla. The cones of Ponderosa Pine are on average smaller (3–6 in.) dull brown instead of yellow-brown, and the prickles curve outward. The bark is yellowish as opposed to the purplish or rosy hue of Jeffrey Pine.

Ponderosa Pine is the most common and widespread conifer in the West. It generally occurs at lower elevations than other pines. In California, Ponderosa Pine reaches its southern limit in the Cuyamaca Mountains of San Diego County. It is most common on the wetter, west-facing mountain slopes, generally at lower elevations than Jeffrey Pine, between 4,000 and 6,500 ft (1,215–1,975 m; Figure 6.22). At lower elevations it is often associated with Coulter Pine (*P. coulteri*), and California Black Oak (*Quercus kellogi*). At higher elevations with more moisture, it occurs with White Fir (*Abies concolor*), Sugar Pine (*P. lambertiana*), Bigcone Douglas Fir (*Pseudotsuga macrocarpa*), Incense Cedar (*Libocedrus decurrens*), and Canyon Live oak (*Quercus chrysolepis*). The understory supports

FIGURE 6.22 Chaparral and Ponderosa Pine converge near Idyllwild at around 5,000 ft (1,520 m). Manzanita and Red Shank dominate in the foreground.

various species of manzanita and occasionally "wild lilac" (*Ceanothus* spp.) along with Mountain Misery (*Chamaebatia* foliosa), Hairy Yerba Santa (*Eriodictyon trichocalix*), and California Buckthorn (*Rhamnus californica*).

Jeffrey Pine is more cosmopolitan in Southern California, tolerating a wider range of temperatures and soils, and replaces Ponderosa Pine completely in Baja California and the mountainous slopes bordering the Mojave. In the latter instance it associates with Singleleaf Pinyon (*Pinus monophylla*), Coulter Pine, Sierra Juniper (*Juniperus occidentalis*), and Mountain Mahogany (*Cercocarpus betuloides*, *C. ledifolius*). At higher, moister localities it often occurs with some of the same coniferous associates as Ponderosa stands, namely Sugar Pine, Incense Cedar, Bigcone Douglas-fir, and White fir.

Pines are the world's most-important source of timber, but in Southern California, they have not been subject to the intense harvesting as in the Pacific Northwest. Still, their economic impact here is huge with respect to recreation and as water sheds. Pine seeds are in the diet of many birds and small mammals, and twigs and needles serve as forage for deer and other browsers.

6.6.1 Montane Forest Vertebrates

Southern California's montane forests support amphibians such as the Arboreal Salamander, California Slender Salamander, Monterey Ensatina, the Large-blotched Ensatina, Western Toad, and Pacific Treefrog. Besides the widespread Western Fence lizard and Sagebrush Lizard, other montane reptiles include the secretive Gilbert's Skink, California Kingsnake, California Mountain Kingsnake (a Southern California endemic, Figure 6.23), Gopher Snake, and the Southern Pacific Rattlesnake.

FIGURE 6.23 The beautiful California Mountain Kingsnake (*Lampropeltis zonata*) is a Southern California endemic restricted to scattered localities in coniferous and pine-oak woodland.

FIGURE 6.24 Common in pine and pine-oak woodland, the White-breasted Nuthatch (*Sitta carolinensis*) forages for insects and seeds by circling head first down the trunks of large trees. Seeds are often cached in the crevices of the bark for later retrieval.

Common birds include Stellar's Jay, Clark's Nutcracker (at the interface with subalpine forests), Western Screech Owl, White-headed Woodpecker, Williamson's Sapsucker, Red-breasted Sapsucker, Northern Flicker, Mountain Quail, Calliope Hummingbird, Pygmy Nuthatch, White-breasted Nuthatch (Figure 6.24), Brown Creeper, American Robin, Western Bluebird, Dark-eyed Junco, Pine Siskin, and Purple Finch. Present, but more restricted are the Spotted Owl, Northern Saw-whet Owl, Townsend's Solitaire, Western Tanager, and Red Crossbill.

Merriam's Chipmunk is a staple of Southern California's montane coniferous forests. Among other mammals are the Broad-footed Mole, Botta's Pocket Gopher, Agile Kangaroo Rat, Piñon Deer Mouse, Western Gray Squirrel, Long-tailed Weasel, and Mule Deer.

6.7 SUBALPINE CONIFEROUS FOREST

The Rocky Mountains, Sierra Nevada, and Cascades each have impressive peaks and ranges consisting of subalpine forests. In the Sierra Nevada they occur from around 9,500–11,000 ft (2,885–3,340 m). Precipitation averages from 30 to 50 in. (750–1,250 mm), mostly as snow, and the brief growing season lasts from 7 to 9 weeks. In Southern California, subalpine forest occurs only at the highest peaks of the San Gabriel, San Bernardino, and San Jacinto Mountains. Primarily composed of Lodgepole Pine (*P. murrayana*) and Limber Pine (*P. flexilis*), they are relatively open forests and of shorter stature then the coniferous forests down slope. Absent are such subalpine Sierran species as Mountain Hemlock (*Tsuga mertensiana*), Western White Pine (*P. monticola*), Whitebark Pine (*P. albicaulus*), and Foxtail Pine (*P. balifouriana*), a close relative of the Bristlecone Pine (*P. aristata*). Aspen, a common successional member in subalpine forests in the north is absent from Southern California as well. Scattered patches of understory plants may include Whitethorn (*Ceanothus cordulatus*), Green Manzanita (*Arctostaphylos patula*), Mountain Mahogany (*Cercocarpus ledifolius*), currants (*Ribes* spp.), and Bush Chinquapin (*Castanopsis sempervirens*).

6.7.1 Subalpine Vertebrates

The only reptiles that venture into the subalpine forests are the Western Fence Lizard, Sagebrush lizard, and the Southern Pacific Rattlesnake. Nesting birds include Williamson's Sapsucker, Clark's Nutcracker, Red-breasted Nuthatch, Ruby-crowned Kinglet, and Cassin's Finch. Also present but equally representative of mixed coniferous forests are the Red Crossbill, Pine Siskin, Dusky Flycatcher, and Mountain Chikadee.

Mammals of the subalpine forests are few, namely the Northern Flying Squirrel, Golden-mantled Ground Squirrel, Lodgepole Pine Chipmunk, and Long-tailed Vole.

6.8 THE DESERTS

More than two-thirds of Southern California is desert—a life zone defined by meager rain and seasonally brutal temperatures, hot or cold. Sometimes they become ferociously windy. Our deserts, the Mojave and Colorado, are contiguous in space but separate in character. Although they share many of the same species, each has its own subtle associations of plants and animals that derive from latitude, topography, the amount and distribution of rain, temperature extremes, and history.

North America claims four major deserts. The *Great Basin* is cold-temperate. It occupies much of the intermountain west between the Rocky Mountains in the east and the Sierra Nevada and Cascades in the west. Some components of Great Basin Scrubland trickle into the northern Mojave and other parts of Southern California. The two warm-temperate deserts are the Mojave and the Chihuahuan. The Chihuahuan Desert extends from extreme southeast Arizona through large parts of southern New Mexico and west Texas and south into northern Mexico east of the Sierra Madre (Figure 6.25). The fourth North American desert is the Sonoran, a biome with historically subtropical and tropical affinities centered about the Gulf of California. It occupies most of northwest Mexico (Sonora state), southwestern Arizona, southeast California, and much of eastern and central Baja California (Figure 6.25). West of the Colorado River, in southeast California, the Sonoran Desert is commonly referred as the "Colorado Desert."

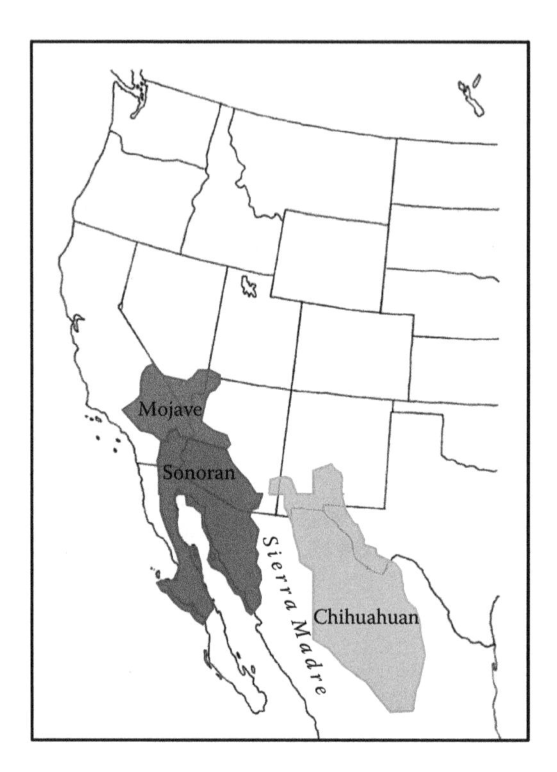

FIGURE 6.25 The three major deserts of the American Southwest. The Mojave and Sonoran Deserts merge with one another in Southern California and western Arizona. The Colorado Desert is the extreme western part of the Sonoran Desert that occupies Southern California. The Chihuahuan Desert is largely isolated from the Sonoran Desert by the Sierra Madre of northern Mexico.

6.9 GREAT BASIN DESERT SCRUBLAND

The Great Basin and Range province of North America consists of a series of north-south trending mountain ranges and valleys between the Rocky Mountains and the Sierra Nevada, mostly north of the 36th parallel. Winters are cold, but daily and seasonal temperature extremes are typical. Rainfall is low and scattered throughout the year. Deep, porous, often saline, soils are the norm. Vegetation in the basins and at lower elevation consists of low-growing, rounded shrubs comprising the sagebrush scrub. The dominant species is Great Basin, or Big Sagebrush (*Artemisia tridenta*). Blackbush (*Coleogyne ramoissima*), rabbit brush (*Chrysothamnus* spp.), saltbush (*Atriplex* spp.), Cotton-Thorn (*Tetradymia spinosa*), and Bitterbrush (*Purshia glandulosa*) also are represented along with various grasses (family Poaceae). Great Basin Scrubland reaches south into Southern California between 2,900 and 5,400 ft (880–1,640 m) along the southern Sierra Nevada foothills (Figure 6.26), and the northern slopes of the San Bernardino Mountains.

FIGURE 6.26 Great Basin Scrub grows along the east slope of the Sierra Nevada south into the northern Mojave Desert as seen here.

6.10 MOJAVE DESERT

Located largely between 34° and 36° latitude, the Mojave is the smallest of the North American deserts, yet still occupies about 48,000 mi^2 (ca. 124,000 km^2) and includes portions of four states. A little more than half of that area is in California, consisting of various parts of Los Angeles, Riverside, San Bernardino, Kern, and Inyo Counties (Figure 6.27). The western edge borders the Tehachapi Mountains and southern end of the Sierra Nevada. Here it extends up Indian Wells Valley into southern Inyo County and the Owens Valley. In California, the Panamint Range and Death Valley comprise the northeast sector of the Mojave where it transitions to the Great Basin. The southern end borders the Transverse Ranges (San Gabriel, San Bernardino, and Little San Bernardino Mountains) and passes through Joshua Tree National Park to the southeast. The Mojave extends east into southern Nevada, mainly Clark County and parts of Nye and Lincoln Counties, as well as to extreme southwestern Utah to St. George. It also covers extreme northwest Arizona to around Kingman.

The Mojave's southeast transition to the Sonoran Desert is rather ill-defined. Numerous "indicator species" comingle in this region, and one can encounter both Mohave and Sonoran representatives. For example, the giant Saguaro cactus (*Carnegiea gigantean*, Figure 6.28) that may reach 30 ft (ca. 9 m) is a Sonoran Desert benchmark conspicuous in much of southern Arizona. It is local in California below 1,500 ft (ca. 450 m) in Creosote Bush Scrub near the Colorado River west of Parker Dam (ca. Whipple Mountains) in eastern San Bernardino County. The southeast transition zone is further complicated by the presence of some Great Basin desert scrubland and Great Basin coniferous woodland. Joshua Tree National Park is an excellent place to experience the transitional flora of the

FIGURE 6.27 The Mojave Desert is Southern California's most expansive feature. It occupies portions of four states although more than half of it lies in California. The western Mojave consists of the triangular Antelope Valley plain, bordered by the Tehachapi Mountains on the northwest and the Transverse Ranges to the southeast. The eastern Mojave is dominated by isolated mountain ranges and gives way to the Colorado Desert southeast of Joshua Tree National Park. (Base map adapted from www.nationalatlas.gov.)

FIGURE 6.28 The impressive Saguaro cactus (*Carnegiea gigantea*) grows throughout much of the Sonoran Desert in southern Arizona and northwest Mexico (above) but occurs in California only in a small transitional area of southeastern San Bernardino County, across the Colorado River from Arizona. Saguaros are favored sites for some cavity-nesting birds like the Ferruginous Pygmy Owl and the Gilded Flicker.

low deserts as well as the shift to upland vegetation, for instance chaparral on the east slopes of the Little San Bernardino Mountains.

The physical geography of the Mojave consists of three regions. Although they transition quietly from one to the other, each is recognizable on its own. The western extreme is an apex of a triangle. One side of the triangle is formed by the Tehachapi Mountains and southern Sierra Nevada. It extends northeast. The other side runs southeast following the base of the north-facing slope of the San Gabriel Mountains. The triangular apex also marks the intersection of the San Andreas Fault and the Garlock Fault along I-5 near Gorman. Within these parameters the western Mojave is composed of the essentially flat Greater Antelope Valley. The average elevation is 2800 ft (850 m), and it is the most populated area of the Mojave Desert. Lancaster and, to its south, Palmdale are the largest communities. Agriculture was once widespread in this area, but it has given way to housing tracts in recent decades. The military also occupies a significant amount of the area in Edwards Air Force Base and the China Lake Naval Weapons Center.

The central Mojave is defined by the Mojave River Valley and associated scattered mountains. It includes the region between Victorville and Barstow east to Soda Dry Lake, and south to Joshua Tree National Park (Figure 6.29). Valley floors vary from around 650 to 3,000 ft (ca. 200–910 m at Joshua Tree National Park). Mountains typically range from about 3,500 to 5000 ft (1,060–1,520 m). Drainages from the local mountains are nearly all annexed by the Mojave River basin. The Mojave River itself drains the north slopes of the San Bernardino Mountains and terminates at Soda Lake and Silver Dry Lake near Baker (Figure 6.30). Surface flow is negligible over most

FIGURE 6.29 Joshua Tree National Park is another crown jewel of Southern California. The boulder-strewn landscape occupies a transitional area composed of elements of our two deserts (Mojave, Colorado) plus chaparral. (Courtesy of Mitch Leslie.)

FIGURE 6.30 Soda Dry Lake near Baker is the underground terminus of the Mojave River. Sand blown off the lake is redeposited at Devil's Playground and Kelso Dunes (background *right*). In the mid-ground is Little Cowhole Mountain with Old Dad Mountain behind it (*left*).

of the Mojave River's length except during particularly wet winters. Most of the year the river flows underground, sand on top. In some areas, however, bedrock constricts the channel and water is forced to the surface, for example at the Mojave Narrows in Victorville and at Afton Canyon between Barstow and Baker (Figure 6.31).

The Mojave Basin was a much larger river system during the pluvial (wetter), latest Pleistocene about 15,000 years ago. At that time, it filled a large basin near Barstow known as Lake Manix. Eventually the water in Lake Manix broke through its natural dam and flowed eastward to create Afton Canyon. From there it flowed into Soda Lake, Silver Lake, and probably north into Death Valley.

Today, good portions of Mojave Basin as well as Death Valley sit atop vast reservoirs of underground rivers, lakes, and aquifers. Springs are thus not uncommon in these deserts. Most are usually no more than a trickle, although in some areas surface water can be substantial, as at Saratoga Springs in the southern end of Death Valley–a haven for migratory birds (Figure 6.56).

Much of the eastern Mojave represents the southern extension of the Basin and Range Province, which is a composition of north-south trending mountain ranges separated by wide valleys. Known as *horst* (meaning "uplifted blocks") and *graben* (meaning "grave") topography, these mountains and valleys are the essence of the eastern Mojave landscape. The valleys are wider, and the mountains are not as high as in the Great Basin, probably because of the thinning crust of the North American plate in that area, coupled with the right-lateral (easterly) slip of the Mojave block brought on by activity of the San Andreas and Garlock Faults.

FIGURE 6.31 The Mojave River flows underground most of the year except where bedrock constricts the channel and water is forced to the surface, as here at Afton Canyon between Barstow and Baker. This riparian community is a lush contrast to the surrounding flats, mesas, and hills.

Telescope Peak in the Panamint Mountains is exceptional in reaching 11,049 ft (3,355 m). (A scant 15 miles to the east, near Badwater in Death Valley, is the lowest point in the Western Hemisphere, at 283 ft [86 m] below sea level). More typical are ranges like the Clark Mountains, New York Mountains, and Providence Mountains, which are fault blocks with elevations of between 7,500 and almost 8,000 ft (2,280–2,430 m). Relatively recent volcanic activity along the thinning crustal faults is evident in the form of interspersed cinder cones and lava fields, as seen around Pisgah, Cima, Amboy, and other areas (Figure 6.32). Faulting can result in trapped magma that heats any water lying underneath it. Where water surfaces, hot springs appear. These springs are surrounded by salt encrusted sediment in which only strongly saline tolerant plants can exist.

6.10.1 Plant Associations

The Mojave is the driest of the North American deserts. The western sector receives around 5 in. (130 mm), the eastern end only about 2 in. (50 mm). Death Valley averages slightly less than 2 in, and some years experience a mere half-inch. Most rain in the western Mojave comes from winter storms that breach the Transverse Ranges, whereas the eastern end faces better odds of picking up summer storms. The Mojave's varied topography, soils, and moisture regimes are responsible for the numerous and varied plant associations in this desert. More than 2,600 species are known below 7,500 ft (2,280 m) in elevation.

FIGURE 6.32 Inside the Mojave National Preserve, creosote and scattered Joshua Trees give perspective to the volcanic cinder cone in the background.

Common families (Table 6.1) include the Asteraceae (sunflower), Brassicaceae (mustard), Cactaceae (cactus), Chenopodiaceae (goosefoot), Ephedraceae (ephedra), Euphorbiaceae (spurge), Fabaceae (pea), Lamiaceae (mint), Liliaceae (lily), Poaceae (grasses), Rosaceae (rose), Scrophulariaceae (figwort), Solonaceae (nightshade), and Polygonaceae (buckwheat), About one-fourth of the plant

TABLE 6.1
Common Mojave and Colorado Desert Shrubs and Trees

Acanthus Family (Acanthaceae)
  Chuparosa (*Justica californica*)—Col
Agave Family (Agavaceae)
  Desert Agave (*Agave deserti*)
  Joshua Tree (*Yucca brevifolia*)—Moj
  Mojave Yucca (*Yucca shidigera*)
  Our Lord's Candle (*Hesperoyucca whipplei*)
Bignonia Family (Bignoniaceae)
  Desert Willow (*Chilopsis linearis*)
Buckwheat Family (Polygonaceae)
  Buckwheat (*Eriogonum* spp.)
Cactus Family (Cactaceae)
  Beavertail (*Opuntia basilaris*)
  California Barrel Cactus (*Ferocactus cylindraceus*)

(*Continued*)

TABLE 6.1 (*Continued*)
Common Mojave and Colorado Desert Shrubs and Trees

Diamond Cholla (*Cylindropuntia ramoissima*)
Fish-hook Cactus (*Mammilaria dioica, M. tetrancistra*)
Hedgehog Cactus (*Echinocereus engelmannii*)
Silver Cholla (*Cylindropuntia echinocarpa*)
Teddy Bear Cactus (Jumping Cholla) (*Cylindropuntia bigelovii*)
Caltrop Family (Zygophyllaceae)
Creosote Bush (*Larrea tridentata*)
Cypress Family (Cupressaceae)
California Juniper (*Juniperus californica*)
Utah Juniper (*Juniperus osteosperma*)—Moj
Ephedra Family (Ephedraceae)
Ephedra (Desert Tea) (*Ephedra* spp.)
Green Ephedra (*Ephedra viridis*)
Four O'clock Family (Nyctaginaceae)
Desert Sand Verbena (*Abronia villosa*)
Goosefoot Family (Chenopodiaceae)
Bush Seepweed (*Suaeda nigra*)
Four-wing Salt Bush (*Atripix canesens*)
Hop-sage (*Grayia spinosa*)—Moj
Iodine Bush (*Allenrolfea occidentalis*)
Shadescale (*Atriplex confertifolia*)
Winter Fat (*Krascheninnikovia lanata*)—Moj
Grass Family (Poaceae, Gramineae)
Big Galleta (*Hilaria rigida*)
Salt Grass (*Distichlis spicata*)
Lizard's-tail Family (Saururaceae)
Yerba mansa (*Anemopsis californica*)
Loasa Family (Loasaceae)
Sandpaper Plant (*Petalonyx thurberii*)
Mint Family (Lamiaceae, Labiatae)
Bladder Sage (*Scutellaria mexicana*)
Desert Lavender (*Condea emoryi*)
Desert Sage (*Salvia dorrii*)
Thistle Sage (*Salvia carduacea*)
Nightshade Family (Solonaceae)
Box Thorns (*Lycium* spp.)
Ocotillo Family (Fouquieriaceae)
Ocotillo (*Fouquieria splendens*)—Col
Palm Family (Palmae)
California Fan Palm (*Washingtonia filifera*)—Col
Pea Family (Fabaceae, Leguminosae)
Blue Palo Verde (*Parkinsonia florida* [*Cercidium floridum*])—Col

(*Continued*)

TABLE 6.1 (*Continued*)
Common Mojave and Colorado Desert Shrubs and Trees

Cat Claw (*Senegalia greggii* [*Acacia greggii*])
Desert Ironwood (*Olneya tesota*)—Col
Indigo Bush (*Prosopis galandulosa*)
Mesquite (*Prosopis schottii*)
Smoke Tree (*Psorothamnus spinosus*)

Pine Family (Pinaceae)
Piñon Pine (*Pinus monophylla, P. edulis*)

Rose Family (Rosaceae)
Apache Plume (*Fallugia paradoxa*)
Bittterbrush (*Purshia tridentata*)
Black Bush (*Coleogyne ramoissima*)
Mountain Mahogany (*Cercocarpus* spp.)

Rue Family (Rutaceae)
Turpentine Broom (*Thamnosma montana*)

Spurge Family (Euphorbiaceae)
California Croton (*Croton californicus*)

Sunflower Family (Asteraceae)
Big Sage Brush (*Artemisia tridentata*)
Brittle Bush (*Encelia* spp.)
Cheese Bush (*Hymenoclea salosa*)
Cotton-Thorn (*Tetradymia axillaris*)
Desert Baccharis (*Baccharis sergiloides*)
Rabbit Brush (*Chrysothyamnus* spp.)
White Bur-Sage (*Ambrosia dumosa*)

Tamarisk Family (Tamarieaceae)
Salt Cedar (Tamarisk) (*Tamarix ramoissima*) Non-native

Col, colorado only; Moj, mojave only.

species are endemic and many are ephemeral—the annuals that respond to seasonal rain. The annuals germinate in either late winter and appear in late spring and early summer or germinate with summer rain and emerge in late summer and early fall.

The vegetation associations outlined below are not detached and separate. Rather, they occur as units that blend into one another or as part of larger associations. One or several dominant shrubs, however, characterize each.

6.10.1.1 Creosote Bush Scrub

Familiar to even the casual desert-goer, whether they know its name or not, Creosote Bush (*Larrea tridentate*—Figure 6.33; Zygophyllaceae) is *the* dominant shrub of the Mojave, as it is in the Colorado and Chihuahuan Deserts. This remarkable plant tolerates a wide range of conditions. It occurs over most of the desert floor, on alluvial fans (*bajadas*), and in dry desert washes. It may grow in

FIGURE 6.33 Creosote Bush (*Larrea tridenetata*) is the dominant shrub of the desert Southwest where it grows in a remarkably wide range of conditions. The narrow black bands on the stems arise from resinous secretions induced by a lac scale insect (*Tachardiella larrea*).

pure stands or as a transitional species in desert grasslands, mesquite woodlands, and around Joshua Trees. Creosote is most commonly associated with White Bur-sage (*Ambrosia dumosa*) when not growing in pure stands. These two plants in fact cover as much as 70% of the Mojave. Although White Bur-sage is the codominant plant in Creosote Bush Scrub, other species with narrower tolerances will be found where conditions are favorable. Some of the more common associates are Brittle Bush (*Encelia farinosa*, *E. actoni*, *E. virginiensis*), Cheese Bush (*Hymenoclea salsola*), Mojave Yucca (*Yucca shidigera*), Silver Cholla (*Opuntia echinocarpa*), and Beavertail (*Opuntia basilaris*).

Mojave Creosote generally displays a rounded form, but it can also be somewhat short and leggy. Chromosome number is the only character differentiating the Southwest populations of *Larrea*. Mojave creosote is hexaploid (six sets of chromosomes), whereas Sonoran populations are tetraploid (four sets). Chihuahua creosote has two sets. In all populations the small, thick and sticky redolent leaves are distinctive. The ½- to ¾-in. (10–15 mm) flowers with yellow petals bloom from November to May. Conspicuous narrow black bands on the stems arise from the resinous secretions induced by a lac scale insect (*Tachardiella larrea*). Creosote also hosts a gall midge that produces black, walnut-sized galls for their larvae.

Summer rains are required for successful sexual reproduction of creosote. The wind-tumbled seeds frequently germinate beneath White Bur-sage, which acts as a nurse plant. Creosote can also reproduce asexually during drought conditions.

FIGURE 6.34 This Sidewinder (*Crotalus cerastes*) all but disappears in the shadows and leafy detritus beneath a Creosote bush (find it in the center of the photo). These rattlesnakes are one of numerous desert vertebrates that find foraging opportunities and refuge in the sandy mounds that form around the base of the plant.

Older branches around the periphery of the plant die back and fall away. With the arrival of rain, new branches sprout from the outer edge of the root crown as clones. Hence, over time the plant grows outward in a ring and may reach several thousand years of age by this process. In addition to their amazing longevity, individuals are widely and evenly spaced because of a mechanism that inhibits the growth of adjacent individuals. In a process known as *root-mediated allelopathy*, the roots of one plant secret substances that inhibit root elongation of its neighbors.

Creosote is the most important canopy plant species of the desert. In that role, numerous birds, reptiles, and small mammals depend on it as a source of food and cover. Sand accumulations around the plant base provide burrowing opportunities and refugia (hibernacula and or aestivation sites) for the desert tortoise and such reptiles as the Fringe-toed Lizard, Western Whiptail, Zebra-tailed Lizard, and certain snake species like Sidewinder rattlesnakes (Figure 6.34). Among mammals, Kangaroo Rats hide in burrows around the base, and Pack Rats (*Neotoma* spp.) and the Black-tailed Jackrabbit forage around creosote for food. Native Americans valued Creosote for its medicinal properties and rendered it into teas and poultices.

6.10.1.2 Blackbush Scrub

Blackbush (*Coleogyne ramoissima*) is the dominant member of this association and may form extensive dense stands that from a distance turn the landscape a dark gray. Blackbush scrub occurs over a wide range in elevation, either on its own or in association with plants such as ephedra (*Ephedra* spp.),

Hop-sage (*Grayia spinosa*), Cheese Bush (*Hymenoclea salsola*), Winter Fat (*Krascheninnikovia lanata*), Turpentine Broom (*Thamnosma montana*), Joshua Tree, and Pinyon–Juniper.

6.10.1.3 Joshua Tree Woodland

Unique to the Mojave Desert, Joshua Trees (*Yucca brevifolia*) are closely related to Our Lord's Candle (*Hesperoyucca whipplei*). They are easily recognizable by their fibrous trunks that can grow over 30 ft (9.1 m) tall (Figure 6.35). They begin life with a straight trunk until the apical meristem (growing tip) is destroyed by insects or by its own flowers, at which time the trunk then starts to branch. These stately images of the Mojave (named by the Mormons for the branches that appear as arms raised in biblical exaltation) grow between about 2,500 (760 m) and 6,000 ft (1820 m) in areas that receive at least 6 in. (150 mm) of rain per year. Some botanists maintain that a line drawn around their distribution would also delimit the perimeter of the Mojave Desert. Plants associated with Joshua Tree Woodland include the Mojave Yucca (*Y. shidigera*, Figure 6.36), Box Thorn (*Lycium andersonii*, *L. cooperi*), Bladder Sage (*Sautellaria mexicana*), various sage species (e.g., *Saliva dorrii*, *S. mohavensis*), and buckwheat (*Eriogonum* spp.). Raptors like Red-tailed Hawks use mature Joshua Trees for nesting sites (Figure 6.35), whereas the toppled, decomposing trunks provide ideal microhabitat for insects, spiders, scorpions, and other arthropods, along with reptiles like the small, reclusive Desert Night Lizard (*Xantusia vigilis*).

FIGURE 6.35 The stately Joshua Tree (*Yucca brevifolia*) of the Mojave Desert occurs from around 2,500 to 6,000 ft (760–1,820 m) where there is at least 6 in. (ca. 150 mm) of rain per year. Notice the nest of a Red-tailed Hawk near the top center.

FIGURE 6.36 Mojave Yucca, or Spanish Bayonet, (*Yucca shidigera*) ranges throughout Joshua Tree woodland and Creosote scrub in the Mojave Desert. It is also grows in Southern California's chaparral and coastal sage scrub communities.

6.10.1.4 Dunes

Massive, windblown deposits of sand can build up into spectacular mounds and hills, like those at Dumont and Kelso. The sand comes from the surface of dry lakes (playas) and erosion. For example, Kelso Dunes, located in the middle of the 2,500 mi^2 (6,475 km^2) Mojave National Preserve (Figure 6.37), developed from sand brought into Soda Dry Lake by the Mojave River, whence it was carried eastward by prevailing winds and settled on the lower slopes of the Granite and Providence Mountains.

Rainwater percolating through the dunes is protected from evaporation by overlying layers of sand. Plants growing on the dunes have deep roots to access the water. Two large, conspicuous dune species are Mesquite (*Prosopis glandulosa*) and Desert Willow (*Chilopsis linearis*—not a true Willow (*Salix* spp.), but it does have narrow, willow-like leaves). Smaller dune plants include California Croton (*Croton californicus*), Desert Sand Verbena (*Abronia villosa*), Sandpaper Plant (*Petalonyx thurberi*), and the grass Big Galleta (*Hilaria rigida*).

6.10.1.5 Desert Washes

The drainages of foothills and canyons transport tons of sand and small rock in ever-widening tracts as they find their way to the desert floor. Thunderstorms

FIGURE 6.37 The Mojave's Kelso Dunes develops from wind-blown sand off of Soda Dry Lake that settles on the lower slopes of the Providence and Granite Mountains. In this view, two Creosote bushes emerge among an expanse of grass (*Hilaria rigida*).

drive the flow, and periodic flooding and scouring require plants that inhabit these potentially unstable conditions to be not only well anchored but also produce seeds that can survive and germinate quickly if they are to take root and survive. Catclaw (*Acacia gregii*) and Smoke Tree (*Psorothamnus spinosus*) are good examples. In fact, these members of the Pea family have seed coats that must be tumbled and abraded before they can absorb water and begin germinating. Additional components of desert washes include Desert Willow (*Chilopsis linearis*), Cheese Bush (*Hymenoclea salsola*), and Arrowweed (*Pluchea sericea*).

6.10.1.6 Alkali Sinks

In valley floors below 2,000 ft (ca. 600 m), subsurface layers of clay or caliche (crusted calcium carbonate) can form an almost impermeable lens in which rainwater collects. The pooling water evaporates in the depressions leaving behind high concentrations of salts. Because salts draw water out of roots, plants that grow in these alkali (high pH) sinks must have counteracting mechanisms. Some will secrete the salts from stems and leaves, whereas others will sequester salts in special organs and tissues until those tissues eventually die back and fall away. Some common halophytes (salt-tolerant plants) found around these sinks are Salt Grass (*Distichlis spicta*), Iodine Bush

(*Allenrolfea occidentalis*), Bush Seepweed (*Suaeda nigra*), Yerba Mansa (*Anemopsis californica*), Mesquite (*Prosopis glandulosa*), and the introduced Salt Cedar (*Tamarix ramoissima*).

Outward from the extreme halophytes, where salt concentrations are not as high, other plants form a *saltbush scrub*. The Goosefoot family, Chenopodiacea, is especially well represented here, for example by Four-wing Saltbush (*Atriplex canesens*) and Shadescale (*A. confertifolia*).

6.10.1.7 Piñon-Juniper Woodland

Evergreens assume dominance at higher elevations from 4,500 to 8,000 ft (1,365–2,430 m) where temperatures are cooler and at least 12 in. (310 mm) of rain and or snow falls annually (Figure 6.38). Piñon Pine is one of the two codominants, of which Single-leaf Piñon (*Pinus monophylla)* is the more common. Two-needled Piñon (*P. edulis*) in the Mojave is restricted to eastern mountains. Piñon pine is slow growing. Nurse plants are required for successful germination because frost and freeze distortion of the soil would otherwise unsettle the delicate roots of seedlings. The other codominant evergreen is juniper. Stands of juniper may occur independently of Piñon pine, and either or both California Juniper (*Juniperus californica*) and Utah Juniper (*J. osteosperma*) may be present. Piñon-Juniper woodland is common in the Mountains of the Great Basin, and in the Mojave, several of the same understory associates can be found, for example, Big Sagebrush (*Artemisia tridentata*), Green Ephedra (*Ephedra viridis*), Apache Plume (*Fallugia paradoxa*), mountain mahogany (*Cercocarpus* spp.) and buckwheat (*Eriogonuum* spp.)

FIGURE 6.38 Piñon and Juniper woodland grow on the high desert slopes above 4500 ft (1,365 m). This view looks northeast from the San Bernardino Mountains into the Mojave Desert. At this particular locality, single-leaf Piñon (*Pinus monophylla*) and California Juniper (*Juniperus californica*) are on display.

6.11 SONORA–COLORADO DESERT

Drier and warmer climates were replacing the cooler, pluvial conditions of the late Pleistocene some 8,000 to 9,000 years ago. Much of northern Sonora was woodland composed of juniper, piñon, oak, pine, and sage. As winter storm tracts shifted northward, the ensuing drier conditions replaced the woodland with a subtropical flora adapted to drier environments. The Sonora Desert today covers 9° of longitude from about 26° to 35° N. Latitudinally, it extends from 109° to 117° W. At its southern extent in Baja California and mainland Mexico, the Sonoran Desert gives way to a semi-arid tropical biome known as *Sinaloan Thornscrub.* To the north it transitions to interior chaparral and the Mojave Desert. To the east it merges with the northern Chihuahuan Desert and semi-desert grasslands. The Peninsular Ranges in Southern California and northern Baja California define the western extent.

Sonora is unique among the North American deserts in its bimodal pattern of rainfall. Most of the annual precipitation comes from summer convection storms of tropical origin (Gulf of Mexico and Sea of Cortez). However, the Pacific storms that bring the bulk of annual rain to Southern California may bring winter rain to the Sonoran Desert when they move east and if they stay far enough south. In other words, the pattern is one of uneven biseasonality separated by dry springs and falls. Such a vast and varied region challenges our ability to coordinate biological diversity.

In looking for unifying features, the eminent desert ecologist Forest Shreve proposed seven Sonoran Desert provinces, each with particular characteristics of seasonal temperature, rainfall, and associated plants. Three of these are inarguably Sonoran Desert Scrub: *Central Gulf Coast* (Sonora and Baja California, Mexico), *Vizciano* (central Baja California), and the *Lower Colorado River Valley.* Most of Shreve's *Magdalena Plain* of southwest Baja California is now regarded as Sinaloan Thornscrub, as is his *Foothills of Sonora.* The remaining two provinces—*Arizona Upland* and *Plains of Sonora*—are a bit problematic in that they both contain a variety of low-growing but tree-like plants and could be regarded as thornscrub as well, at least in part.

Shreve described the Lower Colorado River Valley Province, which includes our Colorado Desert, as a microphyllus ("small leaf") desert. It is the largest and most arid of the Sonoran provinces. Average winter temperatures are about 4°C warmer than they are in the Mojave. Soils in drier areas are often capped by a thin, packed aggregate of small pebbles known as "desert pavement," embellished with a dark, reddish to blackish patina or "varnish" derived from silica and oxides of iron and manganese. Capillary evaporation of soil moisture carries the dissolved minerals to the surface where wind erosion polishes the patina to a gloss. Desert varnish, which can be found as well in the Mojave, also coats boulders and outcrops that Native Americans used for etching petroglyphs. Desert pavement supports no perennial plants, only a sparse seasonal output of short-lived annuals.

On sandier substrates, drainage patterns dictate the spatial distribution of plants. Where relief is negligible, shallow channels form an irregular network that typically leads to a low plain or "playa." Vegetation is scattered along this network, but from a distance, plant distribution appears homogeneous. In areas where topographic relief and rainfall are greater, channels converge into an obvious watercourse, or arroyo, that ultimately leads to some regional drainage. Plants grow linearly adjacent to the watercourse. In either case, the plant species are similar in that some are facultative wash species such as Mesquite (*Prosopis glandulosa*), Blue Palo Verde (*Parkinsonia* [*Cercidium*] *florida*), Desert Ironwood (*Olneya tesota*), and Smoke Tree (*Psorothamnus spinosa*). Others are more obligate wash species like Desert Willow (*Chilopis linearis*), Chuparosa (*Justica californica*), Sandpaper Plant (*Petalonyx thurberi*), White Bur-Sage (*Ambrosia dumosa*), Cat Claw (Senegalia [*Acacia*] *greggii*), Cheese Bush (*Hymenoclea salsola*), and Desert Baccharis (Broom) (*Baccharis sergiloides*).

By far, the most dominant plant community in the Lower Colorado River Valley Province is Creosote Bush Scrub, as it is in the Mojave and Chihuahuan Deserts. Creosote occurs over vast regions of the broad valleys, often in association with White Bur-Sage (*Ambrosia dumosa*) and on the slopes and bajadas in the western sector. Absent from the Colorado Desert are the giant columnar cactus that are so conspicuous in other parts of the Sonoran biome, like the iconic Saguaro (*Carnegiea gigantea*) of southern Arizona and northern Sonora state and the stately Cordon (*Pachycereus pringlei*) in Central Baja California. (A small population of Saguaro occupies the Mojave transition zone near the Colorado River; see above.)

What Californians call the *Colorado Desert*, in reference to the "River," is that portion of Sonora's Colorado River Valley Province restricted to the extreme southeast corner of California (Figure 6.39). The name derives from sixteenth-century Spanish mariners who first encountered the vast river delta at the head of the Gulf (Sea of Cortez). The reddish, silt-laden outflow thus inspired the name *El Río Colorado* (red river). The Colorado Desert encompasses about 11,000 mi^2 (ca. 28,500 km^2) and includes all of Imperial County, portions of eastern San Diego County, southeastern Riverside County, and the southeastern extreme of San Bernardino County. On the east it is bordered by the Colorado River and on the south by northern Baja California. To the north it transitions with the Mojave Desert southeast of Joshua Tree National Park. Its western edge concludes where high desert gives way to chaparral, piñon, juniper, and pine oak woodland of the Peninsular Ranges.

The predominant geological feature of the Colorado Desert is the Salton Trough. This enormous basin is the landward extension of the Gulf of California, specifically the subsiding rift of the East Pacific Rise spreading center that sits between the North American Plate on the east and the Pacific Plate on the west. The spreading center transitions to the San Andreas Fault Zone at the southern end of the modern Salton Sea. Not surprisingly the region is high in seismic and geothermal activity. The trough itself occupies about 8,360 mi^2 (21,650 km^2) or roughly two-thirds the area of the Colorado Desert. The entire basin is at or below sea level. In fact, the lowest point, –278 ft (84 m), is only 5 ft higher than the lowest point in Death Valley. Within this region lie the Coachella Valley, Imperial Valley, and the Salton Sea.

FIGURE 6.39 The Colorado Desert is that part of the Sonoran Desert west of the Colorado River. The Peninsular Ranges (e.g., San Jacinto and Laguna Mountains) define its western margin. The Salton Trough, the defining geophysical feature of the area, is the subsiding rift of the East Pacific Rise spreading center. It transitions to the San Andreas Fault Zone at the northern end of the Salton Sea (in blue). The entire basin, which includes the Coachella and Imperial Valleys, is at or below sea level, and its lowest point, –278 ft (85 m), is only 5 ft (1.5 m) higher than the lowest point in Death Valley. Anza Borrego is California's largest state park. (Base map adapted from www.nationalatlas.gov.)

For the past 3 million years the Colorado River deposited billions of tons of silt and sand as it forged a delta to the Gulf of California. Eventually the sand build-up created a huge levee along the western shore of the Sea of Cortez that effectively diverted the Colorado River from the Salton Trough and kept the Sea of Cortez from inundating the basin.

Lakes large and small filled the Salton Trough during glacial episodes of the Pleistocene. The largest was Lake Cahuilla, the name taken from a native tribe in the area. The present body of saline water called the *Salton Sea*, however, is the product of a man-made accident. In 1905 engineers attempted to bring Colorado River water to the Imperial Valley for irrigation. The soil was fertile and growing conditions were otherwise ideal. Hence a diversion canal was started south of Yuma, Arizona, in northern Baja California. The canal was undersized and inadequately constructed from the beginning. Within 2 years the "Imperial" canal was filled with silt and attempts by engineers to unblock it failed. Then the Colorado River breeched the head gates after particularly heavy rain and snowmelt and began flowing toward the Salton basin. In the process, two watercourses were created by erosion—the Alamo River in the east and the New River in the west. Essentially the entire flow of the Colorado River poured into the basin from these approximately 60-mi-long (96 km) waterways over a 2-year period. By 1907, the Salton Sea had filled the lowest depths of the basin. At 35 miles (56 km) long and 15 mi (24 km) wide, it is the largest inland body of water in California (Figure 6.40).

Creation of the Salton Sea had a twofold consequence on land use and public policy. One of those is that the periodic flooding of the Imperial Valley spurred efforts to dam the Colorado River for flood control. In 1929, construction began on Hoover Dam in Black Canyon. It was completed in 1935. Lake Havasu was created by Parker Dam 155 mi downstream in 1938, and in the middle of this stretch Pyramid Canyon became Lake Mojave by the Davis Dam in 1951. The U.S. Bureau of Reclamation also authorized construction of the All-American Canal along with the Hoover Dam project. This conveyance carries irrigation water 60 mi from the river to distribution points in the Imperial Valley.

On its own, water in the Salton Sea eventually would have evaporated. This inevitability led to the second consequence—a decision to let it serve as a sump

FIGURE 6.40 The Salton Sea was created in 1905 when a poorly constructed irrigation canal from the Colorado River was breached and floodwaters flowed into the lowest reaches of the Salton Trough. By 1907 the inundated basin became the largest inland body of water in California. It presently extends for 35 mi (ca. 55 km) up to the Coachella Valley with a width of 15 mi (24 km) and maximum depth of 52 ft (19 m).

for irrigation wastewater seeping in from the Imperial Valley. That water, in a bit of irony of course, originates from the Colorado River.

For the moment, Salton Sea hydrology is at a tenuous equilibrium. Inflow from the Alamo River, New River, and irrigation more or less matches water lost to evaporation. The surface is 226 ft (68 m) below sea level, and it has a maximum depth of 52 ft (16 m). That will likely change. Southern California's never-ending quest for freshwater led to a 2003 decision between the state and the Imperial County Water District providing for a 200,000 acre-feet transfer to San Diego. After 2019, when the agreement is in full effect, the Salton Sea will lose about 30% of its surface area. Consider also that at its inception in 1907, the salinity was about 3,500 parts per million. It now hovers around 45,000 parts per million, roughly 30% higher than the Pacific Ocean off the west coast. The salts, mostly sodium chloride, come from those naturally occurring in the soil, others leached in from irrigation, and salts from fertilizers. With a decreased inflow of water, salinity of course will increase even more. This has potentially serious consequences for two other rolls of the Salton Sea aside from it being a catchment for Imperial Valley runoff—sport fishing and as habitat for migratory birds.

The initial fish population of the Salton Sea consisted of freshwater species that arrived with floodwater from the Colorado River. They died out over the years because of the increasing salinity. The California Department of Fish and Game in 1930 began stocking numerous saltwater species deemed suitable for sport fishing, but of those intentionally planted only three have managed to adjust to the ever-increasing salinity: Corvina (*Cilus* sp.: Sciaenidae), Sargo (*Anisotremus* sp.: Pristipomatidae), and Bairdiella (*Bairdiella* sp.: Scienidae). Their continued survivorship is uncertain. Corvina, for example, will not spawn in salinities higher than 45,000 parts per million.

Spectacular numbers of shorebirds, ducks, other waterfowl, and even terrestrial species, arrive at the Salton Sea during migration (Figure 6.41). Many spend the winter, attracting bird-watchers from all over the country. Some 400 species of birds have been tallied. We can assume that a diminished and hyper-saline sea will adversely affect avian diversity and abundance. First to be affected will be the fish-eating species like cormorants, gulls, herons, egrets, and pelicans.

Saving the Salton Sea is an epic challenge for California. Left unattended a dying body of water this large will witness environmental and recreational resources drying up like the sea itself. The bare land now encrusted with pollutants and stirred up by winds will become an even worse health hazard for valley residents. Sadly, numerous rescue plans for the Salton Sea have been advanced over the years only to whither from inadequate funding, lack of political will, and indifference.

The Salton Sea aside, much of the Trough is a desert habitat in name only, having been replaced by agriculture and urban development. The Imperial Valley is a renowned food belt, and the Coachella Valley grows everything from citrus to dates. From Mecca in the South to Palm Springs in the north what is not under cultivation in the Coachella Valley has been turned into residential communities and golf courses. What remains of the original Colorado Desert is now restricted to state and federal land east and west of the Salton Trough.

FIGURE 6.41 After its creation, the Salton Sea became a major stopover for large numbers of migratory and wintering birds. This cluster of mostly American Avocets (*Recurvirostra americana*) and Black-necked Stilts (*Himantopus mexicanus*) congregates in the reedy flats of the Sonny Bono Salton Sea National Wildlife Refuge.

Northeast of the trough are the Little San Bernardino Mountains. This range also defines the southwestern margin of the Mojave Desert's Joshua Tree Woodland. Due east of the trough is a series of low-elevation uplifts comprising the Chuckwalla Bench in which the Orocopia Mountains, and the Chocolate Mountains. are the primary components. The Chocolate Mountain Naval Gunnery Range, a live-bombing facility off limits to civilians, occupies a significant portion of the area. Between these two is the main drainage of the bench running west into the trough, the Salt Creek Wash. Eastward of the Chuckwalla Bench, the desert tilts toward the Colorado River Valley. This is a remote area of broad valleys draining toward the river. Dotting the landscape are small but rugged low mountains half-buried in sand and their own alluvium, like the Cargo Muchachos northwest of Pilot Knob.

6.11.1 Sandy Planes and Dunes

Sandy planes and dunes extend over the lower reaches of the Colorado Desert. The Algodones Dunes are the largest and best known (Figure 6.42). These sand deposits cover about 160,000 acres (650 km^2) south and west of the Chocolate Mountains between the Imperial Valley and the Colorado River and are a familiar sight to motorists traveling along I-8 between Gordons Well and the California Inspection Station. Vegetation is sparse. Scattered growth consists of creosote along with the prostate shrub California Croton and woody grasses like Galleta.

FIGURE 6.42 The Algodones are the largest (approximately 160,000 acres; 650 km^2) and best-known of the Colorado Desert sand dunes. In this view looking northeast the Chocolate Mountains are in the background. Some areas of the Algodones are open to off-road vehicle use, but most of the dunes are designated wilderness.

Sand dunes once occupied about 64,000 acres (260 km^2) of the Coachella Valley, but today less than 8,000 acres (32 km^2) remain. Farther west, between the Imperial Valley and Anza Borrego Desert State Park, are the dunes forming the Superstition Hills, also a designated Naval Aerial Gunnery Range.

The west side of the Colorado Desert ascends the slopes of the Peninsular Ranges, sharply so in the north, more gradually in the south. The mountains comprise a series of uplifted parallel ridges separated by basins that were created when geological faulting shifted from one zone to another: Imperial Fault Zone, San Jacinto Fault Zone, and Elsinore Fault Zone (see Figure 1.2). The San Jacinto Mountains are impressively steep on the east side, soaring nearly 2 mi above Palm Springs. The sight of them is arresting from any perspective (Figure 6.43). To the south, the Santa Rosa and Laguna Mountains are less severe, rising to some 6,500 ft (1975 m) at a more protracted incline. This is the place to experience the transition from the flat, cultivated expanse of the Imperial Valley to the upslope environment at the westernmost edge of the Colorado Desert (Figure 6.44). The Bureau of Land Management oversees some of this region. Much of the rest lies within Anza Borrego Desert State Park.

The Anza Borrego Desert State Park is the largest state park in the contiguous United States (633,000 acres [2,560 km^2]). Upslope, deep washes like Bow Willow and Corrizo Gorge with the badlands below it (Figure 6.45) have been cut out by erosion. Large granitic boulders and rock faces anchor the gravely

FIGURE 6.43 The San Jacinto Mountains rise steeply to 10,800 ft (3,280 m) on the west side of the Coachella Valley, as seen here from the Little San Bernardino Mountains on the valley's east side.

FIGURE 6.44 The Lower Borrego Valley consists of sand deposits and Creosote Bush (*Larrea tridenetata*) that give way to the Ocotillo Badlands and the Vallecito Mountains heading west toward the center of the Anza Borrego Desert.

FIGURE 6.45 The Carrizo Badlands in the Anza Borrego Desert State Park attest to the forces of sedimentation and erosion.

alluvium. Anza Borrego is a much more complex environment compared with the open and relatively simple structure of the Colorado River Valley to the east.

As in the Mojave, there are certain plants associated with dunes, washes, alluvial fans, playas, gravelly slopes, and transitional zones into the mountains and chaparral. The Creosote community is again widespread, and other plant types contribute to the impressive Anza Borrego flora. The gangling Ocotillo (*Fouquieria splendens*) grows in thick stands on alluvial fans as well as the flats. Spiny, erect stems up to 20 ft (6 m) high emerge from a compact base. During the wet season, the small green leaves erupt to form new spines, and the red, tubular flowers are unmistakable (Figure 6.46). Ocotillo is represented in the southeast Mojave, but it is otherwise a Sonoran affiliate. Other common associates are Desert Agave, Mojave Yucca, Cat Claw, Desert Willow, Crucifixion Thorn, Ironwood, Smoke Tree, Blue Palo Verde, Mesquite, and at higher elevations, California Juniper. Common cactus include Teddy-bear Cholla (*Cylindropuntia bigelovii*; Figure 6.47), Silver Cholla (*echinocarpa* viz *C. echinocarpa*), Diamond Cholla (*C. ramosissima*), Beavertail (*Opuntia basilaris*), Mojave Prickly Pear (*O. phaeacantha*), Fish-hook Cactus (*Mammillaria dioica, M. tetrancistra*), Engelmann Hedgehog (*Echinocereus engelmannii*), and Barrel Cactus (*Farocactus acanthodes*).

The Elephant tree (*Bursera microphylla*) occurs in scattered localities among rocky slopes and alluvial fans. Its thick elephantine trunks and peeling papery bark are distinctive. *Bursera* is a large genus (about 100 species) with its greatest diversity along the Pacific slope of Mexico as an associate of tropical deciduous forest and thornscrub. The persistence of *Bursera* in Anza Borrego speaks to the area's Sonoran heritage.

FIGURE 6.46 The dense erect stems of the Ocotillo (*Fouquieria splendens*) erupt with small green leaves during the wet season and are followed with terminal, bright red tubular flowers. Ocotillo is essentially a plant of the Sonoran Desert including the Colorado Desert where it grows on alluvial fans and gravelly flats. Dense stands are found in Anza Borrego Desert State Park. (Courtesy of Don Genero.)

FIGURE 6.47 Teddy-bear Cholla (*Cylindropuntia bigelovii*) is common in washes and flats of the Anza Borrego Desert. These upright cacti are a favorite site for nesting Cactus Wrens (*Campylorhynchus brunneicapillus*).

FIGURE 6.48 Desert Fan Palms (*Washingtonia filifera*) like these are ecological oases supporting numerous species of vertebrates, large and small. Fan Palms grow around springs along canyons and rocky slopes of the western margin of the Colorado Desert.

Localized clusters of *Desert Fan Palms* (*Washingtonia filifera)* are scattered about the western and northern edge of the Colorado Desert, growing on mountain slopes and in canyons where groundwater is forced to the surface at springs or along fault lines (Figure 6.48). Palm oases are havens for numerous plant and animal species that are otherwise only encountered in more mesic habitats in the mountains and chaparral. Besides palms, other conspicuous plants may include Western Cottonwood (*Populus fremontii*), Mulefat (*Baccharis salicifolia*), and Narrow-leaf Willow (*Salix exigua*).

6.12 DESERT VERTEBRATES

Coyotes and rattlesnakes are the stereotype of desert wildlife but they are hardly alone. In addition to hundreds of bird species that pass through in migration, spend the winter, or nest, our deserts support around 40 species of lizards and snakes and similar numbers of mammals, about half of which are bats. Some vertebrates exhibit remarkable structural and physiological adaptations for desert life. Others, like coyotes and roadrunners, are ecological generalists and make a living in the desert because they can. The extremes of the desert environment mean that most vertebrates are active only seasonally for a few months during mid-spring into summer. The rest of the year, they are holed up in burrows and other refugia. Many desert vertebrates are also nocturnal, like the snakes and many of the mammals. Yet most lizards are out during daylight. Whatever activity pattern is followed, individual species tend to be habitat-specific except for the widespread generalists. Some lizards and mice, for instance, may favor sandy flats whereas others prefer rocky, brush-covered slopes.

The Mojave and Colorado deserts share many of the same species but not all. The Colorado Desert supports a number of lizards that are northern or northeastern extensions of populations centered in Baja California and the Colorado River basin. They include the Leaf-toed Gecko, Colorado Desert Fringe-toed Lizard, Flat-tailed Horned Lizard, Baja California Collared Lizard, Banded Rock Lizard, Granite and Sandstone Night Lizards, and Switak's Banded Gecko. In turn, the Mojave Fringe-toed Lizard, Panamint Alligator Lizard, and Mojave Rattlesnake are absent from the Colorado Desert.

Desert amphibians are few as would be expected and are restricted to riparian habitats where permanent water is available. The four species are the California Treefrog, Pacific Treefrog, Western Toad, and Red-spotted Toad.

With the exception of the geckos (*Coleonyx*, *Phyllodactylus*) and Night Lizards (*Xantusia*), most desert lizards are diurnal. Some can be more conspicuous than others because of their size or habits, like the Desert Iguana, Chuckwalla (Figure 6.49), Collard Lizard(s), Leopard Lizard, Desert Spiny Lizard, and Tiger Whiptail Lizard. Zebra-tailed Lizards are common on sandy flats and recognizable by their black-banded tail curled up behind them while darting from here to there. Side-blotched Lizards are small, but so numerous in the Southwest that they cannot be avoided. This species is active all year. Horned lizards (*Phrynosoma* spp.) are less easily detected because of their flat bodies, low profiles, and cryptic appearance. The three species of Fringed-toed lizards (*Uma*) possess striking adaptations for living in sandy environments: scales on the toes

FIGURE 6.49 One of our largest lizards, the Common Chuckwalla (*Sauromalus ater*) is an herbivore that inhabits boulder-strewn slopes, rocky outcrops, and lava flows in the Mojave and Sonoran Deserts (including the Colorado). If disturbed, it seeks immediate refuge in cracks and crevices where it wedges itself by gulping air and inflating its body. (Courtesy of Mitch Leslie.)

FIGURE 6.50 The Mojave Fringe-toed Lizard (*Uma scoparia*) looks a lot like its congeners, the Colorado Desert Fringe-toed Lizard (*U. notata*) and Coachella Valley Fringe-toed Lizard (*U. inornata*). Their common name derives from the elongate scales on their toes, which provide traction in the sandy habitats where they live. Note also the flattened body and counter-sunk lower jaw, which are further adaptations for life in the sand.

are elongate to increase the surface area in contact with the sand, thus aiding in traction; body scales are fine-textured to help facilitate rapid shimmy-burrowing beneath the sand; the lower jaw is counter-sunk, which along with the S-shaped valvular nostrils help prevent fine sand particles from entering the mouth and respiratory tract. The Mojave has its own species, *U. scoparia* (Figure 6.50). The Colorado Desert is home to *U. notata*, and a third species, the Coachella Valley Fringe-toed Lizard, *U. inornata*, has lost almost 90% of its former range to agriculture and urban development.

The Gila monster (*Heloderma suspectum*) is a storied image of the Southwest desert (Figure 6.51). Renowned for a powerful, venomous bite, these beautiful animals are actually rather shy and retiring. They range within the Mojave and much of the Sonoran Desert from extreme southeast Utah and southern Nevada through western and southern Arizona south to northern Sinaloa, Mexico. In California, however, they are decidedly rare. Over the past 150 years, 26 credible records (photos, sight records, specimens) have been documented. All but one or two of these are from the mountains of the eastern Mojave, for example, the Kingston Range and Clark Mountains. Providence Mountains, and Piute Mountains. A single record exists for the Imperial Dam on the Colorado River in Imperial County. The appropriate environmental conditions for Gila monsters in California are marginal. Practically all records are associated with seasonal riparian habitats east of 116° longitude. These areas often experience some of

FIGURE 6.51 The exquisitely patterned Gila monster (*Heloderma suspectum*) and its close relative the Mexican Beaded Lizard (*H. horridum*) are the only known venomous lizards. Their bite is quite powerful, although human fatalities are rare if they occur at all. Gila monsters range from southern Utah to northern Sinaloa, Mexico, but they are exceedingly rare in California. Over the past 150 years there have been only 26 credible records, and practically all of them are from isolated mountains in the eastern Mojave.

the summer rains typical of the Sonoran Desert in adjacent southern Nevada and western Arizona, where the Gila monster is fairly common. Gila monsters, much like the Desert Tortoise, are likely late Pleistocene relicts in California, and perhaps were never common in the first place. Nonetheless, new locality records may await, for example in other suitable habitats in the Little San Bernardino, Avawatz, and Granite Mountains, among others.

Most desert snakes are nocturnal and therefore seldom seen by people. Some may be out and about in the morning or at twilight in the late spring and early summer. Others are more likely to be encountered by slowly driving isolated paved roads after dark when snakes come out to soak up heat from the black top and commence their foraging activities. Among these might be the Spotted Leaf-nosed Snake, Desert Patch-nosed Snake, Glossy Snake, Long-nosed Snake, California Kingsnake, Western Shovel-nosed Snake, Spotted Night Snake, Western Lyre Snake, Sidewinder, Speckled Rattlesnake, Mojave Rattlesnake, and Western Diamond-backed Rattlesnake (Figure 6.52, eastern sectors). The Coachwhip, or Red Racer (*Masticophis flagellum*), is an exception to the nocturnal habits shared by most desert snakes. This large, aggressive (but nonvenomous) predator forages during the late morning with its head and neck held up while searching for a favorite prey, the Tiger Whiptail Lizard.

Agassiz's Desert Tortoise (*Gopherus agassizii*) was long regarded as a single species of the Mojave and Sonoran Deserts with a distribution similar to the Gila monster (see previous discussion), although not nearly as rare in California.

FIGURE 6.52 The Western Diamond-backed Rattlesnake (*Crotalus atrox*) in California is restricted to the extreme southeastern part of the state in the eastern Mojave and Colorado Deserts. Outside of California this species, which can grow to more than 7 ft, ranges throughout the Southwest and east as far as Arkansas.

Desert tortoises outside of California recently have been shown to be a distinct species (see discussion under Turtles, Chapter 8). Presently Agassiz's Desert Tortoise (Figure 6.53) ranges mostly in creosote flats from the central and eastern Mojave to the eastern side of the Salton Trough. They require firm but not compacted soils to construct burrows, usually dug into the banks of washes and such. Adequate moisture is essential for incubating eggs and the developing hatchlings. Adults and young feed on herbaceous plants and cactus. On average they spend about 8 months of the year in their burrows. As the official California State Reptile, Agassiz's Desert Tortoise is also a metric of the overall health of our desert ecosystems. Sadly, desert tortoise populations have been declining for the past 50 years. Several factors are responsible including urbanization, military exercises, off-road vehicle activity, predation on young tortoises by ravens who have increased in numbers because of expanded dumps and landfills, and exotic respiratory diseases. The Bureau of Land Management and some non-governmental organizations (NGOs) like the Desert Tortoise Preservation Committee have enacted policies and protocols to help sustain populations and have even established the Desert Tortoise Natural Area, a protected zone that allows only foot traffic. Only time will tell if these efforts yield positive results.

Many of the desert birds are regular seasonal migrants whose presence can span a few days to several weeks in the fall and spring. A few, like the Scissor-tailed Flycatcher, are an accidental occurrence where only one or two may have been recorded as single individuals that strayed well outside their customary migration routes.

FIGURE 6.53 California's State Reptile, Agassiz's Desert Tortoise (*Gopherus agassizii*), inhabits creosote flats in the Mojave Desert. Desert tortoises remain inside sandy burrows for most of the year. California's populations have been declining for decades and are now under careful stewardship of the Bureau of Land Management and the Desert Tortoise Preservation Committee. (Courtesy of Mitch Leslie.)

A small sampling of nesting species includes the Turkey Vulture, Red-tailed Hawk, Gambel's Quail, Great Horned Owl, White-winged Dove, Black-chinned Hummingbird, Costa's Hummingbird, Say's Phoebe, Common Raven, Black-tailed Gnatcatcher, Verdin, Rock Wren, Canyon Wren, Cactus Wren, House Finch, and Black-throated Sparrow. For a full listing consult any of the numerous checklists that are available, such as the one issued by the National Park Service for the Mojave National Preserve, and for Anza Borrego by the California State Parks.

The few mammals active during the day are often common enough under the proper circumstances (habitat, seasonal activity, forage) to be observed by hikers. The Black-tailed Jackrabbit is unmistakable, for instance, and the Antelope (White-tailed) Ground Squirrel often can be seen scurrying across sandy flats with its tail curled forward over its back. The most charismatic of the diurnal mammals is the Desert Bighorn Sheep, *Ovis canadensis* (Figure 6.54). Bighorns are herbivores that live in social groups of from 5–15 ewes, lambs and 1- to 2-year olds; smaller bachelor herds consist of about 2–5 rams. Access to water is critical for bighorn survival, and they are not often found too far from it. They are known to wander onto golf courses around Palm Springs.

Many isolated groups of bighorn are declining. One-third of California's population has died out in the past century. Today they occur in rocky isolated areas from the White Mountains east of Bishop and other ranges east of the Sierra (e.g., Coso, Panamint) south through most of the mountains of the Mojave, the Little San Bernardino Mountains and the eastern Colorado Desert to the Chocolate and Cargo Muchacho Mountains. Isolated populations occur in the San

FIGURE 6.54 Desert Bighorn Sheep (*Ovis canadensis*), like this trio of young males, prefer rocky, isolated habitats with access to water. Much of the year is spent near the desert floor, but as summer progresses and forage deteriorates, they ascend into the mountains. Their numbers have dwindled in California over the past century.

Gabriel's and in the Peninsular Ranges. Numbers of Peninsular Range Bighorn dwindled to around 280 in the mid-1990s, much of that decline attributable to disease acquired through livestock. They have since recovered to around 600.

Nocturnal mammals consist of Leaf-nosed Bats, Vesper Bats, and Free-tailed Bats, along with carnivores and rodents. The carnivores are Coyote, Gray Fox, Kit Fox, Ringtail, Bobcat, and Mountain Lion, along with certain mustelids, for example Western Spotted Skunk, Striped Skunk, and Badger. The rodents include Pocket Mice (*Chaetodipus*, *Perognathus*), Deer Mice (*Peromyscus*), Harvest mice (*Reithrodontomys*), Pocket Gophers (*Thomomys*), Grasshopper Mice (*Onychomys*), Wood Rats (*Neotoma*), and Kangaroo Rats (*Dipodomys*).

6.13 RIPARIAN HABITATS

The word *riparian* means "of, relating to, or living on the bank of a lake, river or stream." In an arid land like Southern California, riparian zones, even where water is present for only a few months of the year, stand out prominently because of their vigorous greenery. Especially in the chaparral-covered foothills, they give evidence to the fact that this "gray scrubland" is in fact alive (Figure 6.55). Riparian habitats are indispensable to the survival of numerous plants and animals. They provide shelter, foraging opportunities, nesting sites, and serve as corridors for dispersal.

Riparian habitats speak to the dynamics of uplift and erosion. Crumbling and abrading of the Transverse and Peninsular Ranges created myriad of rills, creeks,

FIGURE 6.55 Southern California's riparian habitats are the backbone of the region's watershed. They are indispensable for trees like cottonwood, willow, and sycamore, and for such vertebrates as the California Treefrog (*Pseudacris cadaverina*), the Two-striped Garter Snake (*Thamnophis hammondii*), and the Southwestern Pond Turtle (*Actinemys marmorata pallida*).

and ravines, some of which dry up or cease to exist because of geologic forces, whereas others become the principal waterway to the coast. More than a thousand named watersheds are present in Southern California, and many more unnamed watercourses contribute during wet years. Most all of these are associated with the larger river systems, which from north to south are the Ventura River, Santa Clara River, Malibu Creek, Los Angeles River, San Gabriel River, Santa Ana River (the latter three being the largest), San Juan Creek, Santa Margarita River, Escondido Creek San Dieguito River, San Diego River, and the Sweetwater River. The Tijuana River, also a major system, meanders back and forth across the border.

A stream or river is a narrow body of water that moves in one direction. At the headwaters (perhaps from snowmelt at high elevations) water is clear, cold, and carries little if any sediment. Water moves swiftly through a narrow channel over a rocky bottom. Downstream other creeks may feed in, and here the water is more turbid and carries increasing amounts of sediment that has washed in from eroding soils. Near the river mouth, as it approaches the coast, the channel is wide, the current slower, and the bottom is silty from the deposition of sediment over time. Estuaries are thus brackish, with salinity gradients being a function of the physical structure of the site, the interplay of ocean tides, and the volume of incoming fresh water.

Nutrient content is determined by terrain and by the vegetation through which a stream flows. Fallen leaves add substantial organic matter, and inorganic nutrients are added from weathered rocks. The water is oxygenated where the flow is swift and turbulent, but hardly at all in slower, murkier stretches of the stream. Stream-dwelling animals that occupy sections of vigorous flow use various adaptations to prevent themselves from being carried away. Many small arthropods are flat and can attach themselves temporarily to rocks or logs. Others seek out the downstream side of a rock that is sheltered from turbulent flow.

Plants such as Bulrush (*Scirpus microcarpus*), cattails (*Typha* spp.), and rushes (*Juncus* spp.) are obligate aquatic species, growing in the water or along muddy banks. Others take hold on adjacent dry land so long as their root systems have access to reliable water. Periodic flooding is no problem. Among these are the dominant trees we see along Southern California's creeks such as Western Cottonwood (*Populus fremontii*), Willow (*Salix* spp.), Sycamore (*Platanus racemosa*), and at higher elevations Ash (*Fraxinus* spp.), California Black Walnut (*Juglans californica*), and White Alder (*Alnus rhombifolia*). Various oaks could also be present, and such shrubs as baccharis (*Baccharis* spp.), bush mallow (*Malacothamnus* spp.), nightshade (*Solanum* spp.), brickelbush (*Brickellia* spp.), wild grape (*Vitis* sp.), currant, and poison oak. A common exotic that has invaded costal riparian areas is the Giant Reed (*Arundo donax*). This pest has choked out large sections of some creeks and streams and requires aggressive action to keep it in check.

Desert riparian is best expressed along the Colorado River, sections of the Mojave River, and parts of the Salton Sea and its tributaries. Water is present year-round. Elsewhere in the deserts, riparian habitats are found around scattered and often isolated permanent springs like Saratoga Springs in the southern end of Death Valley (Figure 6.56) or as seasonal streams fed by mountain runoff, like

FIGURE 6.56 Saratoga Springs is located at the southern extreme of Death Valley in the northern Mojave Desert. Halophytic plants dominate the shoreline, and the spring supports a population of the Amargosa Pupfish (*Cyprinodon nevadensis*) along with numerous species of migratory birds.

FIGURE 6.57 San Felipe Creek is a primary riparian feature of the Anza Borrego Desert State Park. It begins in the Volcan Mountains to the west, winds through the desert canyons then meanders as a wash across the Borrego Valley to the Salton Sea.

San Felipe Creek in the Anza Borrego Desert (Figure 6.57; see also the Mojave's Afton Canyon, Figure 6.31). Most all desert vertebrates, probably 80 percent, use these unique ecological resources.

Familiar trees and shrubs include cottonwood, various species of willow, ash, and Mulefat. The problematic Tamarisk (*Tamarix* spp.), or Salt Cedar, has invaded many desert riparian areas. This tree, native to eastern Asia, is common in washes and along roadsides in many parts of the West. Salt cedar has a superior ability to extract water thereby depriving native plants, especially cottonwoods, of adequate moisture. Eradication efforts are being applied in critical areas with mixed results.

6.13.1 Riparian Vertebrates

6.13.1.1 Fish

California's comparatively low diversity of 63 native freshwater fish species is the outcome of evolution under California's unique geology and isolation from eastern and mid-western lakes and rivers. Out of the Sierra Nevada flow cold, fast-moving streams fed by snowmelt. Few types of fish have adapted to these rigorous conditions, namely trout, salmon, suckers, minnows, and sculpins. Moreover, as water drains westerly into the Central Valley, it warms, slows, and coalesces into two large rivers, the Sacramento from the north and the San Joaquin from the south. They unite in a braided delta that twists and turns for 50 miles before reaching San Francisco Bay and its narrow opening to the Pacific. The delta network is a vast array of backwater marshes, lagoons, and channels that require special physiological adaptations over multiple salinity gradients that further limit fish diversity.

The southern half of the state is another challenge. Of its 18 native species, 3 occur in isolated aquatic habitats in the deserts. West of the Transverse and Peninsular Ranges, a more-or-less consistent pattern of species diversity is found in coastal streams from south of Monterey to northern Baja. This diversity mainly consists of minnows, stickleback, perch, killifish, suckers, and trout (Table 6.2). Only two species are endemic: the Santa Ana Sucker (*Catostomus santaanae*) is restricted to the Santa Clara, San Gabriel, Los Angeles, and Santa Ana Rivers.

TABLE 6.2
Native Freshwater Fishes of Southern California

Bass, Sunfish, Crappie—Centrachidae
- Sacramento Perch (*Archoplites interruptus*)[a]

Gobies—Gobiidae
- Tidewater Goby (*Eucyclogobius newberryi*)

Killifish—Fundulidae
- California Killifish (*Fundulus parvipinnis*)

Minnows—Cyprinidae
- Mojave (Tui) Chub (*Siphateles bicolor mohavensis*)
- Arroyo Chub (*Gila orcutti*)
- Boneytail (*Gila elegans*)
- Speckled Dace (*Rhinichthys osculus*)
- Colorado Pikeminnow (*Ptychocheilus lucius*)

Mullet—Mugilidae
- Striped Mullet (*Mugil cephalus*)

Pupfish—Cyprinodontidae
- Amargosa Pupfish (*Cyprinodon nevadensis*)
- Desert Pupfish (*Cyprinodon macularius*)
- Salt Creek Pupfish (*Cyprinodon salinus*)

Salmon—Salmonidae
- Coastal Rainbow Trout (*Oncorhynchus mykiss irideus*)

Sculpin—Cottidae
- Prickly Sculpin (*Cottus asper*)

Sticklebacks—Gasterosteidae
- Three-spine Stickleback (*Gasterosteus aculeatus*)

Suckers—Catostomidae
- Santa Ana Sucker (*Catostomus santaanae*)
- Razorback Sucker (*Xyrauchen texanus*)

Viviparous Perch—Embiotocidae
- Shiner Perch (*Cymatogaster aggregata*)

[a] Brought to Southern California from the Central Valley as part of a redistribution effort.

The Arroyo Chub (*Gila orcutti*) is found in coastal streams of Los Angeles, Orange, and San Diego Counties and in the Mojave River where it was introduced. The remaining coastal natives range elsewhere throughout California. The Speckled Dace (*Rhinichthys osculus*) is the most widely distributed stream fish west of the Rocky Mountains. Another, the Sacramento Perch (*Archoplites interruptus*)—a once abundant food fish from central California—is now restricted to the Delta region and Clear Lake. It has been successfully introduced into other suitable habitats including some in Southern California.

In coastal streams, water volume is variable throughout the year. The majority of fish species reside only in low-salinity, upstream waters. Others move between brackish and freshwater conditions of estuaries, such as the California Killifish (*Fundulus parvipinnis*) and the Shiner Perch (*Cymatogaster aggregata*). The Tidewater Gobie (*Eucyclogobius newberryi*) occurs in partially closed lagoons with low salinity fed by freshwater creeks. These conditions are erratically met such that populations of these fish are isolated from one another up and down the coast; populations are federally endangered in Orange and San Diego Counties.

Surely the most intriguing of Southern California's aquatic habitats are the oases and spring-fed ponds of the deserts. The fish inhabiting them are remnants of populations that existed in the Pleistocene lakes and rivers during glacial times. As these large water bodies dried up, most of the resident fish disappeared. But a few survived the challenges and still persist. Springs in Death Valley support small, disjunct populations of the Amargosa Pupfish (*Cyprinodon nevadensis*) and Salt Creek Pupfish (*C. salinus*). A subspecies of the Mojave (Tui) Chub (*Siphateles bicolor mohavensis*) exists only in a couple of spring pools at the California State University Desert Studies Center at Zzyzx. In the Colorado Desert the Desert Pupfish (*C. macularius*) inhabits protected areas in the Anza Borrego Desert State Park, and in a few tributaries of the Salton Sea.

In 1869, California was home to 67 species of native fish. Today, 4 of those are extinct, and 43 species and subspecies are listed as endangered, threatened, or species of special concern. Dams, reservoirs, levees, channels, and other forms of habitat alteration have been big contributors to their decline. Dams are created to restrain flow and, especially in Southern California, to stockpile water. Dams radically alter downstream ecosystems because of changes they bring to the volume and intensity of water flow.

Bigger still is the presence of introduced species. California's freshwater habitats are grossly polluted with non-native fish species. In most of the larger creeks and reservoirs, the number of exotics often exceeds that of the indigenous forms. The first fish to be introduced in California was a massive release of the American Shad (*Alosa sapidissima*) in 1871. This Eastern-seaboard species was fed into the Sacramento River under authorization of the California Fish Commission (now the California Fish and Game Commission). Subsequent and numerous other introductions followed with no thought to the possible consequences. Of course, at that time there was no understanding of the potential harm to native species that might ensue, nor in general was an ecological conscience part of the land ethic as it is today. Most of California's native fish were deemed of little value, certainly for commercial and recreational purposes. Over the years, sanctioned introductions continued. Various species of bass and catfish, for example, were brought in as game fish. Other species were introduced for biological control, like mosquito fish, and Grass Carp. Moreover, sanctioned releases are not the only source of exotics. Unsponsored introductions have resulted from dumped bait-fish, escapees, stowaways, and common aquarium fish released by households. At present, there are at least 50 non-native species inhabiting the state's freshwater habitats. About 30 non-native species occur in Southern California, primarily in warm slow-moving stream courses and reservoirs. They include catfish, bass, bluegill, crappie, tilapia, sunfish, carp, shiners, and goldfish, among others.

6.13.1.2 Amphibians and Reptiles

Most amphibians depend on aquatic resources during some phase of their life-cycle, especially for egg deposition and larval development. A salamander called the Coast Range Newt (*Taricha torosa*) is always found near water, usually along rocky, isolated streams. Among the riparian frogs are the Arroyo Toad (*Anaxyrus* [*Bufo*] *californicus*), California Treefrog, California Red-legged Frog, and Mountain Yellow-legged Frog. The latter two species are now quite rare in Southern California, having been extirpated over most of their historic range in the southland.

Our riparian reptiles consist of turtles and snakes. Only one native freshwater turtle inhabits the west coast, the Western Pond Turtle. This species ranges from Oregon to northern Baja California chiefly west of the Cascade-Sierra crest. Southern California is represented by the Southwestern Pond Turtle, a subspecies (*Actinemys marmorata pallida*) that inhabits streams, creeks, ponds, and reservoirs. It favors areas with cattails, reeds, and other aquatic vegetation along with emergent features like logs and stumps where a turtle can climb on to bask. A few non-native freshwater turtles have been introduced in Southern California, such as the Common Snapping Turtle (*Chelydra serpentina*) but more importantly are two that have become widely established via the pet trade—the Red-eared Slider (*Trachemys scripta elegans*) and the Painted Turtle (*Chrysemeys picta*).

Southern California's riparian snakes are the Two-striped Garter Snake (*Thamnophis hammondii*) and California Red-sided Garter Snake (*T. sirtalis infernalis*). Both occupy permanent to semi-permanent waterways west of the deserts.

6.13.1.3 Birds

Riparian zones attract all manner of birds, too numerous to enumerate here. Resident species capitalize on nesting opportunities, for example among willow thickets by the Downy Woodpecker. Species of migrant songbirds such as warblers and flycatchers forage within the canopy while wintering sparrows are lured to dense entanglements. Bird diversity and concentrations can be spectacular in desert riparian. For example, around 240 species have been recorded at the California State University Desert Studies Center at Zzyzx, where two large ponds serve a broad spectrum of resident and migrant species, from ducks and shorebirds to passerines (Figure 6.58).

FIGURE 6.58 Lake Tuende at the California Desert Studies Center is a huge draw for migratory birds and a dependable source of water for many mammals. Additionally, it is a sanctuary for the state and federally endangered Mojave Chub (*Siphateles bicolor mohavensis*), a type of minnow.

For decades, human activity polluted and degraded U.S. streams by using them as industrial waste disposals. Historically, this has been less of a problem in southwestern states than in the East. The assumption was that the water would dilute by-products and carry them off. Some are, but others settle to the bottom and are taken up by aquatic animals, or drift downstream and collect in estuaries and bays. Nowadays, various kinds of human-engendered debris still foul our creeks and streams. Items like plastic bottles, shopping carts, tires—you name it—are an eyesore as well as an ecological disgrace. Fortunately, today most of Southern California's important riparian habitats are monitored by state and local agencies. Others receive stewardship provided by a host of nonprofit conservancies. Without this oversight, chronic devastation would befall some of Southern California's most environmentally important habitat.

Section III

Vertebrates

7 Amphibians

7.1 THE MOVE TO LAND

Vertebrates began their transition to land during the late Devonian Period about 375 million years ago after the supercontinent Pangea split into the northern landmass of Laurasia and the southern continent Gondwana. The transformation to a terrestrial existence was neither smooth nor abrupt. Over a period of 15 million years, a mosaic of changes remodeled the fish skeletal design to accommodate new modes of locomotion, feeding, and sensory input for life on land. The limb bones and supporting girdlers, derivatives of the pectoral and pelvic fins, became robust and well ossified. The skull flattened and the orbits (eye sockets) moved to a dorsal position from the lateral placement in their fish ancestors. The bony operculum that covers the gills (find it on your goldfish) was lost to accommodate respiration by other means—lungs or the integument. The pectoral (shoulder) girdle of bony fish is fixed to the back of the skull but became detached in early land vertebrates, which, along with the development of a joint between the back of the skull and the first vertebra (the atlas), permitted some independent up and down movement of the head.

Early land vertebrates had a sprawling posture and polydactylous ("many fingers") forelimbs and hind limbs; the number of digits did not stabilize at five until much later. No doubt locomotion was laborious in early land vertebrates. The body had to be elevated off the ground and then advanced in a series of fish like side-to-side lateral bends before plopping back down. Reproduction was still dependent on egg deposition in aquatic habitats followed by a larval stage of varying duration until metamorphosis brought forth the adult.

All land vertebrates are formally called *Tetrapoda* ("four feet") in reference to their two pair of well-developed limbs. It matters not that some later tetrapod lineages, like snakes, reduced or lost the limbs altogether. Limbless forms are nonetheless tetrapods because they descend from limbed ancestors. Tracing the history of the earliest tetrapods and their descendants has always been difficult if for no other reason than the fossil record for that time is fragmentary and subject to varied interpretation with respect to character homology and significance. Vertebral anatomy was once thought to be a relevant and useful device for tracing lineages, for example. However, this approach was eventually shown to be unreliable, but not before a suite of taxonomic names was proposed for various early tetrapod groups based on vertebral structure. Two of these were the *lepospondyls* and *temnosponyls*. Another taxon, the *labyrinthodonts* was originally applied to those early tetrapods with a unique, enfolded tooth structure, but that name, too, does not apply to any monophyletic (natural) group and thus has no formal standing. Like lepospondyls and temnospondys, however, "labyrinthodonts" persists as a term of convenience, often subsuming most of the other early land vertebrates including the lepospondyls and temnospondys. Some labyrinthodonts were surprisingly large, up to 15 ft (ca. 5 m). All were amphibians in the sense that they were tetrapods with a biphasic lifestyle (larva and adult), unlike later land vertebrates that abandoned the larval stage by way of the shelled, amniotic egg that could be deposited on dry land (see discussion on Amniotes).

7.2 FEATURES OF MODERN AMPHIBIANS—THE LISSAMPHIBIA

The amphibians of today belong to a lineage known as the *Lissamphibia*, a name that refers to their skin texture (*liss* means "smooth" from the Greek). Early nonamniote tetrapods (like labyrinthodonts) had skin variously adorned or imbedded with bony scutes. Lissamphibians are commonly known as the salamanders (*Caudata*), frogs (*Anura*), and a more obscure group the caecilians (*Gymnophiona*). Caecilians are a curious, elongate, limbless clade found in the wet tropics.

Lissamphibia are popularly portrayed as transitional animals between fish and other land vertebrates, as if evolution were directional and predetermined. This is an erroneous and unfortunate misunderstanding. They bear little resemblance to the earliest tetrapods, and in their own right, are quite specialized structurally, ecologically, and behaviorally. We have yet to uncover their deep history because the trail runs cold in the upper Paleozoic. A gap of 100 million years exists in the fossil record between them and the early land vertebrates. Some researchers argue for Lissamphibia monophyly and that they derived from the temnospondyls. Other analyzes suggest that salamanders, frogs, and caecilians have independent origins from separate Paleozoic ancestors. *Triadobatrachus massinoti* from the Triassic of Madagascar is regarded as an early or stem anuran that exhibits a suit of basal amphibian anuran-like characteristics. However, the earliest frogs that possess the skeletal attributes for saltatory locomotion (jumping) show up later in the early Jurassic. The first caecilians and salamander are also from Jurassic Laurasian landmasses. Lissamphibians share these derived characters:

- Pedicellate teeth (see below)
- *Papilla amphibiorum*: A sensory apparatus in the inner ear
- *Extracolumella*: A secondary bone in the ear associated with the stapes for sound reception
- Skin structure with mucous glands that facilitate cutaneous respiration
- *Green rods*: Unique retinal cells (absent in caecilians but probably as a secondary loss)
- *Specialized levator bulbi muscle*: A thin sheet of muscle on the floor of the orbit that causes the eyes to bulge outward when feeding, thereby enlarging the buccal cavity
- Most fishes and aquatic amphibians including tadpoles have a series of pressure-sensitive grooves or canals that extend from the head to the tail,

called the *lateral line system*. Lining the canals is a battery of sensory receptors called *neuromasts* that detect changes in current and water pressure. The lateral line is important for orientation and in some instances prey detection. It is absent in amniotes.

Salamanders are the most morphologically generalized lissamphibians. With simple limbs, an evident tail, and an overall basal tetrapod form, they somewhat resemble a lizard. Frogs on the other hand are peculiar in numerous ways because of their hopping mode of locomotion. With a few exceptions the nearly 700 species of salamanders are entirely Northern Hemisphere (about half are in North America) but for a few that range into Central and northern South America. Frogs, which number around 5,400 species are found worldwide except for polar regions and most oceanic islands.

Lissamphibians typically are small, from few to several inches; some West Indian frogs can fit on a dime. An interesting exception is the enormous Japanese salamander (*Andrias*), which may reach 5 ft (1.5 m). As ectotherms, a constant body temperature is not maintained metabolically as it is in birds and mammals. Nor is active thermoregulation a general behavior. Other ecotherms like lizards and snakes often regulate body temperature by basking or seeking shade. The body temperature of salamanders and frogs primarily corresponds to that of their microenvironment. During winter months many species avoid freezing by hibernating and burying themselves deep into the substrate below the freeze line. Others remain closer to the surface and will freeze, but, remarkably, not die. Frozen frogs will contain as much as 48% ice. In these species, glucose and glycerol produced by glycogen in the liver acts as a natural antifreeze.

The skin of lissamphibians contains little of the water-impermeable protein *keratin* that characterizes the epidermis of amniotes. Therefore, the skin is highly permeable to water and gases, so it functions as an important respiratory surface and an accessory site of osmoregulation. The exchange of gases and water with the environment is passive, although the skin plays an active role in sodium transport from the surface to the inner layers. This is important for osmoregulation and the uptake of water in terrestrial species.

During respiration all that is necessary for the transport of oxygen and carbon dioxide in any animal is a thin moist membrane supplied with abundant blood capillaries. The integument of frogs and salamanders is ideal. In one family of salamanders (plethodontids) the integument is the *only* organ of gas exchange. Glandular secretions of mucopolysaccharides keep the skin moist and permeable. Were the mucous coating to dry, the skin would shrivel and crack and the animal would die. The osmotic permeability of the skin also means that Lissamphibia are intolerant of salt water. If exposed for an extended period they would lose internal water to the saline environment around them.

Cutaneous respiration in amniotes is limited because of their heavily keratinized epidermis and a covering of feathers in birds and scales in reptiles. Nonetheless supplementary gas exchange occurs in several groups of fish, snakes, turtles, and mammals. In these instances, the magnitude of gas exchange is not necessarily even. Bats, for example, eliminate as much as 12% of their metabolic carbon dioxide through the vascularized wing membrane but can only obtain 1%–2% of their oxygen needs this way. Pond turtles, though, can sustain their gas-exchange requirements during periods of low metabolic activity by the skin surrounding their cloaca.

Beyond their role as an emollient and moisturizer, mucous secretions may also participate in defense. For example, some of the Slender Salamanders (*Batrachoseps* spp., see below) wrap their tails and bodies around the head and neck of small garter snakes if seized. The salamander is not only difficult to swallow, but the snake also finds itself slimed by the mucous secretions. The skins primary chemical defense, however, are the specialized poison glands that secrete noxious, distasteful, and toxic substances that deter predators. Secretions of some Central American frogs are extremely potent, even deadly, to humans.

Most all Lissamphibians possess a unique dental morphology called *pedicellate teeth*. Separate conical tooth crowns are attached to a base (pedicle) by soft tissue. They are carnivorous, opportunistic feeders with varied diets that include nearly anything: earthworms, small mollusks, insects, grubs, spiders, and even small vertebrates. Frogs and salamanders capture prey by suction feeding or by a protractible tongue attached to the anterior end of the lower jaw.

7.2.1 Reproduction

The general reproductive cycle begins with courtship, the behavioral repertoire that leads to mating. Fertilization may be external or internal. Fertilized eggs are then deposited in nests on land or in water. Eggs develop into embryos, and embryos into free living larvae. Finally, the surviving larvae transform into adults in a process of metamorphosis. Direct-developing species are known in all three orders of Lissamphibia. In these instances, the larval stage is bypassed so that embryos develop into miniature adults at hatching. Overall, salamanders exhibit the most basic patterns of development and have considerably fewer modifications. This is reflected in their more conserved ecological and geographical range compared with anurans. Caecilians are quite derived reproductively. Many are direct-developing and exhibit sophisticated means of transferring maternal nutrients to the embryo. However, their specialized morphology and lifestyle has constrained their phylogenetic and geographic distribution. By contrast, the great diversity of anurans is no doubt associated with reproductive plasticity; they display an amazing array of egg diversity, clutch structure, nest-site utilization, hatching processes, and temperature tolerance.

7.2.1.1 Salamanders

Courtship among salamanders maybe simple and direct or involve elaborate behavior such as dances, tail waving, and other displays. Pheromones released by males are used for species recognition and hormone activation in females. In terrestrial salamanders like plethodontids, for instance males rub specialized glands located on their chins across the snouts of females.

The majority of salamanders, about 90%, rely on internal fertilization although it does not involve an intromittant organ (such as a penis). Rather, there is a unique form of sperm transfer in which males deposit a sperm "packet" called a *spermatophore* atop a stationary gelatinous stalk. Receptive females pinch off the spermatophore with the lips of their cloaca then transfer the sperm packet to a special compartment in the cloacal wall called the *spermatheca*. Sperm is kept in the spermatheca until ovulation at which time they are released into the oviduct. Sperm storage can last a few hours, or several months. One European species (*Salamandra salamandra*) can store sperm for up to 2.5 years. An interesting variant of the spermatophore/spermatheca approach is for the female to deposit eggs on top of a spermatophore previously put down by a male.

Salamanders with internal fertilization have the advantage of distributing their eggs in time and space or depositing them as a single clutch where the female guards them. The possible options are (1) both eggs and larvae aquatic, (2) eggs terrestrial larvae aquatic, (3) both eggs and larvae terrestrial, (4) eggs terrestrial, direct development (e.g., *Aneides, Batrachoseps*, *Ensatina*), or (5) eggs are retained in the oviduct. Oviposition in aquatic sites can take place in static open water (lentic sites) as favored by *Ambystoma* and newts. Female newts (*Taricha*) wrap individual eggs around plant matter with their hind feet to protect them from other adults who will eat them if they detect a moving embryo. Attaching eggs to stalks and stems also prevents them from sinking into the inhospitable muddy bottom of the pond. Eggs deposited in flowing water (lotic sites) are hidden singly beneath stones or cached into the bank of the stream. In terrestrial sites, used for example by many plethodontid salamanders, eggs are attached by stalks to the roof or wall of some cavity in a crevice or log. Normally the female attends the eggs.

External fertilization is the norm in many aquatic salamanders and eggs develop into larvae. The larvae have feathery gills, which are extensions of the integument and not homologous with those of fish. The larvae otherwise more closely resemble the adult than do frog larvae. Where terrestrial eggs are laid, nest guarding is known in some species. The embryos have gills, but gills are resorbed by the time the larva or preadult stage emerges. Most salamander larvae mature into adults by season's end. Some species overwinter.

Paedomorphosis is an evolutionary phenomenon known in many salamander lineages. Simply put, adults of descendent species retain juvenile features of ancestral species. *Neotony* is one expression of paedomorphosis in which the rate of somatic (nonreproductive) tissue development in the embryo is retarded (evolutionarily), while the rate of gonadal development remains unchanged. The result is reproductive maturity in a descendent species that otherwise retains larval features such as gills as an adult.

7.2.1.2 Frogs

Anuran modes and strategies of reproduction are fascinatingly diverse and exceed that of all other vertebrates except for fish (which outnumber them in species diversity by five to one). No specific visual cues are employed in courtship. Rather, *vocalization* ("ribbit") is the primary means that males use to advertise their availability and location. Frog calls range from simple "peeps," to trills, clicks, or complex notes. Some calls might elicit an analogous sound to our ear. For example, a calling bullfrog sounds to me like a loose, deeply resonating banjo string being plucked—*bah-rooom.* Vocalizing begins with air being forced into the vocal sac, which is the floor of the buccal cavity (mouth). The vocal sac inflates. The call is produced when air forced out of the lungs passes through the larynx across paired vocal cords supported by constrictor and dilator muscles that modulate sound production. The inflated vocal sac functions as a resonating chamber. Contraction of muscles on the floor of the buccal cavity reverses the flow of air back into the lungs. The nostrils are closed during this entire cycle to keep air from escaping.

Call characteristics identify a species and are especially important in large, multispecies choruses. Females choose males not only by call specifics, but also the perceived quality of the male's location because this is where egg deposition and fertilization will take place. In some species, males will defend a calling territory from intrusion by other males, particularly in systems with prolonged breeding seasons.

Calling is not without its costs and risks. Air is forced from the lungs by the contraction of trunk muscles, and this action alone can become energetically expensive. The metabolic costs rise as the number of calls made per unit of time increases. The same is true for the length of the call itself. Yet longer calls may be more attractive to females and might be even more advantageous when the breeding chorus is large. Therefore, metabolic and behavioral compromises are usually made. For instance, male frogs often limit the time spent calling during the breeding season to a few hours per night. In addition to the energetic expense, a calling male can invite the attention of an acoustically oriented predator. Several species of bats, for instance, are quite adept at locating a meal by homing in on a calling frog.

Mating strategy among frogs is a function of the length of the breeding season. Prolonged breeding in the tropics can last several months or even all year. In warm temperate zones like Southern California, breeding is synchronized with the rains of late winter and early spring, and mating happens once or a couple of times in a year. In these instances breeding aggregations can be quite large when conditions are favorable. Explosive breeding like this also happens among tropical species that use ephemeral resources. An advantage to eruptive breeding is that a hyper-abundance of larvae is produced and at least some of them will survive predation or an environmental collapse and reach adulthood. When density is high among explosive breeding populations, males experience intense competition with one another for access to females. The most aggressive ones are usually more successful. Territories, however, are only established and defended by prolong breeding species, such that mating success is determined more by female choice rather than by assertiveness among competing males.

7.2.1.3 Fertilization

Fertilization is external in practically all frogs. The typical scheme is for a male to grasp the female from behind and on top in a process called *amplexus.* Some species grasp the female just behind her forelimbs, and others around the pelvic region. In either case, the male simply deposits a sperm cloud over the eggs as they are being exuded. In a few species of frogs, internal fertilization occurs by cloacal apposition—the male and female simply press their vents together. Then there is the interesting case of the Tailed Frog (*Ascaphus turei*) that inhabits rocky, high-energy streams of the Pacific Northwest (including portions of California). Males of this species have a short tail-like organ (a modification of the cloaca) with a small terminal opening. He inserts the organ into the female, thus assuring sperm transfer in the fast-moving waters.

Nests and egg-deposition sites are commonly ponds, lakes, and quiet streams and rivers, which is the case for most frogs in Southern California. Many tropical species take advantage of other resources, especially where the habitat is structurally complex, resources are in short supply, predation risk is high, and competition from other frog species is especially intense. Some examples include depositing eggs in bromeliads or other epiphytic plants or constructing foam nests on the surface of the water or in depressions in the ground or in trees. The males whip the water with their hind limbs during amplexus while eggs, sperm, and fluids are emitted. In the aptly named marsupial frogs (e.g., *Gastrotheca* spp.), the female carries the fertilized eggs in pouches located on her back.

Anuran larvae are commonly call *tadpoles*, a word derived from the Middle English *tadde* (toad) + *pol* (head); hence the toad that seems all head. An appropriate description, for sure. The vernacular term *polywog* derives from *polwygle*: pol (head) + wiggle. The simple morphology of anuran larvae consists of a round, bulbous body followed by a long, laterally compressed tail. Tadpoles face the same physiological and ecological challenges as adult frogs. They respond to changes in water chemistry, light, and temperature. Some graze on algae, others are carnivorous. Predators are diverse and include other aquatic vertebrates and arthropods such as dragonfly larvae. Most tadpoles metamorphose into adults by season's end, although developmental times vary greatly among species. Some may overwinter. Tadpole metamorphosis, the structural transition to the adult, is the end-point of the developmental sequence that begins with the cleaving egg, to the embryo, to the hatched larva. Metamorphosis is regulated in part by thyroxin, a hormone produced by the thyroid gland. The entire tadpole is radically transformed internally and externally. The general sequence begins with the initial appearance of the adult head and anterior region followed by the budding of hind limbs. Next the forelimbs emerge and lastly the tail is resorbed. Numerous frog species bypass the larval stage and undergo direct to development from the egg to the adult.

7.3 CAUDATA—SALAMANDERS

Salamanders abound across eastern North America with exceptional diversity throughout the Appalachians and the Gulf States; the majority of salamanders in the United States (ca. 190 species) are found here. In the arid Southwest, diversity is far lower, and activity is restricted to the late fall, winter, and spring. In Southern California 12 species are distributed among three families: Ambystomatidae, Salamandridae, and Plethodontidae (Table 7.1).

TABLE 7.1
Amphibians of Southern California

CAUDATA—Salamanders
  Ambystomatidae—Mole or Tiger Salamanders
    Ambystoma californiense—California Tiger Salamander
    Ambystoma tigrinum—Tiger Salamander
  Salamandridae—Newts
    Taricha torosa—Coast Newt, California Newt
  Plethodontidae—Lungless Salamanders
    Aneides lugubris—Arboreal Salamander
    Batrachoseps aridus—Desert Slender Salamander
    Batrachoseps gabriele—San Gabriel Mountains Slender Salamander
    Batrachoseps major—Garden Slender Salamander
    Batrachoseps nigriventris—Black-bellied Slender Salamander
    Batrachoseps pacificus—Channel Islands Slender Salamander
    Batrachoseps stebbensi—Tehachapi Slender Salamander
    Ensatina eschscholtzii—Monterey Salamander Ensatina
    Ensatina klauberi—Large-blotched Salamander
ANURA—Frogs
  Pelobatidae—Spadefoots
    Scaphiopus couchii—Couch's Spadefoot
    Spea hammondii—Western Spadefoot
  Bufonidae—Toads
    Cranopsis alvarius—Colorado River Toad
    Anaxyrus boreas—Western Toad
    Anaxyrus cognatus—Great Plains Toad
    Anaxyrus californicus—Arroyo Toad, Southwestern Toad
    Anaxyrus punctatus—Red-spotted Toad
    Anaxyrus woodhousei—Woodhouse's Toad
  Hylidae—Treefrogs
    Pseudacris cadaverina—California Treefrog
    Pseudacris regilla—Pacific Treefrog
  Ranidae—True Frogs
    Rana aurora—Red-legged Frog
    Rana boylii—Foothill Yellow-legged Frog
    Rana muscosa—Mountain Yellow-legged Frog
    Lithobates yavapaiensis—Lowland Leopard Frog
    Lithobates catesbeiana—Bullfrog (Int.)
    Lithobates pipiens—Northern Leopard Frog (Int.)
    Lithobates berlandieri—Rio Grande Leopard Frog (Int.)
  Pipidae—Clawed Frogs
    Xenopus laevis—African Clawed Frog (Int.)

7.3.1 Ambystomatidae—Mole Salamanders

Ambystomatids comprise 2 living genera and 33 species. The family is widespread in North America from extreme southeastern Alaska and Labrador to the Gulf States and the Mexican plateau. All but a few of the species are in the nominate genus *Ambystoma*. Four of these occur in the West.

Adult mole salamanders are stocky, up to about 8 in. (200 mm) in snout-vent length and have broad, rounded heads with small bulbous eyes. Adults of many species are variously marked with spots or blotches. Most are likely to be seen at night migrating overland to breed in quiet streams, lakes, or reservoirs beginning with the onset of winter rains. The Tiger Salamander (*Ambystoma tigrinum*) occurs throughout most of the eastern half of the United States, where numerous subspecies have been proposed and are still being evaluated. It is essentially absent from the Great Basin, most of the Pacific coast, and the Mojave and Colorado Deserts. California populations are problematic, including those in the southern half of the state. All are probably translocations. The natural distribution of Tiger Salamanders has been obscured because of access to stock ponds and irrigation canals and also their transport to lakes and reservoirs where larvae have been used as fish bait.

The California Tiger Salamander (*Ambystoma californiense*) is remarkably similar in appearance to *A. tigrinum* (with which it will hybridize). This species is endemic to the central coast and adjacent areas of the Central Valley. Its southernmost extent is an isolated population near Lompoc in Santa Barbara County that some authorities consider a distinct species. The California Tiger Salamander inhabits grasslands and oak woodland. It mostly breeds in seasonal ponds but will also use slow-moving streams. The present range is so fragmented because of habitat loss from agricultural development that it is now probably extinct over at least 50% of its historic range.

7.3.2 Salamandridae—Newts

The word *newt* derives from Middle English *neuter*, which itself comes from the syllabic merging of *an eute*. Newts are the dominant group of salamanders in Europe and Asia. In North America they are represented east of the central plains by three species of the genus *Notophthalmus*. In the west, Pacific newts include three species of the genus *Taricha* distributed along coastal ranges from southern British Columbia to Southern California, and one, the California Newt, along the western slopes of the southern Cascades and Sierra Nevada in California. All are aquatic breeders. Adults transformed from larvae will make terrestrial excursions, especially during wet periods. Pacific newts have a somewhat rough, unicolor skin (orangish to blackish-orange) that is notorious for its toxic secretions. The California Newt (*Taricha torosa*) is the only species of *Taricha* in Southern California. That population runs along the Coast Ranges from Mendocino County south through San Diego County. It once occurred throughout the western slopes of the Transverse and Peninsular Ranges in suitable drainages, but populations in our region have been in severe decline because of human activity in one form or another. Dams and other forms of stream alteration are significant contributors along with the introduction of non-native fish and excessive fires.

7.3.3 Plethodontidae—Lungless Salamanders

Plethodontids are the largest family of salamanders at more than 350 species. With the exception of 7 species in southern Europe, they are entirely North American plus a few representatives in the New World tropics. Plethodontids are mainly terrestrial, and they lay eggs on land. They lack lungs and rely on cutaneous respiration exclusively.

West Coast plethodontids include five species of the genus *Plethodon*, two of which reach northwest California (Humboldt, Trinity, Siskiyou, and Del Norte Counties). The three rare species of *Hydromantes* are found in granite and limestone crevices in the Sierra Nevada and near Mt. Shasta in northern California. In Southern California nine species of plethodontids are placed among three genera: (1) *Aneides*, (2) *Batrachoseps*, and (3) *Ensatina*.

Aneides are climbing salamanders. Robust limbs, elongate, blunt toes, and a round, quasi-prehensile tail aid them in scaling logs, tree trunks, and rockslides. One of the four western species occurs in Southern California, the *Arboreal Salamander* (*A. lugubris*) that attains 4 in. (400 mm) in snout-vent length. It is the largest salamander in our area. The Arboreal Salamander inhabits coastal forest zones from Humboldt County south into northern Baja California. Outlier populations are known on the western slopes of the Sierra Nevada (El Dorado County to Madera County), on Catalina Island, and on Los Coronado Islands off northern Baja. In Southern California, it favors oak and sycamore woodland up to about 5,000 ft but may also occupy coastal sage scrub and chaparral. From November to April, Arboreal Salamanders hide under rocks, logs, boards, and in damp animal burrows. During the dry season, they take refuge in tree cavities.

The slender salamanders of the genus *Batrachoseps* (Figure 7.1) are a fascinating study in microevolutionary processes. In the mid-1960s only two species were recognized, the Pacific Slender Salamander (*B. pacificus*) and the California Slender Salamander (*B. attenuates*). Two decades later, there were 8 species. Presently 20 species are recognized. All but one of these (the Oregon Slender Salamander [*B. wright*]) is confined to California except for 2 species whose range extends marginally out of state (*B. attenuates* into Oregon and *B. major* into Baja, California). These small (less than 2 ½-in. (ca. 65 mm) snout-vent length) slim salamanders with diminutive limbs are maddeningly similar to one another in physical appearance, which explains why so many discrete lineages were undetected for so long. Biochemical and genetic markers have revealed (and likely will continue to) the remarkable extent of their diversity. One recent study using DNA sequences estimated that *Batrachoseps attenuatus* alone could be composed of 39(!) cryptic (hidden) species.

Batrachoseps inhabit damp places around springs, beneath leaf litter, and under or inside rotting logs. Some species will take up residence in well-watered backyards beneath stones and potted plants, and hence have earned the nickname of “flowerpot” salamanders or “worm” salamanders because the grooves

FIGURE 7.1 The aptly named slender salamanders of the genus *Batrachoseps* are no more than about 2½–3 in. (65–75 mm) in total length. At least six cryptic species occur in Southern California, all with restricted distributions. They inhabit damp places around creeks, springs, and beneath leaf litter and rotting logs. Occasionally they turn up beneath flowerpots in well-watered gardens.

on the sides of their bodies give the impression of segmentation. If you find one by flipping over a rock or other suitable hiding place, it tends to remain tightly curled and motionless when first exposed. If handled, the salamander may flip and gyrate to escape, then bound erratically across the ground.

As currently understood, six species of *Batrachoseps* occur in Southern California. The Garden Slender Salamander (*B. major*) was once common in the foothills and lowlands throughout its range in Los Angeles and Riverside Counties but is now difficult to find there; more robust populations persist in southern Orange County and San Diego County south into northern Baja to El Rosario. They also occur on Catalina Island. The Pacific Slender Salamander (*B. pacificus*) is now applied to the slender salamanders on the northern Channel Islands of San Miguel, Santa Rosa, Santa Cruz, and Anacapa. The Black-bellied Slender Salamander (*B. nigriventris*) also occurs on Santa Cruz Island, as well as on the mainland from southern Monterey County south into the Tehachapi Mountains at Fort Tejon (where it overlaps with the *Tehachapi Slender Salamander* [*B. stebbensi*]), and into the Transverse Ranges (to Cajon Pass) then south to the northern Peninsular Ranges. The San Gabriel Mountains Slender Salamander (*B. gabriel*) is locally distributed from the upper San Gabriel River to the western end of the San Bernardino Mountains. The Desert Slender Salamander (*B. aridus*, a close relative of *B. major*) is confined to mesic canyon bottoms along the east slope of the Santa Rosa Mountains, for example at Hidden Palm Canyon, a state ecological reserve southwest of Palm Desert in Riverside County.

Ensatina is the third plethodontid genus in Southern California. It is a close relative of *Plethodon* that goes by the simple common name Ensatina. Ensatina *eschscholtzii* is divisible into a least six subspecific populations from British Columbia south into northern Baja California. The California populations form a continuous "ring" distribution down the Coast Ranges through Southern California on one side and the Sierra Nevada west of the crest on the other. They are absent from the Great Valley. The *Monterey Ensatina* (*Ensatina e. eschscholtzii*, Figure 7.2) is a uniform pinkish-brown (white below) and ranges from San Luis Obispo County south through the Transverse and Peninsular Ranges in chaparral and oak-coniferous woodland. The *Large-blotched Ensatina* (*E. klauberi*, Figure 7.3) is striking in its

FIGURE 7.2 The Monterey Ensatina, (*Ensatina e. eschscholtzii*) is one of six subspecies of *Ensatina* salamanders found from southern British Columbia south into northern Baja California, Mexico. The Monterey Ensatina ranges from San Luis Obispo County south through the Transverse and Peninsular Ranges in oak-conifer woodland.

FIGURE 7.3 The beautiful Large-blotched Salamander (*Ensatina klauberi*) is less common than its close relative the Monterey Ensatina. These salamanders inhabit pine-oak woodlands in the San Bernardino Mountains south through the Peninsular Range.

large yellow to orange blotches on a dark background. It prefers pine-oak woodlands from the San Bernardino Mountains south through the Peninsular Ranges. Some authorities regard it as a subspecies of *E. eschscholtzii.*

7.4 ANURA—FROGS

Frogs are a structural improbability and an evolutionary prodigy at the same time. Some jump (saltation), others hop, walk, climb swim, or burrow. Skeletal architecture (Figure 7.4) is based on a standard template, modified according to the particular style of locomotion. In all frogs, the trunk is shortened relative to other land vertebrates because the vertebral column has been pared to a mere 5 to 8 presacral vertebrae (24 in humans). The sacrum (where the pelvic girdle attaches) consists of a single vertebral element (the mammalian sacrum is composed of from 3 to 5 vertebrae fused to one another). Although *anura* means "tailless" (from the Greek *an*, without and *uro*, tail), frogs nonetheless have a series of caudal (tail) vertebrae, but they are fused into a single rod called the *urostyle.* The urostyle articulates with the sacral vertebra (as caudal vertebrae do in all tetrapods), but instead of being a distal external appendage the urostyle is directed forward and lies between the long, exaggerated ilial bones that form the V-shaped pelvic girdle.

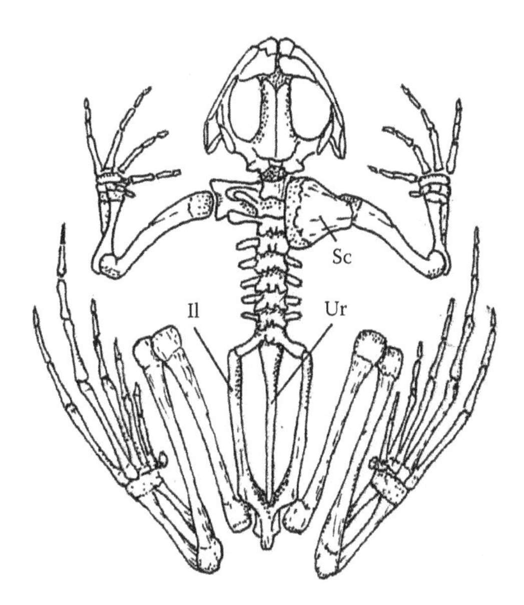

FIGURE 7.4 The powerful hind limbs of a leaping frog consist of three linkage components (thigh, shank, elongate ankle and foot) that extend sequentially to send the compact body airborne in about 100 ms. Note the exaggerated ilial bones (Il) of the pelvic girdle, and the urostyle (Ur) between them. The urostyle is actually the caudal (tail) vertebrae fused into a slim rod. This skeleton is shown in dorsal view, with the left scapula (Sc, dorsal part of the pectoral girdle) removed. (From Porter, K.H., *Herpetology,* W. B. Saunders Company, Philadelphia, PA, 1972.)

The urostyle serves as a site for the attachment of pelvic muscles. The head and trunk comprise a compact projectile that experiences next to no torque or twisting as it flies through the air.

Launch is accomplished by the hind limbs reacting simultaneously from a crouched position. Three components comprise the leg: the proximal femur (thigh) followed by the shank bones, the tibia, and fibula, which in frogs are fused into a single bone (tibiofibula), and thirdly the elongate tibilae (astragalus) and fibulare (calcaneum), which are modified tarsal (ankle) elements. At rest, the three components are folded against one another next to the trunk. Groups of muscles extent each leg section starting with the femur; peak maximal force is transmitted in as little as 100 ms as the foot leaves the ground.

Not all skeletal modifications for locomotion performance are designed solely for jumping, but they do correlate with habitat preferences and foraging tactics. Aquatic and semi-aquatic frogs, like jumpers, also have long hind limbs. Feet are strongly webbed, and the act of swimming is accomplished in a manner similar to saltatory movement. Jumpers like the ranid frogs (see Section 7.4.4) are usually ambush predators that sit and wait for a meal to come into range. They are often cryptically patterned and lack skin toxins or other chemical defenses. If detected by a potential terrestrial predator while sitting pond-side, they escape by a series of rapid, erratic leaps before disappearing beneath the water surface and out of view of the predator.

Frogs that move by short hops or by walking, like the toads (Bufonidae) and the Western Spadefoot (*Spea hammondii*), have stout bodies, short limbs with little or no webbing on their feet, and blunt snouts. Short-legged species are terrestrial except when breeding and typically forage widely within their home range. Because they cannot move rapidly to escape from their own predators, they rely on other defenses and deterrents, especially powerful skin toxins.

Arboreal frogs have long legs, slim waists, and large heads and eyes. They climb, walk, and leap with stunning agility. Most of these species inhabit tropical biomes with complex, multileveled vegetation. Fingers and toes are equipped with expanded discs on their tips. The discs are arrayed with microscopic peglike projections supplemented with mucous glands that combined with water produce a fluid layer between the disc and the substrate. Arboreal frogs can adhere to a surface that is as much as 60° past vertical by a combination of surface tension and viscosity (wet adhesion). They climb in quadrapedal fashion, hand over foot. The toe pads are peeled loose from back to front as the foot is lifted. When at rest, they orient themselves with their head directed skyward to keep their toes pointing up. If they faced downward, their body weight would cause the toe pads to peel away from the surface. For the same reason, when descending vertically they must do so by moving backward. Toe discs are not limited to the arboreal frogs. They also occur in numerous species that clamber over leaf litter on the forest floor.

As with salamanders, the anuran species diversity is rather limited in the arid Southwest. Southern California hosts 17 species of frogs in five families (Table 7.1). They include jumpers, hoppers, climbers, swimmers, and burrowers.

7.4.1 Pelobatidae—Spadefoots

Most of this family of some 80 species is Old World (Eurasia, North Africa, western China) and Southeast Asian (Indo–Australia and the Philippines). North America is home to 6 species allocated to two similar genera, *Spea* and *Scaphiopus*. Five of these are in the western states, two of which reach Southern California—Couch's Spadefoot and the Western Spadefoot. Spadefoots take their name from the black, spade-shaped tubercles on the inner edge of their hind feet. The spades assist in burrowing, which is done backward, feet-first into soft substrates in a rotating fashion. Spadefoots have short legs and a rounded stout body that is reminiscent of toads. They differ from the true toads (Bufonidae), however, in a number of ways such as their smooth as opposed to bumpy skin and in the absence of the poison producing paratoid glands at the back of the head.

Spadefoots breed after heavy rains in temporary pools, ponds, and irrigation canals. In dry years no breeding occurs at all. When rains are adequate, large breeding aggregations assemble hastily and males aggressively compete with one another for access to females. Calling males are audible for a considerable distance, which presumably is advantageous when breeding sites are often temporary and uncertain from one year to the next. Given the ephemeral nature of the breeding sites, the development time from egg to metamorphosis happens remarkably fast in some species, as brief as 1 to 2 weeks.

Couch's Spadefoot (*Scaphiopous couchii*) ranges across southern Arizona, New Mexico, much of Texas, and south through Mexico, including the east side of Baja California. In our area, it is restricted to the eastern sections of the Colorado Desert. Scattered populations exist in Imperial and San Bernardino Counties, typically in creosote bush scrub. Breeding takes place from June to September when the summer rains of the Sonoran Desert have arrived.

The Western Spadefoot (*Spea hammondii*) is nearly endemic to California. This is a lowland species of sandy or loose soils in short grasslands, open chaparral, and oak woodland from the Central Valley and Coast Ranges south of Monterey Bay into northwest Baja California. The loss of habitat to agriculture and urbanization has, unfortunately, eliminated the Western Spadefoot from much of its historic range. In Southern California it has been extirpated from the lowlands of Ventura and Los Angeles Counties. Viable populations persist in parts of Riverside County, coastal Orange County, and San Diego County. The Western Spadefoot remains hidden below ground most the year. It emerges to breed during the warm winter and spring rains. The tadpoles linger anywhere from 1 to 3 months before metamorphosis, depending on the temperature and depth of the water at the breeding site.

7.4.2 Bufonidae—Toads

The toad has been reviled and revered, loathed and admired for much of recorded history. Toads figure into folk literature, witchcraft, and various sorts of artistic expression. Some have been stuffed, varnished, and outfitted with mariachi costumes and sold at Mexican border crossings and curio shops. Bufonids (around 335 species) are found on all continents except for Antarctica and Australia (the Marine or Cane Toad [*Chanus marinus*]) has been introduced in Australia). They occur in the West Indies, but like anurans generally they are absent from most oceanic islands except where introduced, for example Hawaii.

Toads are terrestrial but breed in aquatic habitats and typically lay eggs in paired strings. They tend to walk rather than hop, have a stocky body, and thick, granular skin. All frogs (except one species) lack teeth on the mandible, but bufonids are also absent dentition on the maxilla (upper jaw). The prominent, paired paratoid glands on the back of the head are another hallmark. These structures secrete a whitish, viscous poison as a deterrent to predators. Some predators are apparently immune to it, but others, including small pets, are repulsed or may even succumb to the poison's effects. The active compounds are a class of cardiotoxin steroids (bufotoxin and bufogenin). In some species, like the Sonoran Toad (*Cranopsis alvarius*) the secretions are a powerful hallucinogen. The psychotropic properties are behind the bizarre and unfathomably stupid fad of "toad licking." Whether urban mythology or not, toad licking can be added to the canon of bufonid mystique. People have long had an affinity for fiddling with their minds, so it would not be unlikely that some fool would try to get stoned licking the paratoid glands of a toad. Stomach acids would probably mitigate hallucinogenic effects, but other properties of the secretions could well result in a trip to the emergency department. An alternative approach to actually licking the toad has the secretions dried and pulverized then smoked or snorted.

A decade ago, the cosmopolitan genus *Bufo* contained more than half the species of all bufonid frogs, including the 11 species in the western United States. Rigorous phylogenetic analysis of anurans demonstrates that *Bufo* in the traditional sense was composed of several distinct lineages. These findings necessitated nomenclatural changes. For examples, most North American toads are now placed in then genus *Anaxyrus*. Eight species are found in California, two of which are endemic to the Sierra Nevada (the Black toad [*A. exsul*] and the Yosemite Toad [*A. canorus*]). Six species have ranges that include Southern California. The most common of these are the Western Toad and the Red-spotted Toad of the deserts.

The Western Toad (*Anaxyrus boreas*, Figure 7.5) is the familiar toad of the coastal valleys, foothills, and mountains of Southern California. It ranges into parts of our deserts but is largely replaced there by *A. punctatus*. The Western Toad is remarkably adaptable and occurs in a variety of woodland habitats from southern Alaska to northwest Baja California and east to Idaho, Montana, and the Rocky Mountains of Colorado south into northern New Mexico. Although it tends to avoid arid habitats, isolated populations occur for instance in the Owens Valley and at Afton Canyon in the Mojave Desert. Three subspecies have been proposed of which *A. b. halophylus* is the one in our area. It breeds in ponds, reservoirs, and creeks from late January into summer. The Western Toad lacks a vocal sac but does produce a peeping-like call.

FIGURE 7.5 The Western Toad (*Anaxyrus boreas*) is the common bufonid frog of coastal valleys, foothills, and mountains of Southern California.

The Red-spotted Toad (*Anaxyrus punctatus*, Figure 7.6) occurs throughout the Southwest and well into Mexico. In California, it ranges from the southern Great Basin through the Mojave and Colorado Deserts to around 6,500 ft. Although they have been found far from water, these toads usually associate with springs, streams, palm oases, rocky canyon bottoms, and other places where water is permanent most of the year. The melodious trill of its call is distinctive when it is breeding from March to September.

FIGURE 7.6 Male frogs call by forcing air out of the lungs across regulated vocal cords. The inflated vocal sac acts as a resonating chamber (see text). Shown here is a male Red-spotted Toad, (*Anaxyrus punctatus*) calling from a drainage ditch in Imperial County.

The Arroyo Toad (*Anaxyrus californicus*) is practically endemic to California. It occurs west of the deserts in the mountains, foothills, and valleys from southern Monterey County south into northern Baja to around San Quintin. The Arroyo Toad is a riparian specialist confined to sandy washes, arroyos, and streams, either ephemeral or permanent, particularly where cottonwood, willow, and sycamore are present. It breeds in the quiet waters from April to June. This nocturnal species, closely related to the Arizona Toad (*A. microscaphus*), is considered endangered by U.S. Fish and Wildlife Service and is a species of special concern in California. Probably never widespread in its native range, it has suffered losses from habitat degradation and non-native predators. Non-native bass (*Micropterus* spp.), catfish (*Ameiurus* spp.), and carp (*Cyprinus* spp.) feed on adults, larvae, and eggs; crayfish opportunistically eat frog eggs and tadpoles; and bullfrogs will consume any unwary larva or adult.

Three other toad species recorded in Southern California are known only form the Colorado River drainage in Imperial County: Woodhouses's Toad (*Anaxyurus woodhousii*), Great Plains Toad (*A. cognatus*), and the Sonoran (Colorado River) Toad (*Cranopsis alvarius*). Woodhouse's Toad and the Great Plains Toad are otherwise common in the central plains south into Mexico. The Sonoran Toad, the largest toad in the west (up to about 7.5 in.) is a resident of creosote-mesquite grasslands and sycamore, ash and oak canyons with permanent water. Its present status in extreme southeast California is uncertain, and it may be extinct there.

7.4.3 Hylidae—Treefrogs

This large family of some 800 species occurs throughout North, Central, and South America, the West Indies, across Europe to Western Asia, Australia, New Guinea, and the Solomon Islands. Hylid frogs are especially diverse in the New World tropics.

Seven species of hylids occur in the Southwest, but only two of these are found in California, both in our area. The more common one is the Pacific Treefrog (or Pacific Chorus Frog [*Pseudacrus regilla*]). This small frog (1.5–2 in.) comes in shades of green, brown, grey, and red; a dark eye stripe is almost always present (Figure 7.7). The Pacific Treefrog occupies varied microhabitats from British Columbia south through the Pacific states into northwest Baja California. In Southern California, it occurs from the coast to the mountains and occasionally turns up in desert oases. *Pseudacrus regilla* is the only frog species indigenous to the Channel Islands. These small frogs are active year-round and are often encountered some distance from water if plant cover is sufficiently dense. This is the frog that Southern Californians are most likely to hear calling in their backyards or in nearby canyons. Pacific Treefrogs breed in almost any permanent or temporary water source from late fall into summer.

The California Treefrog (*Pseudacris cadaverina*), California's other hylid species, has a geographic distribution similar to that of the Arroyo Toad. It occurs from around Santa Barbara County south into the Sierra San Pedro Martir of northern Baja. Thus, its range overlaps with the Pacific Treefrog in Southern

FIGURE 7.7 The Pacific Treefrog (*Pseudacris regilla*) is common in Southern California from the coast to the mountains and occasionally around high desert springs. These small frogs tend to remain hidden in dense vegetation near watercourses, so they are more often heard calling than actually seen.

California, but the two species are rarely found together in the same place. The California Treefrog prefers mountainous stream-sides with rocks, boulders, and shady pools. It occasionally shows up in high desert oases.

7.4.4 Ranidae—True Frogs

All families of anurans are "true frogs" by definition. Hence this old, long-in-use common name for ranids should not confuse you into thinking otherwise. Ranids are a large cosmopolitan family of around 700 species. The majority, like the 25 species in North America, are slim, long-legged, have webbed feet, and live in streams, lakes, ponds and other aquatic habitats. Some ranids, however, are arboreal, others terrestrial, and a few burrow. Ranids range from 2 to 3 in. in most instances, but the largest frog in the world, *Conraua goliath* of southern Africa, reaches a prodigious 12 in. snout to vent.

Nine species of ranids are known from California, three of which are introduced and occur in Southern California. Most of California's indigenous species have suffered serious declines in recent decades and are now federally protected. Three native species range into Southern California. Until recently, all North American ranids were placed in the genus *Rana*, a name that still persists in various textbooks and field guides. In the same study that reassigned populations of the cosmopolitan toad genus *Bufo*, *Rana*, too, was found to be composed of several discrete lineages that required generic reassignment. Practically all North American *Rana* were allocated to *Lithobates*. The Southern California natives, however, are retained in the genus *Rana*.

The native Southern California ranids west of the deserts are the California Red-legged Frog (*Rana draytonii*), the Mountain Yellow-legged Frog (*R. muscosa*), and the Foothill Yellow-legged Frog (*R. boylii*). Sadly, these three species have been nearly extirpated from Southern California if they are not already extinct. All three inhabit foothill and montane streams and ponds with shady vegetation and boulder-strewn sandy banks, although each has its own microhabitat preferences. The loss of native ranid frogs has come at the hands of pollution, excessive siltation, poor stream management, and the introduction of exotic predators such as certain non-native fish, and the Bullfrog.

The Red-legged Frog (*Rana aurora*) was once considered a single species that ranged from southern British Columbia west of the Cascades south to northern Baja California, as well as in the central Sierra Nevada foothills; desert populations have been recorded from the Mojave River and San Felipe Creek. Most of the California populations of *R. aurora* are now recognized as a separate species, the California Red-legged Frog (*R. draytonii*). In Southern California, isolated populations survive in sheltered habitats in the San Gabriel, San Bernardino, San Jacinto, and Santa Rosa Mountains. They have otherwise vanished from localities where they were once common.

The Mountain Yellow-legged Frog was recognized historically as two disjunct populations. One of those, in the Central Sierra Nevada, is now regarded as a separate species, *Rana sierra*. The other, *R. muscosa*, inhabited Southern California's Transverse and Peninsular Ranges. Both species are seriously depleted or locally extinct; they are gone from nearly 99% of their historic range. Southern California's Mountain Yellow-legged frogs were once found in rocky streams in the San Gabriel Mountains east to the San Bernardino Mountains then south into the San Jacinto Mountains to Mt. Palomar. Fewer than 10 minuscule populations remain. The Foothill Yellow-legged Frog (*R. boylii*) has fared even worse. It was at one time locally common in the Transverse Ranges, for instance along the upper reaches of the San Gabriel River, but now it has essentially disappeared from its historic distribution in the central coastal ranges south into Southern California.

East of the mountains the distribution of the Lowland Leopard Frog (*Lithobates yavapaiensis*) once included the Colorado River Valley from around southern Utah to the delta in Imperial County. It is now likely extinct in that region. Older records also exist for San Felipe Creek in Imperial County. The Lowland Leopard Frog is primarily an inhabitant of oak woodland and streams with oaks, cottonwood, sycamore, and willow in central and southeast Arizona and northern Sonora, Mexico. A second ranid species, the Rio Grande Leopard Frog (*L. berlandieri*), was introduced to the Colorado River around Yuma, Arizona, then spread into parts of eastern Imperial County. Its current status is unknown there, but it, too, is probably extinct. Its native range is western Texas and adjacent parts of New Mexico.

A second introduced ranid frog has had greater success at establishing itself, the Bullfrog (*Lithobates catesbeiana*). The Bullfrog is indigenous to the eastern half of the United States and is not native west of the Rocky Mountains.

It is now widely established in the Pacific states and occurs throughout Southern California. It is the largest frog in the state; adults range from about 5 to 8 in. Bullfrogs prefer calm waters such as lakes, reservoirs, and ponds grown to reedy vegetation. They are best detected at night by reflected eye shine from a flashlight. These voracious feeders will consume almost anything they can subdue. Their diet includes all manner of arthropods as well as fish, amphibians (even conspecifics), small mammals, turtles, and snakes. Needless to say, the Bullfrog has had an alarmingly negative impact on some of our native aquatic species.

The third introduced ranid is the Northern Leopard Frog (*Lithobates pipiens*), common across southern Canada and the northern parts of the United States as well as the Rocky Mountain States south into northeastern Arizona. It has a spotty presence in California. In our region it became established in, for example, Malibu Creek and the Santa Ana River. Doubtless it occurs in other locations, but its current status in Southern California is poorly understood.

7.4.5 Pipidae—Tongueless Frogs

Pipids are a curious family of 4 genera and 26 species native to sub-Saharan Africa and across the Atlantic in Panama and tropical South America east of the Andes. Strictly aquatic, they have flat, smooth bodies and long legs with fully webbed feet. The toes terminate in keratinized claw-like tips. Forelimbs are relatively short, but the fingers are long and unwebbed. Pipids are the only frogs that lack a tongue. The African Clawed Frog (*Xenopus laevis*) first arrived in the United States during the 1940s and 1950s where it was used in pregnancy testing and later as a laboratory subject in embryology. Others found their way into the pet trade. By the early 1970s, Clawed Frogs began showing up in Southern California reservoirs and other permanent waterways. Clawed Frogs are completely dependent on aquatic environments but will venture over dry land during heavy rains. They are now established west of the deserts from Santa Barbara south into Sand Diego County and the Tijuana River. African Clawed Frogs feed on invertebrates as well as other frogs and native fish, for instance the federally protected Three-spine Stickleback (*Gasterosteus aculeatus*) and the Tidewater Goby (*Eucyclogobius newberryi*). They will even cannibalize their own tadpoles and newly transformed adults. African Clawed Frogs are now banned from the pet trade and their transport or possession is illegal.

7.5 THE WORLDWIDE DECLINE IN AMPHIBIANS

Thirty years ago amphibian biologists working in Central America began noticing something odd. Populations of frogs were disappearing from localities where they were once abundant. A few years later, entire species were edging toward the cusp of extinction. No one was sure why. Fast-forward to the present and the extent of amphibian decline has become global. Populations have plummeted in North America, Central America, South America, Europe, Australia, and Africa. Amphibians are now the most-threatened vertebrates on the planet.

Losses have occurred not just in habitat pressured by agriculture and urbanization but also in wilderness areas and national parks. At least 10 species in the Sierra Nevada and Rocky Mountains have disappeared or are extremely rare. Then, of course, there are the three imperiled species of ranid frogs in Southern California. Isolated populations in other parts of the Southwest have dwindled as well. Amphibians have vanished from the Great Lakes region, and numerous frogs in the upper Midwest develop unusual physical abnormalities such as extra or missing appendages and distorted faces.

The causes behind this collapse appear to be multiple and act individually and collectively. Some are obvious such as deforestation and other loss of habitat, habitat alteration, pollution, siltation, pesticides, and the introduction of exotic predators like non-native fish. Unsustainable collecting by commercial enterprises associated with the pet trade is another. An especially pernicious factor was discovered by the investigations that came out of the research in Central America. An exotic, pathogenic chytrid fungus was infecting amphibians. Chytrids are a primitive lineage of fungi and the only fungus that has flagellated spores. The chytrid in question is *Batrachochytrium dendrobatidis*, or simply Bd. Mortality is near 100%, and at least 200 species are extinct because of it. The fungus has since been reported infecting amphibians in the United States, for example among populations of the Arizona Toad (*Anaxyrus microscaphus*) and the Sierra Nevada Yellow-legged Frog (*Rana sierra*). A new strain of *Batrachochytrium* has been discovered among Asian salamanders, although infected species appear to resist the fungus's fatal affects to some extent. However, European salamanders, contaminated with it through the pet trade, have fatality rates of around 96%. If it reaches North America, home to the world's highest salamander diversity, the consequences could be catastrophic.

Why and how have amphibians become vulnerable to the fungus? Is the fungus itself a recent adventive, or have amphibians become susceptible when in the past they might not have been? There are no clear answers at this point. We do know that it is usually futile to reintroduce captive reared stocks to decimated areas because the fungus stays in the environment. Remember that the physiology of frogs and salamanders closely couples them with their habitat and that their permeable integument leaves them potentially vulnerable to exotic disease. With or without the fungal infestation amphibians are facing tough odds.

8 Turtles and an Overview of the Amniotes

8.1 AMNIOTES

As successful as amphibians are and have been, they nonetheless require access to water, minimally for reproduction, if not also for other phases of life history. Vertebrates would not become a dominant presence on land unless that constraint was alleviated. The solution was the amniotic egg, one with a protected embryo that could be deposited on dry land and eliminated the larval stage. Although energetically expensive for the female parent, the amniotic egg is an evolutionary masterpiece. It underpins the great phyletic and ecological diversity of nonamphibian tetrapods—the turtles, lizards, snakes, birds, and mammals. Glibly put, the egg indeed preceded the chicken, by about 260 million years.

The name *amniote* derives from one of four extra-embryonic membranes associated with the developing embryo. In addition to the *amnion*, the other membranes are the *chorion*, *allantoise*, and the yolk sac. Fertilization is necessarily internal in amniotes, after which reproduction proceeds in either of two ways. In most amniotes an outer leathery or calcareous shell completes the package, that is, the shell, the embryo, and its extra-embryonic membranes. The egg(s) is deposited in a nest on land to incubate until hatching. Known as *lecithotrophy* (or oviparity), embryonic development takes place outside the female's body and nutrition is supplied entirely by the yolk mass. In matrotrophy (viviparity), no shell forms and the embryo develops within the female until birth. It receives most of its nourishment from the female reproductive tract. Matrotrophy characterizes marsupial and placental mammals, although it arose independently among several lineages with lecithotrophic ancestry.

Development among living tetrapods is easily observed directly. But vertebrates long extinct rarely leave evidence of their reproductive biology in the fossil record. Therefore, the early origin and diversification of amniotes must be inferred from other signs, usually discovered by extrapolating backward in time from better-understood fossil and living sources.

Stem amniotes arose in the Late Carboniferous about 320 million years ago within a lineage called the *diadectomorphs*, stout, pig-sized animals that appear to be structurally intermediate to non-amniotes and amniotes. The basal amniotes themselves are referred to as *Cotylosauria*, which literally means, "stem reptile." Cotylosaurs included several putative basal amniote lineages, and, as a taxon, the name is often used rather casually. In any case two major amniote groups emerged—the *Sauropsida* and the *Synapsida*. The former includes the basal and derived groups of reptiles and birds. The latter comprise the mammals and their ancestors.

The traditional approach to representing amniote relationships is based on peculiarities in the temporal region of the skull (the area just posterior to the eye sockets). Specifically, they are the openings (fenestra) and the associated bones on either side of the skull. In amniotes, the muscles closing the jaw are attached to the undersurface of the bones in the temporal region. When muscles contract they bulge, and their action is enhanced if there is an opening in the skull through which the swelling can protrude. Hence, such openings permit an evolutionary increase in muscle size and therefore the force at which the jaws operate.

The anapsid skull lacks temporal openings altogether, which is presumably the ancestral condition. Anapsids include early amniotes and later the turtles and their ancestors. Synapsids (mammals and ancestors) have a single temporal fenestra on either side, where as diapsids have two. The pattern of temporal openings has endured as a diagnostic tool that, along with other evidence, supports the following classification:

Amniota
- Synapsida (mammals)
- Reptilia
  - Chelonia (turtles)
  - Diapsida
    - Archosauria
      - Pterosauria (extinct flying reptiles)
      - Dinosauria (extinct sauropods and theropods)
      - Crocodylia (crocodiles and alligators)
      - Aves (birds and ancestors)
    - Lepidosauria
      - Sphenodontia (e.g., Tuatara)
      - Squamata (lizards and snakes)

In this scheme, successively indented taxa are subordinate to the taxa above them. For example, Archosauria and Lepidosauria are sister groups belonging to the more inclusive group Diapsida. Notice, too, that crocodiles are the closest *living* relatives of birds.

8.2 TURTLES

The trademark shell is unlike anything else in vertebrate life, and for that, turtles are one of evolution's more improbable fancies. An upper component, the *carapace*, houses the soft body tissues and provides a space for retraction of the head and limbs. Ventrally, the plastron serves as a kind of skid plate. Both units are composed of bone overlain by keratinized epidermal scutes. The bony aspect of the carapace includes the vertebrae, expanded and fused ribs, and several peripheral elements. The plastron, which is hinged in many species, receives contributions from the clavicles and interclavicles along with other dermal bone (basically directly ossified skin) from the abdomen. Consider this for a moment and then realize how truly odd the whole apparatus is; the appendicular skeleton now lies *within* the rib cage instead of outside of it.

Proganochelys quenstedi was a swamp dweller of Western Europe (and probably elsewhere) during the late Triassic about 210 million years ago. *Proganochelys* was a perfectly modern turtle in the sense that it had a complete carapace and plastron. Turtles have changed very little since, and in the absence of any transitional evidence to go on, the origin of their remarkable morphology remained a mystery since *Proganochelys* was described in 1887. The plodding, boxy land turtle is easy to understand functionally because the shell provides a safe place to withdraw the head and limbs otherwise vulnerable to predators. But how did this radical design come about? It is difficult to imagine a gradual transformation because the shell is so unique in construction and function. Current molecular genetics hints that it may have happened rapidly, evolutionarily speaking, as a result of changes in a class of nucleotide sequences called a *homeobox*, or *Hox* genes, that affect all animal development. In turtles, *Hox* genes preside over patterning of the axial skeleton. Insights about the embryological development of the carapace reveal the presence of a "carapacial disk," a thickening of the dermis of the skin that initiates carapace construction in concert with vertebral formation and arrested development of the ribs to a horizontal plane. The recent (2008) discovery of an even older turtle (220 my), *Odontochelys semitestacea*, from China supports the idea that the turtle carapace is an emergent property. *Odontochelys* had a fully developed plastron but an incompletely formed carapace, and the shoulder girdle was in front of the anterior-most rib. Rather than the shoulder girdle moving backward, as has been suggested, the anterior ribs apparently grew forward across the shoulder blade.

Turtles are thought to belong to a lineage that differs from diapsid reptiles in respects other than the shell. Temporal fenestrations are absent (the anapsid condition), the squamosal bone instead of the quadrate supports the eardrum, and the foot structure has unique articulations between the anklebones and the digits. Turtles are the only surviving members of this group, or so we thought. Recent morphological evidence that includes fossils argues that turtles might be diapsids after all. The issue is yet unresolved, but if this turns out to be the case, then the "anapsid" condition of turtles must be a secondary feature in which the double fenestra of the temporal region of other diapsids has closed over. Hence, some of what we understand about the evolution of reptile morphology will need to be reconsidered if turtles no longer serve as an ancestral template for some features.

Living turtles belong to either of two groups defined by their method of head retraction into the shell. Both have specialized cervical vertebrae that permit a considerable degree of rotation. *Plerodira*, the side-necked turtles, flex the neck laterally to withdraw the head into the shell. Pleurodirans comprise a group of about 80 species of the Southern Hemisphere in three families (Chelidae, Pelomedusidae, and Podocnemidae), all of which are at least semiaquatic. *Cryptodira*, the hidden neck turtles, flex and retract the neck vertically in an S-shape. Of the two, cryptodirans are the dominant group consisting of around 220 species placed in 10 families. They are present on all continents except Antarctica and Australia. Most of the northern hemisphere turtles are cyptodirans including those found in Southern California and the marine species that pass by off shore.

The shell limits where turtles can live, what they can do, and thus restricts diversity and ecological opportunity. For instance, no turtles glide from tree branches and climbing is restricted to moderately inclined surfaces. Most species are aquatic, others are terrestrial, and a few are marine. However, carapace shape and limb structure usually do correlate with lifestyle. Pond turtles tend toward flattened shells and feet are generously webbed. Snapping turtles (Chelydridae) occupy pond bottoms and slow streams; they have been introduced into Southern California. The plastron is reduced significantly so they rely on powerful jaws for defense (I've seen them crush a ¾-in. dowel like a dried twig). The highly aquatic softshell turtles have flat "pancake" shells in which ossification of the carapace is greatly reduced. Terrestrial species like the Box Turtles (*Terrapene* spp.) and the Tortoises have high, domed shells and stout limbs.

8.3 EMYDIDAE—POND AND BOX TURTLES

These omnivores are freshwater, semiaquatic, or terrestrial turtles of small to medium size (up to ca. 24 in. [600 mm]). In general, they have small heads short tails, and a broad connection between the carapace and plastron. Several species are popular aquarium pets, such as those commonly referred to as sliders (*Trachemys*) or painted turtles (*Chrysemys*) that frequently are released into local ponds and reservoirs outside their native range, as here in Southern California. Until recently Emydids were the most diverse of the turtle families, comprising some 100 species from around the world except Australia. With the assignment of the Asian box turtles to their own family (Bataguridae), Emydidae is now an essentially Western Hemisphere group of around 50 species placed among 10 genera. One of only two native turtles in Southern California is an Emydid, the Southwestern Pond Turtle (*Actinemys marmorata pallida*, Figure 8.1). This relatively small (carapace length up to 8 in. [200 mm]) and drab species is a resident of lakes, ponds, reservoirs, and slow streams throughout our region [see discussion under RIPARIAN HABITATS].

FIGURE 8.1 The Western Pond Turtle (*Actinemys marmorata*) is the only native freshwater turtle of the Pacific coastal states. The southern subspecies (*A. m. pallida*), referred to as the Southwestern Pond Turtle, is found in quiet waters of our region's ponds and reservoirs south into northern Baja California.

8.4 TESTUDINIDAE—TORTOISES

Tortoises are terrestrial species with high, domed shells and columnar legs. They inhabit some of the Earth's most arid places. Most of the 40 or so species are herbivores. Testudinids include the giant tortoises of Galápagos and Aldabra and the indigenous species of the Seychelles that were eaten out of existence by eighteenth- and nineteenth-century mariners. North America's representatives are the gopher tortoises (genus *Gopherus*) of the Southwest, northern Mexico, and the southeast United States. Gopher tortoises are accomplished burrowers that have spade like front feet for digging.

California's state reptile, Agassiz's Desert Tortoise (*Gopherus agassizii*) inhabits the Mojave Desert and Colorado Desert east of the Salton Trough. Compelling evidence now demonstrates that the populations east and south of the Colorado River in the southern Great Basin through southwestern Arizona and Sonora, Mexico, long regarded as *G. agassizii*, are in fact a distinct species described and named Morafka's Desert Tortoise (*G. morafkai*) in 2011. California's state reptile is now referred to as *Agassiz's Desert Tortoise* (Figure 6.53). The fallout from this is that the California tortoises, already imperiled (see Desert Vertebrates), actually occupy only 30% of their presumed former range and that the populations east and south of the Colorado River (*G. morafkai*) can no longer serve as a genetic reservoir for conservation efforts.

8.5 OTHER LAND TURTLES

The Sonoran Mud Turtle (*Kinosternon sonoriense*, Kinosternidae) is native to central and southeast Arizona, southwestern New Mexico, and adjacent regions of Mexico. There are old records for this turtle along the Colorado River in Imperial County, but its current status there is unknown. The Spiny Softshell (*Trionyx spiniferus*, Trionychidae) found its way into the Colorado River around 1900 probably from the Gila River Drainage to the east. This flat "pancake turtle" presently occurs in the irrigation canals of Imperial County, and it has been introduced to numerous ponds, lakes, and slow rivers west of the deserts, for example the San Gabriel River in Los Angeles County, and lower Otay Reservoir in San Diego, among others.

8.6 CHELONIIDAE AND DERMOCHELYIDAE—SEA TURTLES

Marine turtles are well designed for life in the sea. The carapace is moderately flat and tapers posteriorly to reduce hydrodynamic drag. Forelimbs are modified as flippers that stroke in unison. Marine turtles occur around the globe generally favoring warm waters. The Green Turtle (*Chelonia mydas*) reaches the Pacific coast as far north as Bahia San Quintin, Baja California. They do not nest in Southern California, but occasional individuals wander north into our waters. In San Diego Bay, Green Turtles assemble around the warm-water discharge from the San Diego Gas and Electric power plant. The Loggerhead (*C. caretta*) is a large turtle found in most of the world's tropical and temperate waters. Sightings off Southern California are rare and are usually juveniles that have dispersed across the Pacific from nesting grounds in Japan. The Pacific (Olive) Ridley (*Lepidochelys olivacea*) is a relatively small sea turtle that nests along the Pacific Coast no farther north than the cape region of Baja California. Like other marine turtles, some may stray well north into our area. The largest sea turtle, the Leatherback (*Dermochelys coriacea*), has a shell length of 8 ft (ca. 2.5 m) at maturity and can weigh 1,600 lbs. (730 kg) or more. Leatherbacks range worldwide, generally in more temperate waters than other sea turtles and they are often seen far out at sea.

9 Lepdiosaurs—Squamates and Tuatara

Lepidosaurs are the largest group of living tetrapods outside of birds. They include the approximately 5,000 species of lizards and 3,000 species of snakes (which together comprise the squamate reptiles), plus the two species of Tuatara (*Sphenodon*) restricted to small islands off the coast of New Zealand. Lepidosaurs are covered in small epidermal scales (*Lepido* means "scale"). As in all amniotes, the skin is imbued with an insoluble protein called *keratin* and is thus fairly impervious to water. In lepidosaurs, the skin is shed periodically in large pieces. Lepidosaurs are also unique in having a transverse vent, that is, the cloacal aperture extends crosswise at the base of the tail. In other tetrapods, the vent is longitudinal.

Sphenodontids were fairly diverse in the Mesozoic, but survived today by only two species (*Sphenodon punctatus* and *S. guntheri*) commonly called the Tuatara (a Maori word meaning "spines on the back"). Tuatara (no "*s*" is used to form the plural) once inhabited both the north and south islands of New Zealand until the arrival of humans and their commensals (rats, pets, and livestock) brought about their eradication from the main islands. Superficially lizard-like in appearance, adult Tuatara can attain a total length of nearly 2 ft (ca. 600 mm). They are nocturnal, and during the day occupy burrows they share with nesting seabirds. Tuatara feed largely on various arthropods and other invertebrates attracted to the guano deposits and occasionally consume fledglings and food scraps brought in by parenting seabirds.

9.1 SQUAMATES—AMPHISBAENIANS, LIZARDS, AND SNAKES

Amphisbaenians are a group of burrowing squamates that for the most part are familiar only to herpetologists. They are so specialized for burrowing that their phylogenetic position among squamates is not understood with much confidence. All but two species are limbless, and the pectoral and pelvic girdle is variously reduced or absent in other genera. Amphisbaenians are the only squamates to occupy permanent tunnels that they construct with their heavily reinforced skulls, using it as a sort of battering ram. Body scales encircle the body in distinct annuli. Some species are only about 4 in. (ca. 260 mm), whereas others can attain 2 ft (600 mm). About 150 species are recognized in 19 genera and four families. Their primary distribution is sub-Saharan Africa, the West Indies, and South America. In the United States, the Florida Worm Lizard (*Rhinura floridana*) is the only extant representative. The Mole Lizard or *Ajolote*, (*Bipes bipora*) occurs along alluvial western slopes of Baja California from around Guerro Negro south to the Cape. *Bipes* has no hind limbs but retains functional forelimbs, a condition rare among tetrapods. The hands are used for penetrating the surface.

Lizards and snakes, the best-known squamates (from *squama* meaning scale), are a timeless study in ecological diversity. As with arid lands elsewhere in the world, Southern California hosts a diverse assemblage: 10 families and 37 species of lizards and 5 families and 35 species of snakes. Lizards and snakes are popularly thought of as discrete from one another, and snakes indeed have numerous characteristics of their own. However, snakes are embedded within lizards phylogenetically and should not be construed as evolutionarily distant relatives. The ongoing debate is about just where snakes belong as a clade.

Most lizards have well-developed appendages, yet limb reduction and loss have occurred numerous times. This trend toward limb reduction explains in part why squamates have been the subject of such intense phylogenetic study over the past 200 years. The principal sticking points have been the relationships among the burrowing, elongate "snake-like" forms of lizards, and which of these groups might be the sister lineage to snakes themselves, if at all.

Besides the trend toward limb reduction another area of contention involves the relationships among the various clades of carnivorous squamates. In recent decades, molecular character analysis has joined the fray bringing both congruence and discordance to analyses based on traditional phenotypic criteria. This is not to say that our understanding of deep squamate evolutionary history is hopelessly muddled, only that some uncertainty remains. In fact, vertebrate systematics at large has benefited from the scrutiny given to lizard–snake relationships because the effort has probed and refined some of the assumptions behind phylogenetic reasoning generally and how we interpret hypotheses about ancestry.

Unlike sphenodontids, squamates (males) have a unique copulatory organ called a *hemipenis*. This is an admittedly odd name because at first reading one might assume that they only have half a penis. In fact, they have two (pleural is hemipenes), and the female likewise has paired receptacles to receive one or the other. Cranial kinesis is another squamate feature. This means that the skull has one or more moveable joints (especially in snakes) that are important structural adaptations for feeding. The bony arch at the bottom of the lower temporal opening (see diapsids) has been lost in squamates resulting in a condition known as *streptostyly*; the quadrate bone (the articulation site for the lower jaw at the back of the skull) is moveable relative to the braincase.

Squamates range in size from 2 in. (50 mm) or so in some lizards to impressively large species like some of the monitor lizards (*Varanus* spp.; e.g., 9 ft [2.75 m] or more in the Komodo Dragon [*V. komodoensis*]) and a few snakes, for example, the Giant Anaconda (*Eunectes murinus*), a semi-aquatic snake of South America that can reach 30 ft (ca. 9 m). By and large, squamates are terrestrial but some are aquatic, especially snakes, and a few are marine (Marine Iguanas of Galápagos, sea snakes). Some lizards, like iguanas, are herbivores, but most lizards consume small arthropods, such as insects, spiders, and scorpions. Most snakes are carnivores that feed on small birds, mammals, and lizards as well as other snakes.

9.1.1 Squamate Senses

Vertebrates respond to their environment by information received from three *primary sense organs* associated with the brain. Each is specific to a particular class of stimuli: chemical, mechanical, or electromagnetic. The nasal apparatus and tongue are chemosensory organs involved with olfaction (smell and taste). The inner ear and associated auditory structures receive mechanical vibrations transmitted through water or air that are interpreted as sound. In tetrapods, the inner ear is also important in maintaining balance and body orientation. Optical organs—eyes and other photoreceptors—receive visual (electromagnetic) information. In other words, the eyes, ears, tongue, and nose are receptors that capture information, code it, and send it on to the brain.

Besides the primary senses, an array of other receptors, structurally less complex, are scattered throughout the vertebrate body, each of which is stimulus specific. Internal chemoreceptors help regulate hormone balance, mechanoreceptors detect pressure and pain, and other receptors sense heat loss or gain. The *parietal organ* is a dorsal evagination of the vertebrate midbrain that ancestrally was involved with photoreception. The organ was inserted into an opening in the top of the skull and covered by a thin layer of skin. It likely functioned in responding to annual changes in photoperiod. A great deal of structural and functional variation exists with the parietal organ, as does the terminology surrounding it. In amniotes, where it functions with the endocrine system, it is called the *pineal organ* or *pineal gland*. In lizards and tuataras, the parietal organ is so distinctive it is sometimes called the "third eye."

Vertebrates rely on all of their primary senses but typically depend on one more than the other two to navigate and respond to their world. Hearing, for example, is limited in squamates, and except for some geckos, they do not vocalize. Snakes lack a tympanic membrane ("ear drum"), which is the structure that receives airborne vibrations that are then transmitted to the inner ear by a chain of one to three bones. Despite the popular notion that snakes are deaf, experimental evidence demonstrates that auditory centers in the brain do in fact respond to sound waves.

Many squamates are visually oriented. Visual displays are important in courtship and territorial defense in Iguania and Gekkota, the most visually oriented clades. Some arboreal snakes have excellent visual acuity and are able to assess depth of field, an advantageous capability when preparing to strike at prey. All vertebrates have rod cells in the retina that provide visual reception in dim lighting and at night. However, many vertebrates lack the cone cells necessary for color vision. Up to four types of cone cells may be present, each being particularly sensitive to a spectral region of visible light: violet, blue, green, and orange. Vertebrates with tetrachomatic (four-color) color vision include many fish, turtles, lizards, and birds. Snakes have color vision, but the "cones" appear to be modified rod cells adopted for that purpose. Color vision accompanied natural selection for a diurnal lifestyle. (Many amphibians lack color vision, and those that have it are trichromatic.)

Some vertebrates have retinal structures receptive to ultraviolet (UV) radiation, which lies just to the left of visible light on the spectrum of electromagnetic radiation. It has been known for more than a century that insects are UV sensitive, and now that capability is known to occur in some fish, turtles, lizards, and birds. Its full role in tetrapods is not well understood but seems to be related to foraging behavior and territoriality. Even visually oriented lizards like iguanians can be sensitive to UV light, which is absorbed by pheromones released onto rocks and other hard surfaces that the lizards can detect.

Infrared radiation lies to the right of visible light, and some vertebrates have receptors that are sensitive to it. All objects above absolute zero give off infrared radiation from their surfaces, and cooler objects can be warmed by it. For that reason, it is sometimes called "heat radiation," but this is incorrect because strictly speaking infrared is a narrow wavelength of electromagnetic radiation, not heat. Still, infrared receptors are stimulated by the warming effect that infrared radiation produces. Such receptors are known in some bats but are especially keen in boas, pythons, and pit vipers where they are located around the anterior end of the face.

Less visually oriented squamates rely more on olfactory cues by collecting chemical signals with their tongue. The slender, terminally bifurcated tongues of scincomorphs, anguimorphs, and snakes contrast with the fleshy tongues of Iguania and Gekkota. The degree of forkedness correlates with foraging mode. Deepest forks are found in the most widely foraging species. Most snakes depend on chemical cues for locating prey, potential mates, and for general exploratory behavior. The constant flicking of the tongue, the sampling phase, collects olfactory information (molecules adhering to sublingual pads) from the environment, and the bifurcated tip is sensitive enough to detect gradients in the strength of the chemical cues. When the tongue is withdrawn into the mouth, the tips transfer the chemical stimuli to a vomeronasal apparatus called *Jacobson's Organ* (present in all squamates), where the information signal is transmitted to the brain.

9.1.2 Ectothermy, Foraging, and Feeding

Beyond the generalization that fence lizards, for instance, are insectivorous or that desert iguanas are herbivorous, the full range of dietary preferences is known only for a few species. Most species are opportunistic feeders because food abundance and density will always vary in space and time. For endotherms like mammals and birds, much of the energy converted from food is dedicated to maintaining a constant body temperature. The rest is allocated to cell and tissue maintenance and to the reproductive effort if the individual is a female. Ectothermy, on the other hand, is energetically inexpensive because there is no metabolic requirement for a constant body temperature. Ectotherms can convert food energy to other needs when they are active and can sustain long periods of no or minimal energy intake and expenditure when they are not. Similarly, the greater the energy content per food item, the less often the animal needs to forage and feed. Snakes that eat relatively large, soft-bodied prey like small mammals or birds need only feed every few weeks during the season and not at all during the winter. Smaller lizards and snakes forage and feed more regularly, especially if they are active thermoregulators.

Many ectotherms regulate their body temperature behaviorally. An individual can maximize the radiant heat received directly from the sun by orienting its body perpendicular to the incoming rays. Watch basking fence lizards when they first emerge in the morning. They position themselves at right angles to the sun. Notice, too, that they are very dark, almost black. Imbedded within the integument are packets of pigment crystals called *chromatophores*. Chromatophores give vertebrates their color and pattern. The blackness of the lizard comes from the concentration of dark pigments (melanin) that absorb the sun's heat. When the lizard achieves a desired temperature the dark pigment granules scatter within the chromatophores and its coloration lightens. An ectotherm can also gain heat by *conduction*, which is heat exchange by direct contact with a warm surface like a rock or a sandy surface. *Convection* is the transfer of heat by air or liquid moving across the body surface, which can thus either heat up or cool down the body. Lizards and snakes dissipate heat by taking shelter in a cool retreat away from direct sunlight. By these means, active thermoregulation can achieve remarkably stable body temperatures over a 24-h period during the active time of the year. Most thermoregulators operate within a specific temperature range referred to as the *critical thermal minimum* and the *critical thermal maximum*, above which the temperature becomes lethal. Both of these limits will affect when, where, and for how long an individual will forage.

Foraging strategies and tactics vary from those species that hunt over a wide area to those that are ambush predators who sit and wait for prey to come into range. Ambush feeders such as Collared Lizards and phrynosomatids are visually oriented, generally have stocky bodies, and consume fewer but larger prey items per foraging episode. They have high sprint capability but low endurance. Most of the day is devoted to finding food in this way. Predation risk is lower because they are not calling attention to themselves by moving widely about. By contrast, widely foraging species, like whiptail lizards and many snakes, rely on chemosensory signals to locate prey, have slender, gracile bodies, and encounter smaller but more numerous prey items. They have high endurance but weaker sprint capabilities, and generally are active for only part of the day. Widely foraging species face greater predation risk than sit-an-wait foragers and will expend as much as 150% more energy. However, they experience as much as 200% greater food intake.

9.1.3 Squamate Defenses

A predator relies on its primary senses to determine if the potential prey is an object distinct from the background and whether it is edible or not. Of course, the best defense is not to be detected in the first place. Avoiding detection is often accomplished by a cryptic pattern and color so that the body disappears into its background so long as the prey remains motionless. Other means of avoiding detection include limiting activity time and, as a species, being rare or uncommon. If it is detected and the predator is approaching, the prey's only option is to flee. An attempt to escape may happen at the last instant, when a sudden burst of speed briefly startles the predator. If possible execute some evasive moves—feint left and cut right until safety is reached. Many lizards avoid the challenge by simply remaining close to a retreat while keeping a wary eye on their surroundings. They rely on speed to escape and usually take off before the predator is in striking distance.

If a predator succeeds in making contact the prey's options are limited. The strength to escape or some structural or mechanical defense are about all that it has left. For example, many lizards will detach their tail. Involuntary muscular contractions keep the tail wiggling diverting the predator's attention while its owner takes off. Tail autotomy is possible because of the vertical fracture planes that pass through the body of the individual vertebra. Hence a tail separates not between successive vertebrae but at point within a single vertebra. Usually only the anterior third or so of caudal vertebrae have fracture planes. The tail will regenerate at the point of detachment, but the regenerated tail consists of a shorter, cartilaginous rod that may not even ossify. Regenerated tails are easy to spot because they taper rather abruptly, and the external scales are darker and somewhat irregularly configured. Tails can be lost and regenerated more than once so long as whole vertebrae with fracture planes remain from the original caudal sequence.

Other forms of physical defense involve various forms of noxious behavior. These can include threatening postures, a favorite response by rattlesnakes, or hissing, biting, and expelling feces, and other foul-smelling secretions. Anguimorphs, in particular, are noted for this type of behavior.

Many species of horned lizards squirt blood from the corner of their eyes, either one or both. This remarkable defense is rarely used, and in some quarters, the behavior is thought to be a myth. It is not. Blood pressure increases significantly in the head by shunting venous blood back to the heart while maintaining arterial flow into the ocular sinuses. When sufficiently agitated some horn lizards can expel a thin stream of blood up to 6 ft. Interestingly, the behavior seems to occur mainly when they are molested by canids—dogs, foxes, and coyotes for instance. No chemical substances are added to the blood as an added defense, and it is unclear why these particular mammals are more vulnerable. One suggestion is that the horn lizard diet of ants increases the concentration of formic acid in the blood, which canids would find especially distasteful.

9.1.4 Squamate Reproduction

Lizards and snakes typically employ either visual or chemical signals to locate mates, as they do when foraging. Iguania are visually oriented and males use a repertoire of displays to solicit females. Fence lizards perform sequences of push-ups and head bobs, for example. Other species will inflate their dewlap (throat fan) or assume particular postures and orientation. And males of some species are marked in bold, contrasting colors. The vivid blue belly patches of fence lizards and the strong throat and facial pattern of collared lizards are good examples.

Where females are similarly patterned, the colors are duller and not as crisply defined. However, among phrynosomatid and crotaphytid lizards, gravid females of some species will take on bright coloration.

In advertising for a mate, a single, primary stimulus does not necessarily exclude other indicators. For instance, chemical signals can also be important between Iguania even though as a group they are largely sight oriented. Secretions from glands located on the ventral surface of the thigh (*femoral glands*) are left behind on the substrate by a number of species (males), both as a territorial marker and as a means of alerting females. These secretions also absorb UV light, to which lizards are sensitive, so the secretion can also act as a visual stimulus.

Most Scincomopha and Anguimorpha have a limited field of view, and they rely mainly on pheromones, a class of chemical signals, to locate potential mates. Pheromones are species-specific. Both sexes can recognize the signal as from one of their own kind as well as its sex. Generally males will follow pheromone trails of females, but not of males, and females will not follow that of either sex. Pheromones have the advantage over visual displays in that they can persist in the environment for days.

Embryogenesis follows successful courtship and mating. Oviparity is the ancestral condition among squamates and the female deposits shelled eggs. The embryo is completely yolk-dependent and develops entirely outside the female's body. In viviparity, eggs are retained and develop within the oviducts; the female gives birth to live young. Embryos are yolk-dependent in some cases (leicithotrophic), rely on maternal tissue in others (matrotophic), or use some combination of the two. Viviparity has arisen at least 45 times in lizards and 35 times among snakes. Viviparity is common in skinks, although Southern California's two species (Gilbert's Skink and Western Skink) are egg layers. The live-bearing lizards in our area include the Night Lizards and the California Legless lizard. The region's live-bearing snakes are the Boas, the Garter Snakes, and all rattlesnakes. Amuse yourself but don't be duped by the novelty box of "rattlesnake eggs" sold in some truck stops across the Southwest.

The advantage of viviparity often expressed is that thermoregulatory behavior can be employed by the mother to incubate her eggs. Hence, viviparity often has evolved in species that live in short breeding season environments such as cold climates or high elevations. This is an appealing idea. Active thermoregulation could hasten the developmental period and reduce the amount of time the female is gravid. Viviparity can also be viewed as a form of parental care. Eggs are not exposed to possible predation or some environmental casualty. The down side is that viviparity increases energetic demands on the female, reduces her agility, and potentially limits reproductive output.

Fecundity refers to clutch size or the number of live-born offspring, and the frequency of mating. It is a function of both phylogenetic history and opportunity. All geckos, for example, produce one or two eggs. However, females can store viable sperm and may deposit up to 3 clutches in a season under appropriate conditions. Multiple clutches per year are not uncommon among Southern California's smaller lizard species and even some snakes. Ample winter rains prior to the spring/summer breeding season and a healthy bloom of flowering plants and insects are important to that end. Female Side-blotched lizards may have up to 7 clutches of from 1 to 8 eggs. By contrast a large lizard like the Chuckwalla typically deposits a single clutch of up to 16 eggs every 1 to 3 years.

Parental care is limited within squamates; typically the eggs are deposited, or the offspring are born and the job is done. Nonetheless nest guarding is known in at least 100 species. Females of some snakes and many lizards remain with their clutch. Some pythons encircle the clutch and actually brood the eggs. Beyond that, there is practically no contact between the young and parent after hatching. (Some parental–offspring interaction is known in juvenile Green Iguanas.)

Parthenogenesis is a form of reproduction without male involvement because there are no males in parthenogenetic species, only females. This curious phenomenon in some respects muddies our concept of just what a species is. Parthenogenesis likely comes about through hybridization events between two bisexual species, or sometimes between a unisexual species and a bisexual one. Hybridization alters meiosis (nuclear division of gametic cells resulting in half the number of chromosomes of the parent) by inhibiting nuclear and cytoplasmic divisions. Hybridization is most likely to take place at the edges of preferred habitats where the ranges of two species overlap. For that reason, parthenoforms are sometimes referred to as "weed" species because they occupy habitats that are marginal to their bisexual parental species.

Among tetrapods, parthenogenesis is known to occur only in squamates. It has been identified in six families of lizards and one family of snakes. Parthenogenesis is particularly common in Whiptail lizards. Pairing females undergo pseudocopulation, which triggers the release of sex hormones and then egg development. In other words, unisexual species reproduce by cloning. Because there is no genetic recombination (contributions by X and Y chromosomes from two parents), unisexual species do not evolve in the Darwinian sense, and from an evolutionarily perspective, all-female species are probably quite young in geologic time. Unisexual species do not look any different than bisexual ones, and their discovery is usually by accident when field studies for completely different reasons reveal that a population is entirely female. There are likely others that have yet to be recognized.

9.2 LIZARDS OF SOUTHERN CALIFORNIA

Lizards occur on all continents except for Antarctica and on practically all oceanic islands. Families are placed into one of the following major clades: *Iguania, Gekkota, Scincomorpha, Anguimorpha*, as shown in Figure 9.1. Notice that Scincomorpha + Anguimorpha comprise the *Autarchoglossa*, and that the Autarchoglossans + Gekkota form the *Scleroglosss*. Some squamate families are cosmopolitan whereas others are geographically restricted but correlated, for instance Old World versus New World or Southern Hemisphere (Gondwanan) versus Northern Hemisphere (Laurasian). We will survey the families and species of lizards that occur in Southern California. A key to the common genera is presented in (Table 9.1). A complete list of Southern California reptiles can be found in Table 9.2.

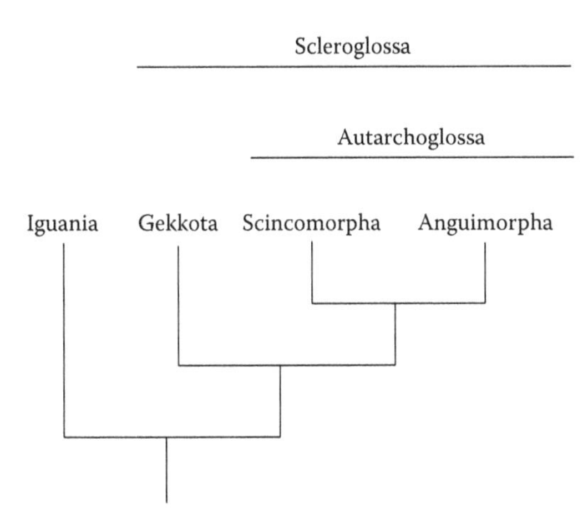

FIGURE 9.1 A phylogeny of the four major clades of squamate reptiles (lizards and snakes): Iguania, Gekkota, Scincomorpha, and Anguimorpha. Snakes are embedded in the Anguimorpha. Together, the Scincomorpha and Anguimorpha comprise the Autarchoglossa. The Autarchoglossa plus the Gekkota comprise the Scleroglossa (see text).

TABLE 9.1
Key to Some Common Lizard Genera of Southern California

1A. Limbs present	2
1B. Limbs absent, eyelids present	*Anniella*
2A. Body elongate, limbs reduced, and proportionately short	3
2B. Limbs robust, not reduced	4
3A. Scales smooth, cycloid, shiny all over body	*Eumeces*
3B. Scales square, plate-like; distinct lateral fold present	*Elgaria*
4A. Eyelids absent, eye covered by transparent spectacle	*Xantusia*
4B. Eyelids present	5
5A. Dorsal scales fine and granular abruptly distinct from large flat scales on belly arranged in rows	*Aspidoscelis*
5B. Scales not as above	6
6A. Body scales distinctly keeled, mucronate	7
6B. Scales weakly keeled or not at all	8
7A. Body robust limbs stout	*Sceloporus*
7B. Body flat oval and compressed; spinous projections from back of head	*Phrynosoma*
8A. Hind limbs conspicuously longer than forelimbs	9
8B. Hind limbs and forelimbs more or less equal in length	10
9A. Toes of hind feet with posteriorly projecting fringe-like scales	*Uma*
9B. Toes not as above; head large, distinct	*Crotaphytus*
10A. Body large, somewhat compressed	*Sauromalus*
10B. Body small, less than 3 in.; pupils vertical	11
11A. Toes expanded distally	*Phyllodactylus*
11B. Toes not expanded distally	*Coleonyx*

TABLE 9.2
Reptiles of Southern California

Turtles

Cheloniidae: Sea turtles
- *Caretta caretta*: Loggerhead
- *Chelonia mydas*: Green turtle
- *Lepidochelys olivacea*: Pacific Ridley

Dermochelyidae: Sea turtles
- *Dermochelys coriacae*: Leatherback

Trionychidae: Softshells
- *Trionyx spiniferous*: Spiny Softshell

Kinosternidae: Mud turtles
- *Kinosternon sonoriense*: Sonoran Mud Turtle

Emydidae: Pond and box turtles
- *Actinemys marmorata [pallida]*: Southwestern Pond Turtle

Testudinidae: Tortoises
- *Gopherus agassizi*: Agassiz's Desert Tortoise

Lizards

Iguanidae: Iguanids
- *Dipsosaurus dorsalis*: Desert Iguana
- *Sauromalus ater*: Chuckwalla

Phrynosomatidae: Sand lizards, Fence lizards and allies

I. Sand lizards
- *Callisaurus draconoides*: Zebra-tailed Lizard
- *Uma inornata*: Coachella Valley Fringe-toed Lizard
- *Uma notata*: Colorado Desert Fringe-toed Lizard
- *Uma scoparia*: Mojave Fringe-toed Lizard

II. Spiny Lizards, Brush lizards, Rock lizards
- *Sceloporus magister*: Desert Spiny Lizard
- *Sceloporus occidentalis*: Western Fence Lizard
- *Sceloporus orcutti*: Granite Spiny Lizard
- *Sceloporus vandenburgianus*: Southern Sagebrush Lizard
- *Urosaurus graciosus*: Long-tailed Brush Lizard
- *Urosaurus nigricaudus*: Black-tailed Brush Lizard
- *Urosaurus ornatus*: Ornate Tree Lizard
- *Uta stansburiana*: Side-blotched Lizard
- *Petrosaurus mearnsi*: Banded Rock Lizard

III. Horned Lizards
- *Phrynosoma blainvillii*: Coast Horned Lizard
- *Phrynosoma mcallii*: Flat-tailed Horned Lizard
- *Phrynosoma platyrhinos*: Desert Horned Lizard

Crotaphytidae: Collared and leopard lizards
- *Crotaphytus bicinctores*: Great Basin Collared Lizard
- *Crotaphytus vestigium*: Baja California Collared Lizard

(*Continued*)

TABLE 9.2 (*Continued*)
Reptiles of Southern California

Gambelia sila: Blunt-nosed Leopard Lizard
Gambelia copeii: Cope's Leopard Lizard
Gambelia wislizeni: Long-nosed Leopard Lizard
Eublepharide: Geckos
Coleonyx switaki: Barefoot Gecko
Coleonyx variegatus: Banded Gecko
Gekkonidae: Geckos
Phyllodactylus xanti: Desert Leaf-toed Gecko
Xantusiidae: Night lizards
Xantusia gracilis: Sandstone Night Lizard
Xantusia henshawi: Granite Night Lizard
Xantusia (Klauberina) riversiana: Island Night Lizard
Xantusia vigilis: Desert Night Lizard
Scincidae: Skinks
Eumeces gilbert: Gilbert's Skink
Eumeces skiltonianus: Western Skink
Teiidae: Whiptails
Aspidoscelis hyperythrus: Orange-throated Whiptail
Aspidoscelis tigris: Western Whiptail
Anguidae: Alligator lizards and allies
Anniella pulchra: California Legless Lizard
Elgaria multicarinata: Southern Alligator Lizard
Helodermatidate: Gila monster, Beaded Lizard
Heloderma suspectum: Gila Monster

Snakes
Leptotyphlopidae: Blind snakes
Leptotyphlops humilus: Western Blind Snake
Boidae: Boas
Charina bottae: Rubber Boa
Lichanura trivirgata: Mexican Rosy Boa
Lichanura orcutti: Desert Rosy Boa
Colubridae: Colubrid snakes
I. Ring-necked and Leaf-nosed snakes
Diadophis punctatus: Ring-necked Snake
Phyllorhynchus decurtatus: Spotted Leaf-nosed Snake
II. Racers and whipsnakes
Masticophis flagellum: Coachwhip, Red Racer
Masticophis fuliginosus: Baja California Coachwhip
Masticophis lateralis: California Whipsnake, Striped Racer
Masticophis taeniatus: Striped Whipsnake
Coluber mormon: Western Yellow-Bellied Racer

(*Continued*)

TABLE 9.2 (*Continued*)
Reptiles of Southern California

III. Patch-nosed snakes, Rat snakes, Kingsnakes
Arizona elegans: Glossy Snake
Bogertophis rosaliae: Baja California Rat Snake
Lampropeltis getula: Common Kingsnake
Lampropeltis zonata: California Mountain Kingsnake
Pituophis catenifer: Gopher Snake
Rhinocheilus leconti: Long-nosed Snake
Salvadora hexalepis: Western Patch-nosed Snake
IV. Garter snakes
Thamnophis elegans: Western Terrestrial Garter Snake
Thamnophis hammondii: Two-striped Garter Snake
Thamnophis marcianus: Checkered Garter Snake
Thamnophis sirtalis infernalis: California Red-sided Garter Snake
V. Ground snakes
Sonora semiannulata: Ground Snake
Chionactus occipitalis: Western Shovel-nosed Snake
Tantilla hobartsmithi: Southwestern Black-headed Snake
Tantilla planiceps: California Black-headed Snake
VI. Night and lyre snakes
Hypsiglena torquata: Spotted Night Snake
Trimorphodon biscutatus: Western Lyre Snake
Hydrophiidae: Sea snakes
Pelamis platurus: Yellow-bellied Sea Snake
Viperidae: Pit Vipers; Crotalinae—Rattlesnakes
Crotalus atrox: Western Diamond-backed Rattlesnake
Crotalus cerastes: Sidewinder
Crotalus helleri: Southern Pacific Rattlesnake
Crotalus mitchellii: Speckled Rattlesnake
Crotalus ruber: Red Diamond Rattlesnake
Crotalus scutulatus: Mojave Rattlesnake

9.3 IGUANIA

Iguania is a diverse assemblage of around 1,450 species, and two of the three principle lineages are Old World. These are the Agamids and Chamaelonids that are often referred to collectively as the Acrodontids because of their unusual dentition; teeth are fused to the jaw. Agamids consist of 380 species that range from the Middle East, through Africa, Australia, and the Indo-Australian Archipelago. Most are generalized and terrestrial lizards that range in size from around 4 in. (100 mm) to 3 ft (0.9 m). The chamaeleonids are a bizarre-looking group of 160 mostly arboreal species found in Africa, Madagascar, and southern Spain. Specializations for life in trees include laterally compressed bodies, prehensile tails, eyes that can pivot 360°

independently, a projectile tongue for catching prey, and zygodactylous or yoke-toed feet in which two pairs of toes oppose one another for grasping branches and twigs.

The third group of Iguania is the *Iguanians*, composed of around 900 species. They are exclusively new world except of a few species in Madagascar and in the Fiji Islands of the southwestern Pacific. The large, arboreal Green Iguana (*Iguana iguana*) of Middle America and the Marine Iguana (*Amblyrhynchus cristatus*) of Galápagos are familiar examples. However, these are but two species of Iguanians belonging to one (Iguanidae) of seven lineages (families). Southern California supports 23 species placed in three of those: Iguanidae, Phrynosomatidae, and Crotaphytidae.

9.3.1 Iguanidae—Iguanas and Allies

Our two species of iguanids, the *Chuckwalla* (*Sauromalus ater*—formerly *S. obesus*, Figure 6.49) and the *Desert Iguana* (*Dipsosaurus dorsalis*, Figure 9.2) have nearly identical distributions in the desert Southwest from southern Nevada south through the Southern California deserts, western Arizona, and down the gulf coasts of Sonora and Baja. Within their range, Chuckwallas favor boulders and rocky outcrops with crevices. Desert Iguanas are generally found in sandy areas of creosote scrubland. Both of these are among the largest lizard species in the Southwest. Chuckwallas will reach a snout-vent length of 9–10 in. (ca. 230–255 mm). Like other iguanids, the Desert Iguana and Chuckwalla are herbivores with multicuspate teeth and a valvular intestine for processing plant matter. Both have high thermal tolerances. Desert Iguanas, for example, are active in an average ambient temperature of 107°F and will remain active to 115°F.

FIGURE 9.2 The Desert Iguana (*Dipsosaurus dorsalis*) has a geographic distribution similar to that of the Chuckwalla (Figure 6.49) in the Mojave and Sonoran Deserts but prefers sandy, Creosote scrublands with scattered rocks. Like the Chuckwalla, it is also herbivorous. (Courtesy of Mitch Leslie.)

9.3.2 Phrynosomatidae—Sand Lizards, Horned Lizards, and Allies

This is the largest family of Iguanians in Southern California. The phrynosomatids are arranged into three groups: (1) Sand Lizards (*Calisaurus, Uma, Hobrookia*), (2) the Spiny (*Sceloporus*), Brush (*Urosaurus, Uta*), and Rock (*Petrosaurus*) Lizards, and (3) the Horned Lizards (*Phrynosoma*).

1. Sand Lizards possess a suite of adaptations for living in arid, sandy habitats like flats, dunes, and gravelly washes. They range from the central and southwestern United States to central Mexico. The Zebra-tailed Lizard (*Callisaurus draconoides,* Figure 9.3) occurs in our area deserts where it is often seen running with its tail curled over its back exposing the black and white ("zebra") markings. The Earless Lizards (*Holbrookia*, *Cophosaurus*), found outside California in the west-central and southwestern United States and Mexico, are close relatives that, unlike Zebra-tailed Lizards, lack external ear openings. All three genera have slender bodies with long legs and toes.

   The five sand lizards of the genus *Uma* are quite similar to one another. They have flattened bodies, elongated feathery scales extending back from their toes, a dorsal pattern of fine ocelli (small red dots on small white circles), and an adeptness at shimmy burrowing beneath the sand. Each of the three California species is restricted in distribution as indicated by their common names: The Mojave Fringe-toed Lizard (*U. scoparia*, Figure 6.50), Colorado Desert Fringe-toed Lizard (*U. notata*), and the Coachella Valley Fringe-toed Lizard (*U. inornata*). The latter species is

FIGURE 9.3 With its long legs and slender body, the Zebra-tailed Lizard (*Callisaurus draconoides*) is a familiar site darting across desert hardpan and other firm, flat surfaces. These speedy runners curl their black and white banded tail forward up over the body while sprinting from one place to another. (Courtesy of Mitch Leslie.)

imperiled because of habitat loss (see discussion under Deserts). A fourth species, the Yuman Fringe-toed Lizard (*U. rufopunctata*) is confined to extreme southwest Arizona and adjacent northwest Sonora, Mexico; it was until recently in synonymy with the Colorado Desert Fringe-toed Lizard. A fifth species, *U. paraphygas*, is restricted to the sand dunes of Bolsón de Mapimí, Chihuahua, Mexico.

2. Spiny Lizards, genus *Sceloporus*, are so named because of their acute, keeled body scales. This is a large group of about 80 species found throughout the United States (mostly the west) south through Mexico to Panama. Of the four species found in California, all occur in our area. The Granite Spiny Lizard (*S. orcutti*) is restricted to the Peninsular Ranges from near San Gorgonio Pass down through Baja to near the Cape. It favors granite outcrops in chaparral and oak woodland on the coastal slopes, and upper elevation rocky canyons in the Colorado Desert. The Desert Spiny Lizard (*S. magister*, Figure 9.4) ranges throughout the arid Southwest in a variety of habitats in the northern Great Basin south through the Mojave and Sonoran Deserts (including the Colorado). One of the most familiar lizards to Southern Californians is the Western Fence Lizard (*S. occidentalis*, Figure 6.17). In fact, it is one of the most common lizards in the West, ranging from Washington, Idaho south through Nevada and most of California to northern Baja, and from sea level to nearly 11,000 ft. It occupies multiple types of habitat except for the extreme deserts. It does well in urban areas and often can be seen basking on rocks, tree trunks, the sides of buildings, and fence posts. As kids we called them "blue-bellies" because of the vivid blue patches found on males along the ventral sides of the trunk and the throat. In females, the blue markings are less intense and

FIGURE 9.4 The Desert Spiny Lizard (*Sceloporus magister*) is found in a variety of habitats throughout the arid Southwest and northern Mexico. Look for these lizards in our Mojave and Colorado Deserts around large rocks, boulders, and toppled vegetation.

may even be absent. "Blue-bellied Lizard" was an early common name for this species. Blue markings like theirs, however, are present on many species of Spiny Lizards and Brush Lizards. The Southern Sagebrush Lizard (*S. vandenburgianus*) is closely related and similar to the Western Fence Lizard. It differs in its slightly smaller size and weakly keeled scales. Until recently, this species was regarded as a subspecies of the Sagebrush lizard (*S. graciosus*). The Southern Sagebrush Lizard is usually found at higher elevations in upper chaparral and montane habitats of the Transverse and Peninsular Ranges.

Brush (*Urosaurus*) and Side-blotched Lizards (*Uta*) resemble Spiny Lizards in general proportions, but they are typically smaller and lack the distinct keeled dorsal scales. The Common Side-blotched Lizard (*Uta stansburiana*, Figure 9.5) is, like the Western Fence Lizard, one the most abundant lizards in the West. Primarily a ground-dweller, "Utas" occupy a multitude of microhabitats from below sea level to around 9,000 ft. These are small lizards (up to 2 ½ in. [65 mm]) with a blue to turquoise patch (blotch) on each side of the body behind the armpit, although it is sometimes difficult to see. In females and even some males, the patch is feint or ill defined.

As a desert species (Mojave, Colorado, northern Sonora), the Long-tailed Brush Lizard (*Urosaurus graciosus*) cryptically perches on branches and twigs among small trees and shrubs like creosote. The Ornate Tree Lizard (*Urosaurus ornatus*) occurs in portions of Utah, Arizona, New Mexico,

FIGURE 9.5 The Common Side-blotched Lizard (*Uta stansburiana*) is abundant throughout much of the western United States. Dazzling are the brightly colored males like this one. The side "blotch" is located just behind the armpit.

West Texas, and parts of northern Mexico where it can be found climbing about tree trunks and brush piles, especially in riparian areas grown to Cottonwood and Mesquite. It reaches California near the vicinity of the Colorado River. The Black-tailed Brush Lizard (*Urosaurus nigricaudus*) inhabits Baja California thornscrub and chaparral. Its range barely extends into Southern California near Anza Borrego on the desert side and in chaparral/riparian areas on the west in southeast San Diego County.

The Banded Rock Lizard (*Petrosaurus mearnsi*) is a handsome, agile rock climber (up to 4½ in. [115 mm], snout to vent) in the boulder regions of the desert slopes of the Peninsular Ranges, south into the upper third of Baja California. Two other species of Rock Lizards are endemic to the Baja peninsula: *P. repens* and *P. thalassinus*.

3. Horned Lizards (*Phrynosoma*) are close relatives of Sand Lizards. Nine species occur in the western half of the United States and Mexico, and an additional five species are found in Mexico alone. Four species include California in their range, of which three of these (all but the Pygmy Short-horned Lizard [*Phrynosoma douglasii*]) can be found in Southern California. Horned Lizards ("horny toads") are crowd favorites. Unmistakable are their oval, flat bodies and bony spikes protruding from the back of the skull. Blainville's (or Coast) Horned Lizard (*P. blainvillii*) ranges throughout much of California west of the Sierra Nevada south through Baja. It inhabits coastal sage scrub, chaparral, oak woodland, and piñon-juniper forest. Sadly, many populations have been extirpated as a result of urbanization and agriculture (see discussion under Coastal Sage Scrub). Many readers may know this species as the Coast Horned Lizard (*P. coronatum blainvilli*). The Desert Horned Lizard (*P. platyrhinos*, Figure 9.6) replaces Blainville's Horned Lizard on the desert side of the Sierra Nevada and the Transverse and Peninsular Ranges, east through Nevada and western Arizona. Our third Southern Californian species, the Flat-tailed Horned Lizard (*P. mcallii*), is restricted to the sandy regions of the lower Coachella Valley southeast to the head of the Gulf of California including extreme southwest Arizona. Human impact of various sorts (off-road-vehicles, agriculture, and mining) has taken a toll on their long-term viability.

9.3.3 Crotaphytidae—Collared Lizards and Leopard Lizards

Crotophpytids are among the most striking vertebrates of the Southwest. There are 12 species in this family of 2 genera, the Collared Lizards *Crotaphytus* (9 species) and the Leopard Lizards *Gambelia* (3 species). Five of them can be found in Southern California. Collard Lizards are characterized by often-bright colors and patterns, large heads, and robust limbs. These ambush predators lie in wait for large insects and even other lizards that they leap after and seize with powerful jaws. The Great Basin Collared Lizard (*C. bicinctores*, Figure 9.7) is chiefly a rock dweller that occurs throughout the Great Basin south into the Mojave and Colorado

FIGURE 9.6 The Desert Horned Lizard (*Phrynosoma platyrhinos*) is one of three species of the genus in Southern California. It occurs throughout the Great Basin and the Mojave and Sonoran Deserts. The Desert Horned Lizard dwells in a variety of habitats, although sandy substrates are generally present. Another desert species, the threatened Flat-tailed Horned Lizard (*P. mcallii*) is restricted to sand dunes and flats in southeast Imperial County and adjacent areas of the lower Colorado River basin. (Courtesy of Mitch Leslie.)

FIGURE 9.7 Collared Lizards (*Crotophytus* spp.) are among the most striking of the desert vertebrates. They generally inhabit rocky hillsides but occasionally occur on gravelly slopes and adjacent flats. The Great Basin Collared Lizard (*C. bicinctores*) shown here occurs throughout the Great Basin south through the Mojave Desert to southeast California and western Arizona. The Baja California Collared Lizard (*C. vestigium*) occupies desert slopes of the Peninsular Range. (Courtesy of Mitch Leslie.)

FIGURE 9.8 A relative of the collared lizards, the Long-nosed Leopard Lizard (*Gambelia wislizenii*) inhabits much of the arid Southwest and prefers sandy or gravely surfaces among open areas of low shrubs. This one appears to be half-nodding off while catching some early morning sun.

Deserts. The Baja California Collared Lizard (*C. vestigium*) also favors arid rocky hillsides, but along the eastern slopes of the Peninsular Ranges south into Baja.

The Long-nosed Leopard Lizard (*Gambelia wislizenii*, Figure 9.8) is also an arid land inhabitant (Great Basin, Mojave, Sonoran, and Chihuahuan Deserts) but prefers open areas of low shrubs and substrates that may include sand dunes, gravel, and hardpan. These lizards take refuge inside burrows, or beneath rocks, and other surface debris such as toppled Joshua Trees. The Blunt-nosed Leopard Lizard (*Gambelia sila*) once occupied large areas of the San Joaquin Valley and adjacent foothills south to Santa Barbara and Ventura Counties. Agriculture, industry, and urbanization have eliminated it from more than 90% of its former range. Cope's Leopard Lizard (also called the Baja California Leopard Lizard [*Gambelia copeii*]) is a resident of peninsular Baja and barely reaches the United States in southern San Diego County.

9.4 GEKKOTA

Granular scales, fragile tails, large eyes covered by a transparent spectacle (most), and expanded, adhesive toe discs (lamelle) characterize the most familiar species of Geckos. Most are nocturnal and excellent climbers. Geckos are a large cosmopolitan group of more than 1,100 species especially well represented in tropical and subtropical regions and also in some arid temperate zones. Southeast Asia, Australia, and many islands in the Pacific and the West Indies have a particularly diverse gecko fauna. Three species belonging to two families are present in Southern California.

FIGURE 9.9 Soft, delicately granular skin is a hallmark of geckos. Moveable eyelids and thin, unexpanded toes further differentiate species of the Eublepharidae, like this Western Banded Gecko (*Coleonyx variegatus*).

9.4.1 Eublepharidae—Banded Geckos and Allies

Eublepharids stand out among the Gekkota because they are the only ones with moveable eyelids. In addition, they have slender toes that lack expanded terminal discs. The Western Banded Gecko (*Coleonyx variegatus*, Figure 9.9) is primarily a resident of the deserts and arid regions of the Southwest where it usually associates with rocky areas and occasionally sandy dunes. The Banded Gecko does occur west of the deserts along the slopes of the Peninsular Ranges in rocky areas of chaparral, but very little is known about these populations.

The Barefoot Gecko, or Switak's Banded Gecko (*Coleonyx switaki*), was discovered in 1974, in central Baja, California. Subsequently it was found in San Diego County near Yaqui Pass and Borrego Springs. Only a smattering of individuals has been discovered in the area since then, but presumably the population is continuous south into Baja. At the time of its discovery, Switak's Banded Gecko was placed in its own genus, *Anarbylus*, which was later synonymized with *Coleonyx*.

9.4.2 Gekkonidae—Gekkonid Geckos

The third gecko in our area is the Peninsular Leaf-toed Gecko (*Phyllodactylus xanti*). This is another species whose distribution is primarily peninsular Baja. It is a secretive rock/boulder inhabitant that in Southern California occurs mainly along the eastern slopes of the Peninsular Ranges southward. An outlier population is known from Joshua Tree National Park. Leaf-toed geckos get their common name from their flap-like toe pads. The skin is covered in granular tubercles.

9.5 SCINCOMORPHA

Three families of scincomophans are present in Southern California and other areas of the Southwest. These are the night lizards, the skinks, and the whiptail lizards. The families are distinct from one another in appearance yet share structural details that unite them with this clade.

9.5.1 XANTUSIIDAE—NIGHT LIZARDS

The night lizards are an elegant but inconspicuous group of mostly small crepuscular and nocturnal species found in the Southwest and in Mexico. One peculiar outlier is known from Cuba. Night Lizards are live-bearing and lack eyelids. The Granite Night Lizard (*Xantusia henshawi*, Figure 9.10) occurs along the Peninsular Ranges south into northern Baja California. These handsome flat-bodied lizards are denizens of granite boulder crevices in coastal sage scrub, chaparral, and chaparral–creosote communities of the high desert. For many years now, a few have occupied the stucco exterior of our house, apparently finding it an adequate substitute for a granitic surface; they take refuge behind the downspouts and emerge during summer evenings.

The Sandstone Night Lizard (*X. gracilis*) was described in 1986 as a subspecies of the Granite Night Lizard (*X. henshawi*). It is similar in appearance but distinctive in several ways, including its geographic restriction to the Truckhaven Rocks in the Anza Borrego Desert State Park.

The Desert Night Lizard (*X. vigilis*) once thought to be extremely rare because of its small size and secretive habits is quite common in many arid parts of the Southwest. It ranges from south central and Southern California to southern Nevada

FIGURE 9.10 The flat profile of the Granite Night Lizard (*Xantusia henshawi*) is adapted for moving among cracks and exfoliating sheets of granite boulders. This species is found along the Peninsular Range south into northern Baja California.

and Utah, central and southeastern Arizona, and Baja California. These lizards favor microhabitats with toppled and decaying trunks and branches such as those of Joshua Tree, agaves, and large cactus. They can also be found under dried cow pies.

The Island Night Lizard (*X. riversiana*) is a large species (up to 4½ in. [115 mm], snout to vent) confined to the Channel Islands of San Clemente, San Nicholas, and Santa Barbara.

9.5.2 Scincidae—Skinks

At more than 1,300 species, this group comprise one of the largest of all lizard families. Skinks are found on all continents except Antarctica and are diverse on many oceanic islands. Many species are small, quick, and agile and have flat cycloid, shiny scales. Others are quite large like the Australian Blue-tongued Skink (*Tiliqua*) and the Arboreal Skink *Corucia zebrata* of the Solomon Islands. Only two skink species occur in California and are members of the diverse and rather unwieldy genus *Eumeces*. Recent studies indicate that *Eumeces* as currently construed is fraught with taxonomic problems. Some researches advocate placing all North American *Eumeces* (ca. 14 species) in the genus *Plestiodon*. Additional work on Southern California's two species, *E. gilberti* and *E. skiltonianus*, indicate nomenclatural problems at several levels.

Gilbert's Skink (*Eumeces gilberti*) is distributed primarily in California from around the latitude of Yuba County south along the foothills of the Sierra Nevada and Coast Ranges, the Central Valley, and south into Southern California. Outside of California its range extends into extreme southern Nevada and west central Arizona. Gilbert's Skink frequents a variety of habitats, except the low deserts. Because adequate ground cover is an essential requirement, these lizards are seldom seen in the open. In Southern California, juvenile Gilbert's Skinks are distinguished by their red tails, hence the subspecific designation, for example, *rubricaudatus*.

The Western Skink (*E. skiltonianus*, Figure 9.11) is the smaller of the two species. Its distribution in California is similar to that of Gilbert's skink but extends north into British Columbia as well as into Idaho, most of Nevada, and the western half of Utah. Adults tend to resemble the juveniles of Gilbert's Skink in their dorsal stripes (adult Gilbert's are unicolor). Like Gilbert's Skink, it avoids the deserts but is otherwise common in many habitats with adequate ground cover and places to shelter.

9.5.3 Teiidae—Whiptail Lizards

Whiptail Lizards are included in this large, New World family of about 130 species. Teiids are especially diverse in Middle and South America and the West Indies (e.g., *Ameiva* spp.). Until 2002, the 40 or so species of whiptails were all placed in the genus *Cnemidophorus*. At that time systematic studies concluded that all species north of Mexico belonged to a separate lineage, *Aspidoscelis*. This recommendation has largely been accepted for those

FIGURE 9.11 Skinks comprise one of the largest families of lizards (Scincidae). The adult Western Skink (*Eumeces skiltonianus*) shown here is one of two species in Southern California. The other is Gilbert's skink (*E. gilberti*). Juvenile Western Skinks have bright blue tales that fade with age. In our area, juvenile Gilbert's have red tails. Both species associate with rotting logs, rocks, and surface debris in a variety of habitats except for deserts.

16 species. The name *Cnemidophorus* is still used in some field guides, as you might discover, and it has a deep history in the literature.

All but one species of *Aspidoscelis* (*A. sexlineatus*) is confined to the western half of the United States. Whiptails are terrestrial long-bodied, diurnal lizards that forage widely, moving about with jerky, hesitant motions. Most have stripes, spots, or some combination of the two, and individual species can be difficult to identify. About a third of the species are unisexual, being all-female parthenoforms. Knowledge of a particular species geographic range is especially useful when attempting to identify whiptails in the field.

Southern California supports two species, both bisexual. The Orange-throated Whiptail (*Aspidoscelis hyperythrus*) is a small, striped, peninsular Baja species that ranges into Southern California as far north as San Bernardino and Riverside Counties. It occupies undisturbed areas of coastal sage scrub and lower Chaparral where buckwheat (*Eriogonum*) and sage (*Salvia*) are often predominate. Unfortunately, habitat loss and degradation have been seriously detrimental to this species' survival in Southern California.

The larger Western Whiptail (*A. tigris*, Figure 9.12) occurs in varied, open habitats throughout the west from as far north as Oregon and Idaho and south into Baja and mainland Mexico east to west Texas. Currently eight subspecies are recognized. In Southern California, Western Whiptails occur on both sides of the Transverse and Peninsular Ranges. To the west, they inhabit coastal sage scrub, chaparral, and pine-oak woodland (*A. t. stejnegeri*). To the east, they roam in the upper desert slopes to the low desert creosote communities (*A. t. tigris*).

FIGURE 9.12 At least 16 species of the genus *Aspidoscelis* (formerly *Cnemidophorus*) occur across the Southwest. Variously known as Racerunners or Whiptails, only two of these are found in California. The Western Whiptail (*A. tigris*), shown here, is found over much of the inter-mountain west in deserts, sagebrush communities, and chaparral. Western Whiptails are widely foraging lizards usually seen mid-morning moving purposely from one shrub to another in search of food. (Courtesy of Mitch Leslie.)

9.6 ANGUIMORPHA

The Anguimorpha consists of the Monitor Lizards (Varanidae) of Africa and Indo-Australia, the Gila monster and Beaded Lizards (Helodermatidae) of North America as well as the Alligator Lizards and their allies (Anguidae). Because Anguimorpha also includes the snakes along with several other limbless to near limbless taxa, it has been and remains the most problematic group of Squamates to resolve phylogenetically.

9.6.1 Anguidae

This small family of around 100 species has a far-flung distribution that includes Europe, North Africa, Asia, parts of Indonesia, and the Americas including the West Indies. In the eastern United States, anguid lizards are represented by four species of the legless Glass Lizards *Ophisaurus*. The West, including Baja, California, is represented by seven species of Alligator Lizards (*Elgaria* spp.) and two species of the Legless Lizard (*Anniella* spp.).

Three species of Alligator Lizards occur in California. The Northern Alligator Lizard (*E. coerulea*) is distributed along the northern Coast Ranges and central Sierra Nevada north to British Columbia. The Panamint Alligator Lizard (*E. panamintina*) is restricted to the desert mountains of Mono and Inyo Counties. The Southern Alligator Lizard (*E. multicarinata*, Figure 9.13) ranges from Washington south in to northern Baja. This is the species in our area. Alligator lizards have elongate

FIGURE 9.13 In the same family (Anguidae) with the California Legless Lizard (Figure 9.14), the Southern Alligator Lizard (*Elgaria multicarinata*) is found in shady, moist areas in coastal sage scrub, chaparral, and oak woodland. It is also common in urban gardens. Alligator lizards do not take well to handling. If molested they will writhe, attempt to bite, and expel feces.

bodies and tails, short limbs, and a distinct skin fold on either side of the body. They prefer shadier, moister areas of chaparral and oak woodland and may be common around urban gardens where ample cover is available. Alligator Lizards have a cantankerous disposition. When handled they are prone to writhing, biting, and defecating.

Legless Lizards are about the length and diameter of a pencil. At first look they can be mistaken for a small snake. Their pointed heads have moveable eyelids (unlike snakes), and they lack the sinuous, coordinated muscle movement of most serpents. Seldom seen because they burrow in soft soil and forage beneath leaf litter and plant cover, they are most likely found when turning loose soil or raking debris. The California Legless Lizard (*A. pulchra*, Figure 9.14) ranges from the Bay Area south through the west side of the Central Valley and Coast Ranges down through the Transverse and Peninsular Ranges into northern Baja, generally avoiding the deserts. The Baja California Legless Lizard (*A. geronimensis*) is endemic to the Baja Peninsula along a narrow coastal stretch from El Rosario north to Colonial Guerrero where it overlaps with the California Legless Lizard.

9.6.2 Helodermatidae

This family is represented by the Gila monster and Beaded Lizard. These two large (up to 14 in. [355 mm], snout to vent), strikingly colored animals are the only known venomous lizards. Venom glands are located alongside the jaws and grooved teeth deliver the venom during a bite. Helodermatids have a fossil record extending to the Early Cenozoic of North America, a history that demonstrates

FIGURE 9.14 The California Legless lizard (*Anniella pulchra*) is a burrowing species about 4–6 in. long that spends most of its time underground. It may emerge on the surface during warm evenings. Legless lizards prefer loose soils in coastal sage scrub, chaparral, oak, and piñon-juniper woodlands.

that the primary morphological specialization has been their powerful, stoutly constructed jaws capable of crushing large prey. Living helodermatids may consume proportionately larger prey than any other squamates aside from certain snakes. Venom capabilities may have arisen in helodermatids as a defense mechanism, given that widely foraging species like these are also more likely to encounter predators of their own. However, it is also possible that venom delivery initially evolved in a feeding context and was later coopted for defense. In any case, human fatalities from a bite are very rare.

The Gila monster (*Heloderma suspectum*, Figure 6.51) occurs in the Southwest (east to extreme southwestern New Mexico) and northern Mexico, whereas the Mexican Beaded Lizard (*H. horridum*) ranges along mainland Mexico's Pacific Coast from Sinaloa south into Guatemala. Gila monsters are terrestrial and spend much of their time in underground burrows. They are agile climbers (and good swimmers), however, and have no problem ascending into branches in search of bird eggs and fledglings. The Gila monster tends to frequent mesic habitats like canyon bottom with permanent streams. People are often surprised to learn that they occur in California. They do, but only in the eastern Mojave where they are extremely rare (see Desert Vertebrates).

9.7 SNAKES

As we noted in our general discussion of squamates, there are numerous forms with elongate bodies and reduced or missing limbs. Limbs have been lost independently at least a dozen times. Morphological changes in various phylogenetic lineages appear to have started with body elongation followed by reduction of limb size and the loss of distal elements of the appendages, like fingers and toes. Limblessness probably evolved in response to several selective forces. Nonetheless, limbless bodies have common structural constraints that are imposed on locomotion and feeding, and these typically lead to similar lifestyles. Most limbless or near limbless lizards are fossorial; they either burrow directly into the substrate or move about beneath leaf litter and surface entanglements. They consume small prey items that do not exceed the width or gape of the open mouth. In contrast, most snakes have surmounted the constraints of limblessness and its ecological limitations. They dwell in nearly every conceivable structural habitat, and use several different modes of locomotion to transit over irregular ground, to climb, and to swim. Nor is prey size a limiting factor compared with limbless lizards because many species can and will consume prey considerably larger than the size of their open mouths.

I mentioned the contentiousness of snake relationships within Squamata. That dispute results from the troublesome task of trying to identify character similarity that is based on shared evolutionary ancestry versus similarity because of the convergences resulting from a limbless lifestyle. The fossil record of snakes extends back to the Cretaceous, yet these early snakes present almost as much ambiguity as they do clarity with respect to their origin and relationships. Current evidence supports a relationship with anguimorph lizards, but less clear is how this came about. Unclear indeed. Recent molecular studies have uncovered a clade (named Toxicofera) consisting of Iguania + Anguimorpha + Snakes. If true, this finding completely upends our understanding of basal squamate phylogenetics and possibly the origin of snakes themselves.

One hypothesis for snake origins posits a marine ancestry. Evidence for this argument comes from 90-million-year-old, marine, varanoid-like fossils with certain cranial features suggestive of snakes. Other options favor a fossorial origin for snakes, that they diverged early from specialized, burrowing anguimorph lizards. This scenario points out that both specialized burrowing snakes and lizards have degenerative eyes. Nonburrowing, aboveground lizards visually focus by changing the shape of the lens, much like we do. Snakes, on the other hand, focus by moving the lens forward or backward relative to the retina. This suggests that the eyes of snakes experienced a period of degeneration early in their evolutionary history, only to be re-elaborated in a different way later on.

9.7.1 Characteristics of Snakes

Limbs and girdles are absent except in a few snakes like the boas and pythons that retain vestigial hind limbs in the form of a visible "cloacal spur" on either side of the body just in front of the base of the tail. The spur is a tiny remnant of the femur that articulates with a miniscule pelvis.

Snakes have no moveable eyelids; each eye is covered by a transparent scale, or spectacle. Although they have middle and internal ears, external ear openings are absent.

The left lung is reduced and nonfunctional or absent altogether. A modified "tracheal lung" is usually present.

Snakes have a highly modified vertebral column. The move onto land by vertebrates required a remodeling of the vertebral architecture. Fish propel themselves through water by bending the body from side to side, with lateral undulations increasing in amplitude toward the tail. On land, locomotion is accomplished with the appendages, which means that the body needs to be supported off the ground. That requirement and others, for example independent movement of the head (which fish cannot do), led to the regionalization of the vertebral column. In amniotes especially, distinct cervical (neck), thoracic (anterior trunk), lumbar (posterior trunk), sacral (supports the pelvic girdle), and caudal (tail) vertebrae emerge with their own structural peculiarities and functional roles. Most lizards, for example, have 24 or fewer presacral vertebrae; a pair of ribs is present on each thoracic vertebra, which meet the sternum midventrally and surround the lungs and heart. Some lizards retain vestigial ribs on the cervical vertebrae. The elongate body of snakes derives not from lengthening individual vertebrae but by an increase in the number of vertebrae themselves. Snakes have from between 120 and 150 vertebrae in the body and tail. In some, however, that number is as high as 585. Other than the atlas and axis, the first two vertebrae of the neck, regionalization of the vertebral column is essentially lost in snakes. Body vertebrae are pretty much identical and each has a pair of ventrally curving ribs, one pair per segment that corresponds with the number of ventral scales. Ribs are important in serpentine locomotion (see discussion below). Tail length varies according to function. Finally, unlike in other tetrapods, where each vertebra makes contact with adjacent vertebrae at three points (a single ball-and-socket-joint and a pair of facets called *zygopophyses*), snakes have two additional points of contact, an accessory pair of anterior facets, the *zygosphenes*, and posterior *zygantra* that receive the zygosphenes of the vertebra behind it. This is important because the extra points of articulation help minimize torque, the twisting around the long axis of the body. Torque is a mechanical issue confronting any cylindrical, elongate object. Garden hoses, for example, are variously reinforced internally to prevent torque (kinking), according to manufacturers, although that claim is seldom met. The solution for minimizing torque among limbless or near limbless lizards is to keep the body small (short) relative to cross-sectional area. The tradeoff is that short bodies limit locomotor options and the potential to exploit other habitats. Hence, most limbless lizards are fossorial, headfirst burrowers with rounded heads. The axial design of snakes releases them from these constraints and some species can reach up to 30 ft in length.

Axial musculature is also modified. In amniotes, the muscles supporting the vertebrae comprise a complex of short units between one vertebra and the next. Their primary function is to help stabilize the vertebral column during appendicular locomotion. In snakes, the axial muscles have tendons that span as many as 30 vertebrae before they reach their insertion point, with tendons becoming shorter as they approach the head. Thus, rather than functioning in support, axial muscles of snakes are the prime movers of locomotion. The body behaves more like a cable than a hinge. If you've ever had the opportunity to hold a snake, then you've experienced that gracefully coordinated muscular action. The motion really is poetry and there is nothing like it elsewhere in nature.

9.7.2 How Snakes Move

A snake moves in a smooth asymmetric fashion in which the loops formed by the body change proportionately as it glides over the surface. Yet the very motion itself would seem just as likely to propel it backward or leave it thrashing in place. Still, snakes move headfirst. The mechanics of motion are difficult to analyze because the particular properties of the musculoskeletal system must be specified along with the output forces exerted by points along the body as the snake progresses. Numerous elegant studies have examined these issues using sophisticated laboratory equipment and film analysis. For our purposes we need only consider how *styles* of locomotion in snakes correlate with predatory modes, body morphology, and the surface or substrate over which a snake customarily travels.

Most generalized terrestrial snakes traversing irregular surfaces use serpentine locomotion, also called *lateral progression*. By modifying simple lateral undulations, the body is thrown into a series of irregular curves with pressure exerted backward against an object that the body contacts. *Rectilinear locomotion* is characteristic of heavy-bodied snakes during routine cruising and by some while stalking prey. In this mode, alternating sections of the ventral skin are lifted off the ground by muscle slips originating from the ends of the ribs and insert on the ventral scales internally. The in-between sections that remain in contact with the ground support the body. The snake moves more or less in a straight line in response to the waves of contractions. Rectilinear locomotion is slow but efficient. *Concertina locomotion* is employed in movement through narrow passages, such as rodent burrows. The action is like that of an accordion. A snake will stabilize the posterior part its body by pressing loops against the side of the passage. Thus anchored, it then extends itself anteriorly. After it is fully extended, new loops secure the body anterior to the previous pressure point and the posterior part of the body is pulled forward. *Sidewinding* is a mode of locomotion practiced by desert species that inhabit substrates composed of windblown sand. The sand would give way under the lateral pressure of serpentine locomotion, so an alternative means of travel is required. In sidewinding, the body is raised in two or three short loops while the rest of the body is in contact with the sand, supporting its weight. The loops are thrown outward then placed on the sand in successive, synchronous waves with the points of contact moving smoothly along the body. The body extends perpendicular to the snake's line of travel, and because the force is exerted downward, there is no lateral slippage. The distinctive tracks left behind in the sand depict a series of parallel lines (Figure 9.15).

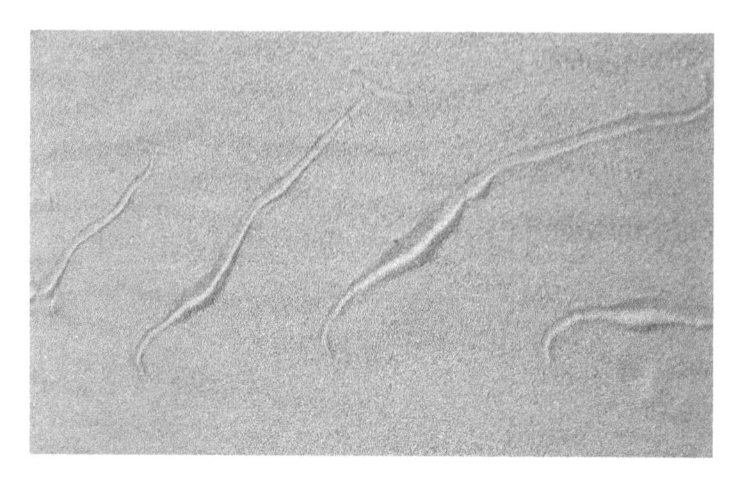

FIGURE 9.15 The Sidewinder Rattlesnake's unique style of locomotion is exquisitely suited for traversing loose, windblown sand, leaving behind this distinctive track (see text).

9.7.3 Modifications to the Skull

Limblessness in squamates imposes limits on prey size and modes of prey acquisition and ingestion. Among snakes, this limitation is evident in the Scolecophidia or blind snakes (see discussion below). In advanced snakes like the Macrostomata ("*large mouth*"), the skull is highly kinetic and variously modified to accommodate ingesting large prey items. The number of bones is reduced, and the upper and lower temporal arcades are absent. Rather than being fused to one another, the anterior tips of the lower jaws (mandibles) are joined by elastic tissue. Unlike in lizards, the braincase is fully enclosed. Other bones form a series of linkage components that can be moved and rotated about their joints, which collectively provide an astounding array of possible actions. To further complicate the mechanics, the paired links act independently on either side of the skull, in particular the tooth-bearing elements, which include not only the upper and lower jaws but also the bones of the palate (palatines and pterygoids).

Snakes usually swallow their prey headfirst, although smaller items may be consumed tail first or even sideways. During swallowing, teeth on the maxilla and the adjacent palatal bones anchor the prey while the snake turns its head so that the maxilla and palatal bones on the opposite side advance forward over the prey. This process is repeated where one side anchors and the other advances until the prey is fully drawn into the esophagus. At this point the snake bends its head sharply side to side to allow neck muscles to assist in pushing the prey into the stomach.

Teeth are sharp and recurved. Those on the maxillae are functionally categorized by their size and arrangement. *Aglyphous* snakes are nonvenomous and have no enlarged teeth; teeth are continuous along the maxilla. Venomous snakes have one of three arrangements of enlarged teeth on the maxillary bones. Proteroglyphous (*proto* means "first"; *glyph* means "knife") snakes have short, hollow teeth (fangs) at the front of the maxilla followed by a series of solid teeth behind. The fangs are permanently erect, that is, they cannot be folded backward. Examples are found among the elapid snakes such as coral snakes, cobras, and mambas. The viperid snakes (e.g., rattlesnakes) are solenoglyphus (*soleno* means "pipe"), meaning that the only teeth on the maxilla are the enlarged, hollow fangs in the front. These fangs can be folded backward and therefore may be quite long. Ophisthoglyphous (*opistho* means "behind") or "rear-fanged" snakes have one or more moderately enlarged teeth at the back of the maxillary bones. They are usually solid, but some may have grooves that assist in venom transfer, which is done typically by chewing motions. The primary prey consists of lizards and birds that are held in the mouth until the prey is immobilized before it is swallowed. Most rear fang snakes are not dangerous to humans, for example the California Lyre Snake that occurs in our area. However, the African Boomslang (*Dispholidus typus*) is a notably deadly rear-fanged species.

During prey capture, some snakes simply grab the item and consume it while the prey struggles. Injury to the snake is possible in these instances, especially when the prey is being swallowed. Unlike in lizards, however, the braincase is fully enclosed, so the brain itself is afforded some protection. The grab-and-swallow tactic probably constrains the type of prey the snake will eat.

Prey constriction probably appeared early in snake evolution and today is a trademark of pythons and boas. Some colubrid snakes, for example kingsnakes and gopher snakes, also use constriction. The snake seizes its prey and rapidly entwines it in body coils tight against one another so that friction prevents the struggling prey from forcing the loops apart. Despite the folklore, the prey is not crushed during constriction. Rather, with each of the prey's exhalations the snake tightens its coils and takes up slack. The prey dies from suffocation or heart failure as a result of the pressure.

Constriction requires the ability to create tight, small diameter coils. This is accomplished in pythons and boas by relatively stout vertebrae and shorter axial muscles. The tradeoff is that locomotion is comparatively slow in these snakes relative to colubrids such as whipsnakes and racers that achieve speed by throwing their body into broad loops.

Fast-moving snakes in North America appeared in the early Miocene as grasslands were expanding. They probably sacrificed prey constriction for greater speed, which would then necessitate alternative means of prey capture. Some of them relied on the delivery of toxins to immobilize prey. Many colubrid snakes possess a pair of structures at the back of the upper jaws called *Duvernoy's gland.* The glands deliver mildly toxic secretions by way of a duct adjacent to the posterior maxillary teeth. The secretions can also partially lubricate food and help maintain oral membranes. Additionally, they may neutralize toxins that could be carried by the prey itself and perhaps initiate early chemical stages of digestion. Modern colubrids like kingsnakes and gopher snakes lack the venom capability. In these species they have become constrictors secondarily.

The venom glands of viperid and elapid snakes are homologous to the Duvernoy's glands of colubrids. Venom glands produce a cocktail of different chemicals of varying function. Some are highly toxic and quickly dispatch prey. Other enzymes break down tissues. Thus, when a snake strikes defensively (at a person) medical treatment must not only neutralize the toxic compounds but also deactivate the proteolytic enzymes to minimize tissue damage. During the strike, the compressor glandulae muscle contracts and forces venom in the gland out through a duct into the fang and then into the prey. A snake does not usually expend its entire reservoir of venom during a strike. Full venom replacement takes about 2 days. Envenomation is a highly efficient means of prey capture because there is almost zero danger for the snake. The prey may wander off a short distance before it dies, but the snake easily tracks the prey's scent to obtain its meal.

9.8 SNAKES OF SOUTHERN CALIFORNIA

Of the 2,900 species of snakes found around the world, Southern California hosts 35 species assigned to 5 families (Table 9.2). Each of the 3 major snake lineages is represented: Scolecophidia, Alethinophidia, and Caenophidia (Colubroidea).

9.9 SCOLECOPHIDIA

This group of small, burrowing snakes has shiny scales and vestiges of the hind limbs. Eyes are reduced in scolecophidians, hence they are often called *blind snakes*. The 3 families are the Anomalepididae of Central and South America (16 species), the Leptotyphlopidae of southwest Asia, Africa, Central America, South America, and the American South West (about 90 species), and the Typhlopidae the largest family (ca. 230 species) and found worldwide except in North America.

9.9.1 Leptotyphlopidae—Blind Snakes

Two species of Leptotyphlopid snakes occur in the United States. The Texas Blind Snake (*Leptotyphlops dulcis*) is found from extreme southern Kansas south through most of Texas and on into eastern Mexico. It ranges west into parts of eastern and southern New Mexico and extreme southeast Arizona. Its close relative the Western Blind Snake (*L. humilis*) is known in our area. This snake may reach 15 in. (380 mm) and has a slender body and blunt head. It occurs from south central California and southern Nevada south through all of Baja, western mainland Mexico, and east to west Texas. The Western Blind Snake can be found in almost any habitat within this range that has loose soil for burrowing or rocky slopes and crevices. It is seldom seen above ground and then usually only after rains.

9.10 ALETHINOPHIDIA

Alethinophidians are a diverse assemblage made up of about 170 species of large and small snakes assigned to 13 families. Approximately 95% of them are macrostomatans. The most familiar of these are the Boidae, which includes the pythons (26 species) of Africa, Asia, and Australia, and the Boas (33 species) in western North America south through subtropical South America and the West Indies. Boids include the largest snakes in the world, for example the Giant Anaconda of South America (up to 30 ft [ca. 9 m]). The largest snake ever known is the extinct boid, *Titanoboa cerrejonensis*. Estimated at 42 ft (13 m) and weighing perhaps 2,500 lbs. (1,135 kg), the fossils were discovered in 2009 in coal mine deposits of Colombia. The site dates to the Paleocene epoch (60–58 million years ago) not long after the last dinosaurs had vanished. This colossal snake was probably the largest nonmarine vertebrate of the era.

9.10.1 Boidae—Boas and Pythons

In the United States, only one family of Alethinophidians is represented today, the Boidae, and by just three native species, the Mexican Rosy Boa (*Lichanura trivirgata*), the Desert Rosy Boa (*L. orcutti*), and the Rubber Boa (*Charina bottae*). All three are restricted to the far west. Of the two genera, *Lichanura*, is the more common in Southern California. Rosy Boas have experienced a long, convoluted taxonomic history. The earliest descriptions from the 1880s identified three species *L. trivirgata*, *L. roseofusca*, and *L. myriolepis*. A century later only a single species, *L. trivirgata*, was recognized that included multiple subspecies differentiated largely by color pattern. In 1993 *Lichanura* was placed in the synonymy of *Charina* until it was resurrected a decade later. The current taxonomy followed here is based on mitochondrial DNA (mtDNA) analysis of samples taken from 76 localities throughout California, Arizona, and Mexico.

The Mexican Rosy Boa (*L. trivirgata*) ranges down the Baja California peninsula, along northwestern Sonora, Mexico, and in Arizona mountain ranges south of the Gila River. It reaches Southern California only in extreme southern San Diego County near the Otay and Tijuana watersheds. The name Desert Rosy Boa (*L. orcutti*, Figure 9.16) replaces the Southern California populations formerly called

FIGURE 9.16 The docile Desert Rosy Boa (*Lichanura orcutti*) has suffered population declines in Southern California from overcollecting and habitat loss. Notice that the head scales are small, not large and flat as they are in colubrid snakes, such as the California Kingsnake shown in Figure 9.18.

L. trivirgata roseofusca and *L. t. gracia*, taxonomic names that may still be used in contemporary field guides. This small (rarely more than 3 ft) gentle snake occurs along the coastal and desert slopes of the Peninsular Ranges northward into the Mojave Desert and eastward into Arizona. Chiefly nocturnal, these snakes are popular in the pet trade, and in some areas, habitat loss and excessive collecting have reduced populations to endangered status.

The Rubber Boa (*Charina bottae*) has a blunt tail and stout body that does in fact feel like rubber. The Rubber Boa is the northern most of all boid snakes, ranging through northern California into British Columbia. It prefers moister habitats in broken woodland and forests. In Southern California, a disjunct assemblage of populations is known around rocky outcrops in the Tehachapi Mountains Mt. Pinos, the San Bernardino Mountains, and the San Jacinto Mountains. These outliers may represent a species distinct from their northern relatives.

9.11 CAENOPHIDIA

The Caenophoidia, or Colubroidea as it is sometimes called, is a large clade of 5 families encompassing some 2,450 species, or 85% of all the world's snakes. One of those families alone, the Colubridae, accounts for about 75% of them (63% of all snakes). Some of the caenophidian lineages are venomous like the coral snakes and cobras (Elapidae), the sea snakes (Hydrophiideae), and the pit vipers (Viperidae). However, the Colubridae, or simply colubrid snakes are for the most part non-venomous, including those in Southern California.

9.11.1 Colubridae—Colubrid Snakes

Southern California's 24 species are all harmless to humans; most are rather docile and account for about 73% of all snakes in the region. Colubrid snakes are morphologically advanced and intrafamilial relationships continue to be tested. Table 9.2 and the following discussion organize Southern California's colubrid snake fauna into 6 groups by morphological similarity and lifestyles. Most of the species are nocturnal and secretive, but some are abroad by day, especially in the spring months. Opportunities for finding them include driving slowly at night on isolated roads during the warm months of the year. Be aware, of course, that permits are required for possessing many native vertebrate species, snakes included. None of our regional species is restricted to Southern California. Most are broadly distributed across the Southwest or range up and down the west Coast. Others have extensive ranges elsewhere that barely include Southern California from the south, the north, or across the Colorado River.

1. The Ring-necked Snake (*Diadophis punctatus*) ranges from the Atlantic to the Pacific Coasts but is absent or spotty in much of the arid west. In Southern California, it occurs along the Peninsular and Transverse Ranges typically in moist situations. This small snake responds to threats by coiling its tail upward to reveal a bright red undersurface.

The Spotted Leaf-nosed Snake (*Phyllorhynchus decurtatus*) is a nocturnal desert species associated with creosote in sandy to gravelly substrates in the Mojave, Colorado, and Sonoran Deserts. Like its sister species the Saddle Lead-nosed Snake (Arizona south through Sonora, Mexico) it takes its name from the prominent, enlarged rostral scale that assists in burrowing.

2. Racers and Whipsnakes are fast-moving snakes distinguished by their slender bodies, diurnal habits and keen eyesight. The California Whipsnake, or Striped Racer (*Masticophis lateralis*), ranges from northern California south through the Sierra Nevada and Coast Ranges down into Baja. In Southern California, it occurs from the coast to the mountains, particularly favoring chaparral habitats. Whereas they generally search for prey over the ground, I've seen these snakes effortlessly climb up and around manzanita. The Coachwhip or Red Racer (*M. flagellum*, Figure 9.17) is the largest of the whipsnakes in our area (to 55 in. [1,400 mm]), and also the most aggressive. Wide-ranging in the southwestern United States, it is found in many habitats, but generally prefers open areas and flat ground. It has a high thermal tolerance and thrives in the deserts. Like other species of *Masticophis*, it hunts for prey by moving deliberately with its head elevated off the ground. If harassed or cornered, they will strike and bite quickly. Although not venomous, wicked lacerations can result from the bite.

   A close relative, the *Baja California Coachwhip* (*M. fuliginosus*) reaches Southern California only in extreme southern San Diego County within about 6 mi of the border. The Striped Whipsnake (*M. taeniatus*)

FIGURE 9.17 An aggressive and efficient diurnal predator, the Coachwhip, or Red Racer, (*Masticophis flagellum*) ranges across the Southwest in open habitats. In Southern California it is found from the coast to the deserts but absent from elevations above 6,000 ft (1,820 m).

is a Great Basin species that extends southeast into the Chihuahuan Desert. In California it extends as far south as the eastern Mojave. The Western Yellow-bellied Racer (*Coluber mormon*) inhabits open areas of the Pacific Northwest south into the Great Basin, and along the coast to Southern California where its distribution is spotty in open areas from the coast to the high desert, generally preferring riparian habitats. Like the Coachwhip, it has an aggressive disposition if molested.

3. Kingsnakes, Gopher Snakes, and Allies: The Common Kingsnake (*Lampropeltis getula*) and the Gopher Snake (*Pituophis catenifer*) are the most familiar species in this group, and among the commonest snakes in Southern California. Both are wide-ranging in the United States and each has been the subject of intensive subspecific analysis. Out West, the Common Kingsnake is called the California Kingsnake (*L. g. californiae*, Figure 9.18), whereas the Gopher Snake is represented in Southern California by four subspecies: the San Diego Gopher Snake (*P. c. annectens*, west of the deserts, Figure 9.19), the Santa Cruz Island Gopher Snake (*P. c. pumilus* Santa Cruz Island, Santa Rosa Island), the Great Basin Gopher Snake (*P. c. deserticola*, Mojave Desert), and the Sonoran Gopher Snake (*P. c. affinis*, Colorado Desert). Gopher Snakes and Kingsnakes typically are gentle animals and common in the pet trade. California Kingsnakes are distinctive in their bold pattern of white or cream-colored bands (occasionally striped instead of banded) across a dark brown background. Gopher snakes are variably patterned in multihued blotches on a tan to yellowish background. Because of their

FIGURE 9.18 The California Kingsnake (*Lampropeltis getula californiae*) is a common resident in much of Southern California and throughout the state. They are usually banded or striped, but some like this individual can be a mosaic of both pattern types. (Courtesy of Kurt Spanos.)

FIGURE 9.19 A Gopher Snake (*Pituophis catenifer*) takes advantage of a water source on a warm, late spring afternoon. Climbing into the birdbath was no challenge. Gopher snakes are one of Southern California's most common serpents.

size (up to 72 in. [1.9 m] in California *P. c. affinis*), color, and pattern, people often assume they are dangerous. Sadly, like many snakes, their first encounter with a human is likely as not to be their last. Several years ago in a moment of despair and anger, I lost my composure with a neighbor who shot a particularly impressive Gopher Snake with his police service revolver, on the dirt road in front of my property as I was returning home one spring afternoon. He showed some remorse after I explained to him that they were not just harmless but beneficial. It was too late for that particular snake, of course. Its skeletal remains became part of the teaching collection for my students.

The California Mountain Kingsnake (*Lampropeltis zonata*) is an exquisite sight with its glossy black, red, and white bands (Figure 6.23). This snake ranges from southern Washington to northern Baja and from sea level to high mountains except in the desert. In Southern California, disjunct populations occur in the San Bernardino Mountains and parts of the Peninsular Ranges, primarily in coniferous and pine-oak woodland.

Vaguely resembling the California Mountain Kingsnake, the Long-nosed Snake (*Rhinocheilus lecontei*) is a resident of the south central and southwestern United States. In our area, it is most common in the deserts, although it has been found in coastal sage scrub and grassland. The slender, Western Patch-nosed Snake (*Salvadora hexalepis*) occurs from the Great Basin south through coastal Southern California and throughout the Mojave, Sonoran, and Chihuahuan deserts where it associates with creosote communities. This snake is an active diurnal predator that hunts in a similar fashion to whipsnakes. Coastal populations in Southern California are cryptic and rarely seen by people. The (Desert) Glossy Snake (*Arizona occidentalis*) is primarily a species of the Mojave and Sonora Deserts (including the Colorado Desert). One subspecies, the California Glossy Snake (*A. o. occidentalis*) ranges from the Central Valley south through coastal Southern California, but now it is rarely seen and is probably extirpated from much of its historic range in coastal habitats.

The most enigmatic snake in Southern California is the Baja California Rat Snake (*Bogertophis* [*Elaphe*] *rosaliae*). The only record of it north of Baja is a single, road-killed specimen found off I-8 along the San Diego-Imperial County line near Mountain Springs. In the Baja peninsula it occurs mainly south of Bahia de Los Angeles, although a record from Guadeloupe Canyon, about 40 mi south of Mountain Springs, suggests that this snake may yet be found again north of the border.

4. Garter Snakes (*Thamnophis*) usually associate with aquatic habitats where they feed on fish, salamanders, frogs, toads, and tadpoles About 30 species are found in North America, and they have long been a source of befuddlement to taxonomists. Many have a single mid-dorsal stripe and another narrower stripe on either side of the body. In this fashion they are thought to resemble an old-fashioned garter. Some species lack dorsal or lateral stripes and instead are spotted or checkered. The tongue is bright red to orange with a black tip. Most Garter snakes defecate and emit a foul-smelling musk from cloacal glands when handled.

   Four species of *Thamnophis* have distributions that include Southern California in their range. The Two-Striped Garter Snake (*T. hammondii*, Figure 9.20) occurs down the coast from around Monterey south to northern and central Baja. It favors riparian situations with permanent to semi-permanent water from the coast to the deserts and can be found at elevations up to 8,000 ft. This species lacks a mid-dorsal stripe. The Common Garter Snake (*T. sirtalis*) is distributed over much of the United States and Canada but is absent from the Great Basin and much of the Southwest. Along the California coast it is represented by the California Red-Sided Garter Snake (*T. s. infernalis*), whose southern extent is just north of San Diego in marshes and wet upland habitats. Unfortunately, human impact has eliminated many populations from our area. The Western Terrestrial Garter Snake (*T. elegans*) ranges over the intermountain west outside of the deserts. In Southern California, it is restricted to populations isolated along the Armargosa River in the Mojave, and in the San Bernardino Mountains where it is called the Mountain Garter Snake (*T. e. elegans*). The Checkered Garter Snake (*T. marcianus*) is the fourth species to reach Southern California. It inhabits of lowland river systems from southwestern Kansas south to Costa Rica. Its range extends west through southern New Mexico and

FIGURE 9.20 The Two-striped Garter Snake (*Thamnophis hammondii*), like this juvenile, occurs along the Pacific coast from around Monterey to central Baja California. It typically prefers riparian habitats.

Arizona to extreme southeast California along the Colorado River, and to some irrigation canals in Imperial County.

5. Ground Snakes: These are secretive, nocturnal snakes from about 14–18 in. (ca. 350–360 mm) in total length. The Western Shovel-Nosed Snake (*Chionactus occipitalis*) is a small, banded serpent of the deserts found mainly in sandy areas from the southern Great Basin through the Mojave and upper Sonoran Deserts including extreme northeast Baja. It is an excellent burrower that easily swims through sand, assisted by a tapered snout with a countersunk lower jaw. It is often encountered on quiet roadways on warm nights. The Western Ground Snake (*Sonora semiannulata*) comes in variable color patterns and is found in the arid regions of the west and east as far as Kansas and Texas. It also ranges throughout much of the Baja peninsula. In California, the Western Ground Snake is restricted to the eastern parts of our deserts.

   Black-Headed Snakes (*Tantilla* spp.) are so named because of a distinct black to dark brown cap usually separated from the neck by a narrow whitish band. These snakes may be common in some areas but are seldom seen because they spend much of their time beneath flat rocks, logs, and other surface debris. Eleven species occur in the United States; 2 of these include Southern California in their distribution. The Southwestern Black-Headed Snake (*T. hobartsmithi*) is found in a variety of habitats throughout its spotty distribution in the Southwest. In California, it occurs along the southern Sierra Nevada foothills and adjacent valleys, and to the east side of the Sierra from the southern Great Basin into the Mojave Desert. The California Black-Headed Snake (*T. planiceps*) is a resident of chaparral and oak woodland along coastal mountains from the Bay Area south to the cape region of Baja. In Southern California, it extends through numerous habitats to the desert edge of the mountains.

6. Night Snakes and Lyre Snakes: These two secretive, nocturnal snakes prey on small vertebrates by injecting toxic secretions through moderately enlarged teeth at the back of the mouth. Neither is dangerous to humans. The Spotted Night Snake (*Hypsiglena torquata*) is a small to medium serpent (up to 25 in. [635 mm]) found in numerous habitats in much of the western United States well into Mexico. They have flat heads and two to three large dark blotches on the neck and smaller lighter spots down the back. In Southern California, they occur from the coast to the desert, often in crevices and under rocks and surface rubble. Spotted Night snakes feed on small frogs, lizards (e.g., Side-blotched Lizards), and some snakes. The Western Lyre Snake (*Trimorphodon biscutatus*) is a larger species (up to 48 in. [1,220 mm]) of the Southwest from central California south through Baja and east to western Arizona on into Sonora, Mexico. The common name derives from the lyre or V-shape marking on the head. It is primarily a rock and crevice dweller. In Southern California, lyre snakes range from coastal canyons to desert foothills. Like Spotted Night Snakes, they feed on small vertebrates such as lizards and snakes but also rodents and even bats that they catch at roosting sites. Symptoms of Lyre Snake bites on humans include mild edema and itchiness. Because their toxic saliva must be delivered by chewing-like motions envenomation is exceedingly unlikely. Repeated harassment is the only reason that a person would be bitten in the first place.

9.11.2 Viperidae—Pit Vipers

Viperids include approximately 260 species of small to large venomous snakes found on all continents accept Australia and Antarctica. Two lineages are recognized: the 60 or so species of "true" vipers (subfamily Viperinae) of Eurasia and Africa and the pit vipers (subfamily Crotalinae) of the New World and Asia. Among the pit vipers are the rattlesnakes and their relatives, the copperheads and cottonmouths. Pit vipers derive their name from the infrared heat sensitive pit located on either side of the face below and to the rear of the nostril but in front of the eye. The sophisticated venom delivery system consists of an enlarged hollow tooth (fang) situated on each cube-shaped maxillary bone at the front of the skull. Replacement fangs sit alongside. Glands at the back of the skull deliver venom to the maxilla by a duct. When the mouth is closed, the fangs fold backward to a horizontal position. A venom strike is delivered when the mouth opens to full gape and causes a linkage of palatal bones to push on the maxilla, which then pivots upward exposing the fangs. The snake stabs its victim with blinding speed.

Copperheads and cottonmouths (*Akistrodon* spp.) are restricted to the central and southeastern United States and a relative in Mexico, the Cantil (*A. bilineatus*).

Rattlesnakes (*Crotalus* spp.) are found from southern Canada to northern Argentina. Their heavy bodies, thin necks, triangular heads, and the rattle at the end of the tail distinguish them. No other snakes have this appendage. The rattle consists of a series of interlocking hollow, horny segments serially added to the base each time the snake sheds its skin. Juvenile rattlesnakes may shed three or four times a year, older individuals maybe once. Newborns have a small button at the end of the tail that anchors the rattle as new segments are added. Twitch muscles in the tail cause the rattle to vibrate when the snake is alarmed, and initially the rattle may be just a few clicks. Rattlesnakes are wary animals and often are heard before they are seen. Interestingly, when provoked, some harmless species like gopher snakes will vibrate their tails in dry leaves, mimicking the sound of a rattle.

Rattlesnake venom is 90% protein that varies in its toxicity by species. Individual snakes also can adjust the amount of venom they deliver. Most rattlesnake venom expresses hemotoxic properties that affect the circulatory system and the tissues it serves. Some species, such as the Mojave Rattlesnake, produce neurotoxins. Neurotoxins are also characteristic of coral snakes and other elapids, but none of them occur in California. The venom of the Gila monster is also neurotoxic.

Rattlesnakes administer venom to immobilize prey; defense is a secondary adaptation. Some folks are unfortunately bitten while hiking or innocently tending to their gardens. That said, many people are struck when attempting to pick up a rattlesnake, often after consuming a couple of beers. In California, around 250–300 people on average are bitten annually. The California Poison Control System (CPCS) maintains these statistics, but they depend on the clinics that dispense antivenin for timely and accurate reporting. Hence, the numbers are probably an underestimate. If you are bitten, forgo the cut-and-suck remedies. Grab your phone and seek immediate medical attention.

About 33 species of *Crotalus* are recognized. Their relative the Massasaugua (*Sistrurus catenatus*) occurs from the Great Lakes region southwest to extreme southeastern New Mexico. Pygmy Rattlesnakes (*S. miliarius*; adults to about 20 in.) are found mainly in the Gulf States and have a preference for riparian habitats. The greatest diversity of rattlesnakes is in the Southwest and northern Mexico. The largest species are the Eastern Diamondback (*C. adamanateus*) and Western Diamondback (*C. atrox*). These two species reach on average an impressive 72 in. (1.9 m). The record for the former is 96 in. (2.5 m), and the latter 84 in. (2.2 m). Most species of large rattlesnake achieve a maximum length of 50–60 in. (c. 1.5 m). Others are smaller still. Be skeptical of the tales about some enormous rattlesnake seen by a friend or relative. Like fish stories, they seldom hold up under measurement or scrutiny. Because of their linear bodies snakes always appear longer than they actually are. The longest snake native to North America is the nonvenomous Eastern Indigo Snake (*Drymarchon corais couperi*) with a length of around 103 in. (2.6 m).

California is home to seven species of rattlesnakes, of which six occur in our area. Four of these are desert dwellers. Two are found on both sides of the mountains, and one is primarily restricted to the coastal side of the Transverse and Peninsular Ranges. The latter, the Southern Pacific Rattlesnake (*Crotalus helleri*, Figure 6.18) is therefore the species that most people in Southern California encounter. It ranges from around Monterey County south through the northern half of Baja and occurs in most all habitats from sea level to 11,000 ft, for example, in coastal sage scrub, chaparral, and pine-oak woodland. Individuals have even turned up on local beaches where they have wandered in from adjacent coastal canyons. Once on a field trip with my class we came upon four individuals in a meadow below campus, all within 20 ft of one another.

The Southern Pacific Rattlesnake snake will reach about 44 in. (ca. 1,100 mm). Venom is primarily hemorrhagic with some neurotoxic potential, and it is virulent. Coupled with its often-irritable disposition, this snake is potentially quite dangerous. It is closely related to the Northern Pacific Rattlesnake (*C. oreganus*; central California to British Columbia). Until recently, both were considered subspecies of the *C. viridis* complex (Western Rattlesnake).

The Speckled Rattlesnake (*C. mitchelii*, Figure 9.21) is an exquisite and variously hued serpent (up to 52 in. [1,320 mm]) of the Mojave and Colorado Deserts, west into the coastal foothills, and south to the cape region of Baja. It typically associates with rocky slopes where its color and pattern keep this snake well camouflaged. Venom of the Speckled Rattlesnake is potent, although bites to humans are uncommon.

The Red-Diamond Rattlesnake (*C. ruber*) is a species of the Peninsular Ranges from San Bernardino County south to the tip of Baja, and mostly below 4,000 ft. Although it occurs in some desert habitat it is more likely to be found

FIGURE 9.21 A Speckled Rattlesnake (*Crotalus mitchelii*) shows off its pinkish hue in the late afternoon sun. Accented with salt and pepper flecks these snakes blend in superbly with granitic outcrops.

in pristine chaparral (including coastal sage scrub) and piñon/juniper woodland. This is Southern California's second-largest rattlesnake and can reach 65 in. (ca. 1.6 m). Although the venom is low in toxicity, an adult is capable of delivering a large amount of it. The Red-Diamond is not especially aggressive, and it seldom rattles. Nonetheless, it is frequently implicated in rattlesnake bites to Southern Californians.

In our deserts, the Mojave Rattlesnake (*Crotalus scutulatus*) occurs in a variety of mostly open habitats throughout the Mojave and east into the Sonoran and Chihuahuan Deserts, south to Puebla in the Mexican plateau. It is absent from our Colorado Desert. Adults may reach 50 in. (1,270 mm). This species may be the most dangerous snake in the United States. Its neurotoxic venom is extremely virulent, and one bite is sufficient to kill a person; left untreated, death is a high probability.

The Sidewinder (*C. cerastes*) is a comparatively small rattlesnake (adults to 33 in. [840 mm]) most frequently found in areas of windblown sand. It often hides tightly coiled at the base of creosote bushes (Figure 6.34). The enlarged supraocular scales that resemble a small horn above each eye distinguish this species. They take their name from the corkscrew, sideways locomotion that they employ across sandy substrates. Sidewinder venom is the least toxic of Southern California's rattlesnakes, and together with their small size, the bite is not nearly as severe. Nonetheless envenomation can cause considerable pain and local tissue damage.

The Western Diamond-Backed Rattlesnake (*C. atrox*, Figure 6.52) is the sixth species found in Southern California. However, in our area it is restricted to the eastern extremes of the Mojave and Colorado Deserts. It otherwise ranges east into Arkansa and south to central Mexico, from sea level to around 8,000 ft. This big, bold snake with fangs more than one-half inch long is capable of injecting a large quantity of venom that has a mix of hemorrhagic, neurotoxic, and proteolytic enzymes. The Western Diamond-Backed Rattlesnake is probably responsible for more human deaths than any other snake in the United States.

9.11.3 Hydrophiidae—Sea Snakes

Marine snakes are probably relatives of the venomous elapid land snakes of Australia and New Guinea. Around 50 species inhabit the tropical and subtropical waters of the western Pacific and Indian Oceans. They have vertically compressed tails that they use to paddle about in an eel-like fashion. Sea snake venom is among the most lethal of all biological toxins. Fortunately for people, they produce small quantities of it and their injection system is not especially efficient. Consequently, most bites do not result in serious poisoning. Sea Snakes either give birth to live young at sea, or come ashore to lay eggs. One species, the Yellow-bellied Sea Snake (*Hydrophis* (=*Pelamis*) *platurus*) has the broadest range of any snake. It occurs across the warmer areas of the Pacific to the west coast of Mexico. Stray individuals occasionally reach the waters off San Diego and Orange Counties.

10 Birds

Most birds are diurnal and easy to observe, making them an obvious choice for exploring an array of natural history topics, many of which are applicable to terrestrial vertebrates in general, how they make a living and survive. Their pure elegance and ability to fly have forever inspired humans. Paleolithic people had no sense that bird-watching would become a billion-dollar industry, but our human fascination started early.

Around the globe, roughly 10,000 species of birds range from the tropical rainforests to the edge of polar ice, and from the lowest and driest deserts to the peaks of some of the tallest mountains. Flight is the fundamental attribute of birds, the foundation of all their biology. It underpins structure, foraging, feeding, and reproduction. Flight is a failsafe escape mechanism from nonvolant terrestrial predators and makes long distance travel possible. Under special circumstances some birds have opted out of the airborne way of life and have become flightless. Others, although physically capable of flight, lead essentially ground-based lives.

10.1 BIRDS AS ARCHOSAURS

Like mammals, birds (Aves) are endothermic amniotes, and for a time, the two were thought to be closely related because of their "warm-bloodedness" and a four-chambered heart. The eminent nineteenth-century British paleontologist Richard Owen suggested as much and declared that the reptilian heyday, in his opinion, had long ago reached its zenith, leaving behind it only a negligible assortment of lizards. Owen believed that species were immutable, so his views on birds received scant attention in the post-Darwinian era. Modern birds, of course, are distinct from mammals and all other living amniotes if for no other reason than they possess feathers. In addition, birds have (Figure 10.1):

- An upright bipedal posture
- Keeled sternum (carina) in most
- Furcula—the fused clavicles with the unpaired interclavicle (the "wishbone")
- Elongate, mobile S-shaped neck
- Fused foot bones (tarsometatarsus); ankle joint is intertarsal rather than between tarsals and lower leg (tibia and fibula)
- Tridactyl foot: typically with three toes forward, one backward
- Fused hand bones (carpometacarpus)
- Pygostyle (fused terminal tail vertebrae)
- Toothless bill (toothed in most Mesozoic forms)
- Oviparity—all lay and brood eggs

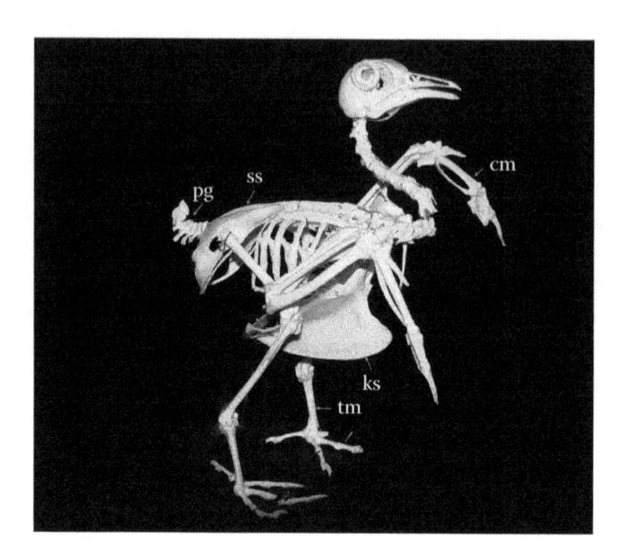

FIGURE 10.1 The skeleton of a bird like this pigeon retains the essential components of its tetrapod ancestry, only severely modified for flight. The bones themselves are pneumatic ("hollow") for weight reduction, the cervical (neck) vertebrae are highly mobile; the massive synsacrum (ss) incorporates the sacral, lumbar, and posterior thoracic vertebrae; the tail is reduced to a few bony elements called the *pygostyle* (pg) for the attachment of tail feathers (rectricies); bones of the hand and wrist are fused into a carpometacarpus (cm) for attachment of the primary feathers; the sternum (breast bone) is equipped with an exaggerated median keel (ks) that serves as the origin for the primary flight muscles, a supracoracoideus that functions as a wing elevator and a massive pectoralis that powers the down stroke; the tarsometarsus (tm) is a hind limb element unique to birds. It incorporates anklebones and some from the foot.

Much of the nineteenth-century interpretation of bird origins relied on the earliest known fossil bird, *Archaeopteryx lithographica*, first discovered in 1861 and persisting as one of the most fortunate finds in the history of vertebrate paleontology. Seven specimens, variously complete, are now known and all have come from deposits in Bavaria (southern Germany) dating from 145 to 150 million years ago (upper Jurassic). *Archaeopteryx* (Figure 10.2) had feathers, clearly evident in the rock impressions, but the rest of the skeleton lacked most of the characters listed for modern birds. It retained teeth, had an unfused hand, lacked a keeled sternum, and had a long bony tail. Without the feather impressions, *Archaeopteryx* would easily have been regarded as a reptile, albeit an odd one because of the elongated forelimb.

The comprehensive study by Gerhard Heilmann (*The Origin of Birds*), first published in 1927, concluded that the similarity to certain reptile lineages in numerous structural aspects leaves scant doubt that birds are archosaurian diapsids. Crocodilians and birds are the only archosaurs to survive past the Mesozoic, and despite their superficial differences, they share similar bone and muscle structure and lay shelled eggs. Crocodiles and birds are each other's closest

FIGURE 10.2 *Archaeopteryx* is the oldest known bird. Feather impressions are evident on this reproduction of a fossil from Bavaria dating from 150 million years ago, but the skeleton bears scant resemblance to modern birds. Note the unfused hands, lack of a keeled sternum, the long tail, and the presence of teeth.

living relatives. Indeed, the evolutionist Thomas Huxley noted the archosaurian affinities of avian structure 150 years ago and famously observed that birds were "glorified reptiles."

In addition to crocodilians and birds, archosaurs also consist of a varied assemblage of stem groups, plus the more familiar dinosaurs and pterosaurs (the flying reptiles of the Mesozoic). Archosaurs tended toward a bipedal posture in which the hind limbs were drawn underneath the body for load bearing and locomotion. Besides the diapsid condition of the skull (Chapter 8), archosaurs have an additional opening on the face, the anteorbital fenestra, and a mandibular fenestra in the lower jaw.

Pterosaurs appear in the late Triassic and were the first vertebrates to develop flight capabilities. They exhibited several body types and ranged in size from the diminutive *Pterodactylus* (about the size of a sparrow) to the outrageous *Quetzalcoatlus* that had a wingspan of around 40 ft. The pterosaur wing was constructed from the elongate fourth finger that acted as a scaffold for the actual airfoil, which was a membrane of skin attached to the side of the body and in some cases the tail. Pterosaurs persisted through the end of the Cretaceous.

Dinosaurs comprise two lineages: (1) Saurischia and (2) Ornithischia. These two groups diverged early from a bipedal ancestor and evolved as independent radiations. Both had some secondarily quadrapedal representatives. Salient differences between saurischians and ornithischians are found in the arrangement of the bones of the pelvic girdle. Saurischians include the theropods, which are the bipedal carnivores like *Allosaurus* and *Tyrannosaurus*, and the enormous, mostly quadrapedal herbivorous sauropods ("Alley Oop" dinosaurs) like *Diplodocus* and *Camarasaurus*. Among the familiar ornithischians are the armored ankylosars, the bipedal duck-billed dinosaurs, the horned ceratopsians, and the pachycephalosaurs.

10.2 THE BIRD WARS

That birds are Archosaurs is not in dispute among paleontologists. What has not been universal is a consensus on which group of reptiles is most closely related to birds. On this point the exchange in the scientific literature has been, shall we say, vigorous, if not hostile, with neither camp displaying much civility toward the other. The politics of science can get ugly at times, but as a spectator sport, the bird wars have been wildly entertaining.

Heilmann proposed that birds arose from a basal group of archosaurs known as pseudosuchians. He did not favor dinosaur affinities as Huxley, and later Henry F. Osborn (1900), had; Huxley's views were more or less overlooked at the time he proposed them. Except for Heilmann's work and that of a few other authors like Osborn, little attention henceforth was paid to the matter of bird origins other than occasional reference to it being problematic. Beginning in the 1970s things heated up. New studies on the origin of bird flight, the accumulation of more fossils, and a phylogenetic approach to systematics renewed the interest in bird origins. The most recent contributions to the issue are the wealth of archosaur fossils coming out of China bearing "feathers," "protofeathers," and filamentous coverings referred to as "dino-fuzz." Most of these Chinese fossils are assigned to various taxa of proto-birds or theropod dinosaurs. Soon "feathered dinosaurs" made their way into popular media, in print, as museum exhibits, and nature specials on television. Today most paleontologists hold the view that birds descend from dinosaurs, and even some ornithologists studying contemporary birds trumpet "ornithology is extant dinosaur biology."

Birds appear to be closest to a clade of theropods known as dromaeosaurs (commonly called "raptors" like those featured in the *Jurassic Park* films).

Numerous details of skeletal morphology have been marshaled in support of the birds-are-dinosaurs position, as for example in the hand and forelimb, the ankle and foot structure, and the pelvic girdle.

Despite the popularity of and evidence for the birds-are-dinosaurs thesis, it is not without its opponents. Critics contend that birds arose from a previous group of archosaurs, not dinosaurs. They maintain that the problem is more complex, and aside from some philosophical differences in methodology, opponents argue that much of the skeletal evidence has been misinterpreted, that flight arose from the trees down rather than the ground, tooth morphology is incongruous between proto-birds and putative theropod outgroups, early feathers or dino-fuzz are not homologous with avian feathers, and lastly that practically all of the theropod fossils exhibiting bird-liked characters are Cretaceous in age and as much as 80 million years younger than *Archaeopteryx* and therefore do not speak directly to the origin of birds.

The origin of avian flight is central to this debate. An arboreal or "trees-down" hypothesis had long been favored. Several other arboreal vertebrates glide or parachute from branch to branch, generally losing altitude with distance. The Borneo Flying Frog (*Rhacophorus*) is one example as are certain neotropical frogs (e.g., *Agalychnis*), the gliding lizards of southeast Asia (*Draco*), and our own North American flying squirrel (*Glaucomys*). In birds, wing beats would have sustained the glide and eventually produce capabilities for a host of aerial maneuvers, including, finally, the ability to take off from the ground. By contrast, if modern birds descended from dromaeosaurid theropods, then flight must have evolved from bipedal runners, in other words from the ground up. In this scenario proto-wings might have first appeared as devices for social signaling. Subsequently they were employed as planar surfaces to create lift and lighten the load while running. Further, according to the ground-up proponents, these incipient wings could have been used to ensnare insect prey (although grasping seems incompatible with flight) while also being suitable for controlling pitch and yaw while running and leaping. Another mechanical problem here is that cursorial activity (running), especially acceleration, requires traction, which would be lost when the body becomes airborne for even a short distance. Alternatively, wings could have been flapped much like a chicken running across a barnyard. Flapping in this fashion is used in modern terrestrial birds like chukars when ascending steep slopes.

In either case, from the trees down or the ground up, powered flight in birds must have passed through numerous evolutionary stages in which increased skeletal specializations and neuromuscular control achieved the range of flight modes that birds now possess. Equally, birds and their ancestors experienced a marked decrease in body size from that of their Archosaur relatives as a precondition to powered flight.

10.3 PHYLOGENY AND CLASSIFICATION

Modern birds arose on the heels of two Cretaceous lineages, neither of which survived the end of the Mesozoic. One of these was the *Enantiornithes*, the so called "opposite birds" because of the peculiar fusion of their metatarsal bones. Most retained teeth and ranged in size from a sparrow to a turkey vulture. Some appeared to have been long-legged waders, whereas others had powerful hawk-like talons. The other Cretaceous lineage was the *Ornithurines*. These were an ecologically diverse group from small arboreal forms to specialized foot-propelled divers later on, such as the toothed *Ichthyornis* and *Hesperornis* of Cretaceous seas.

The modern birds of today, the *Neornithes*, putatively have fossils extending into the Cretaceous but that record is sparse and, for the most part, modern birds should be regarded as a Cenozoic radiation possibly descended from the Ornithurines. All living birds are toothless. At one end of the phylogeny are the Tinamous and Ratites. These birds have a primitive palatal structure and accordingly are referred to as the *Paleognathae*. Tinamous are mostly chicken-sized running birds but capable of short, rapid bursts of flight. The 40 or so species today are found from southern Mexico to Patagonia. Ratites lack a keeled sternum, are flightless, and specialized for running. All living ratites are Southern Hemisphere. They include the ostrich (Africa), rheas (South America), emu (Australia), and cassowaries (New Guinea, Australia). The giant herbivorous moas, ratite birds of New Zealand, became extinct following first human contact around 800 years ago. The enormous elephant birds (Aepyornithidae) of Madagascar met a similar fate.

All living birds with a "modern palate" are referred to as *Neognathae*, or sometimes as "carinate" birds because they have a keeled sternum. All of them fly or descended from ancestors that did. Table 10.1 lists the 18 major clades (Orders) with representatives in Southern California. A phylogeny based on Table 10.1 is shown in Figure 10.3. The tree is incomplete in the absence of several Old World and Southern Hemisphere lineages, such as penguins, which in part accounts

TABLE 10.1
Major Orders of Living Birds with Representatives in Southern California

Anseriformes	*Waterfowl* (ducks, geese, swans); good swimmers, strong fliers; 151 spp.; cosmopolitan
Galliformes	*Grouse*, megapodes, pheasants, quail, turkey, etc.; fowl-like, mainly terrestrial, stout feet, wings short; 253 spp.; worldwide
Phoenicopteriformes	*Grebes*; swimmers, divers; 4 webbed, lobate toes; 8 spp.; fresh water; worldwide
	Flamingos; long-necked, long-legged wading birds, gregarious; 4 spp.; Africa, Madagascar, southern Europe, India, Caribbean, South America
Columbiformes	*Pigeons*, doves; plump, short-legged, unwebbed; frugivorous, granivorus, large crop; 300 spp.; worldwide
Caprimulgiformes	*Poor-wills*, nightjars, and allies; soft-feathered, long pointed wings, small bill, large gape; crepuscular, nocturnal; 93 spp.; temperate to tropical; worldwide
	Hummingbirds and swifts; strong wings, weak-footed; feed on the wing; 388 spp.; hummingbirds temperate and tropical Americas; swifts cosmopolitan
Cuculiformes	*Roadrunners*, cuckoos, and allies; upper bill hooked, never movable; 2 toes front, 2 behind; temperate, tropical; 143 spp.; worldwide
Gruiformes	*Rails*, cranes, and allies; variable in form; either strong flyers of open marshes, prairies, or poor flyers (some flightless) of swamps, marshes; 177 spp.; all continents
Gaviiformes	*Loons*; strong fliers, excellent divers; 4 spp.; Northern Hemisphere
Procellariiformes	*Albratrosses*, shearwaters, and petrels; tubular nostrils extend onto bill; ocean-going, rarely visit land; 90 spp.; worldwide
Pelecaniformes	*Pelicans*, frigatebirds, cormorants, anhingas, boobies, and tropicbirds; fully webbed divers, fish-eaters; 52 spp.; temperate, tropical
	Herons, egrets, and storks; long-legged, long-necked wading birds; 110 spp.; cosmopolitan

(*Continued*)

TABLE 10.1 (*Continued*)
Major Orders of Living Birds with Representatives in Southern California

Charadriiformes	*Gulls*, shorebirds, and allies; diverse group of waders, swimmers, divers; coastal waters, beaches, inland marshes, lakes; 314 spp.; cosmopolitan
Accipitriformes	*Hawks*, eagles, osprey, and vultures; diurnal birds of prey, or scavengers; bill sharply curved, toes long, sharp (talons); 270 spp.; cosmopolitan
Strigiformes	*Owls*; soft-feathered, round face, eyes forward; nocturnal birds of prey; 131 spp.; cosmopolitan
Piciformes	*Woodpeckers* and allies; 2 toes in front, 2 behind, unique leg muscles; 374 spp., worldwide except Madagascar, Australia, New Zealand, oceanic islands
Coraciiformes	*Kingfishers* and allies; large prominent bill, usually brightly colored; forests; 192 spp., temperate, tropical; worldwide
Falconiformes	*Falcons* and caracaras; diurnal birds of prey or scavengers; strong hooked bill, toes long, sharp, wings pointed; 63 spp.; nearly cosmopolitan
Psittaciformes	*Parrots* and allies; large head, strong hooked beak, 2 toes in front, 2 in back; gregarious, raucous, mostly forests; 317 spp., tropical, subtropical; worldwide [Introduced in Southern California]
Passeriformes	*Perching birds*: Nearly 6,000 species in ca. 80 families; placed in 2 groups by structure of vocal apparatus: suboscines, oscines (songbirds); worldwide

for the unresolved position of pigeons and doves (Columbiformes) and of owls (Strigiformes). The higher-level relationships of birds is a work in progress, which in recent years has been fairly substantial, and the tree (Figure 10.3) reflects the most recent consensus of these 18 orders of birds (Hackett et al. 2008, Yuri et al. 2013, Dickinson and Remsen, Jr. 2013, Dickinson and Christidis 2014).

The Paleognathae (Tinamous, Ratites) are shown on the tree because they comprise the sister group to all other modern birds (Neognathae). Readers might recognize (Table 10.1, Figure 10.3) that grebes are now allied with flamingos into a single clade, the Phoenicopteriformes. Hummingbirds and swifts are grouped with the poorwills and nightjars as the Caprimulgiformes, and Herons (formerly Ciconiiformes) are now part of the Pelecaniformes with the pelicans, cormorants, and their allies. The traditional taxon Falconiformes is here restricted to the falcons and caracaras, whereas the hawks, vultures, and relatives are assigned to a separate order, the Accipitriformes. The Falconiformes is the sister group to the parrots (Psittaciformes) plus perching birds (Passeriformes). Note that parrots are not native to Southern California, but populations of introduced birds have become established in San Diego and Los Angeles.

All of the aquatic birds—ducks, shorebirds, and their relatives—that are known from Southern California can also be found in other areas of North America. Many of the land birds likewise may occur in different parts of the west in suitable habitat during appropriate times of the year. A few species, however,

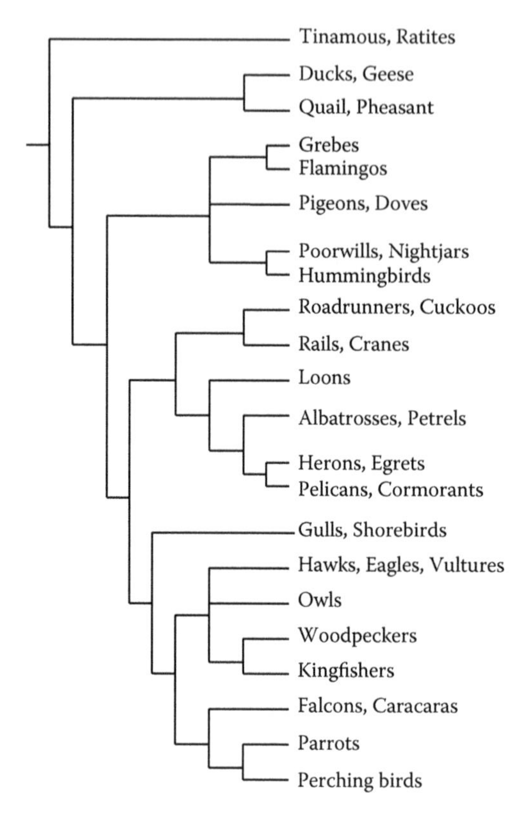

FIGURE 10.3 A phylogenetic tree of the major clades (Orders) of birds that are represented in Southern California (Table 10.1). The Paleognathae (*Tinamous, Ratites*) do not occur in North America but are included in the tree because they represent the sister group to all other modern birds (*Neognathae*). Ducks and geese (*Anseriformes*) plus quail and pheasant (*Galliformes*) are the first branch off the tree, whereas the parrots (*Psittaciformes*) plus perching birds (*Passeriformes*) are their most distant living relatives. Note that parrots are non-native introductions in Southern California. (Data from Hackett, S.J. et al., *Science,* 320, 1743–1767, 2008.)

are West Coast (including Baja) endemics like the Wrentit (*Chamaea fasciata*), California Towhee (*Pipilo crissalis*), California Thrasher (*Toxostoma redivivum*), Tricolored Blackbird (*Agelaius tricolor*), and Nuttall's Woodpecker (*Picoides nuttalli*). The California Gnatcatcher (*Polioptila californica*) is local in coastal sage scrub communities from Southern California south into Baja. The Island Scrub Jay (*Aphelocoma insularis*) is restricted to Santa Cruz Island, and about 25 subspecies of other forms are also confined to our region.

Scanning the taxonomic diversity in Table 10.1, the *Passeriformes* are immediately impressive; they account for over one-half of all bird species. Passerines are found on all continents (except Antarctica) and on almost all oceanic islands as well. Southern California alone has representatives from 27 of the 70 or so passerine families (Table 10.2). Passerines are arboreal birds diagnosed by their

TABLE 10.2
Passerine Bird Families in Southern California

Tyrannidae	Flycatchers and Kingbirds
Vireonidae	Vireos
Laniidae	Shrikes
Corvidae	Jays, Crows, and Ravens
Motacillidae	Pipits
Fringillidae	Finches
Passerellidae	Towhees and Sparrows
Parulidae	Wood Warblers
Icteridae	Blackbirds, Grackles, and Orioles
Cardinalidae	Grosbeaks, Buntings, and Tanagers
Paridae	Chickadees and Titmice
Remizidae	Verdin
Alaudidae	Horned Lark
Hirundinidae	Swallows
Aegithalidae	Bushtits
Sylviidae	Wrentit
Regulidae	Kinglets
Bombycillidae	Waxwings
Ptilogonatidae	Phainopepla and Silky Flycatchers
Certhiidae	Brown Creeper
Sittidae	Nuthatches
Troglodytidae	Wrens
Polioptilidae	Gnatcatcher
Mimidae	Mockingbirds and Thrashers
Sturnidae	Starlings (Introduced)
Cinclidae	Dipper
Turdidae	Thrushes

perching feet (four unwebbed toes on a single plane, three in front, and one in back), a distinct palate, reduced number of cervical (neck) vertebrae, and unique sperm ultrastructure. Young are helpless, almost naked or nearly so, and are reared in the nest. *Oscines* (the song birds) are the dominant group of passerines, distinguished from *suboscines* by a more complex vocal apparatus (*syrinx*). Suboscines have less complex sets of muscles operating the syrinx and therefore perform less intricate vocalizations. Suboscines include, for example, tyrant flycatchers.

The sequence of passerine families occurring in Southern California (Table 10.2) departs from the traditional arrangement of families as they appear in most field guides. Like the ordinal taxonomy already discussed, the list presented here is based on the latest consensus among avian systematists (e.g., Dickinson and Christidis 2014); we might anticipate this scheme to be followed in future editions of field guides of North American birds.

10.4 WHY ARE PASSERINE BIRDS SO DIVERSE?

The earliest known passerine is from the Eocene of Australia. In the Northern Hemisphere, however, passerines were relatively unimportant until the Miocene about 20 million years ago when diversity exploded. At one time, some ornithologists argued that the diversity is simply an artifact of poor phylogenetics and classification, that is, there are not as many passerine species as assumed. Not true.

Speciation is the outcome of genetic isolation. In the majority of cases, it begins when a population no longer breeds freely with other populations that comprise the species as a whole. The reasons for that are varied. Some are obvious like the formation of a physical barrier that isolates one population from others, preventing contact and the exchange of genetic information. For continental birds (and bats), physical barriers like rivers or mountain ranges usually are not an impediment because they can fly across or over them. However, isolating mechanisms can be subtle and not necessarily involve spatial separation of populations. For instance, the appearance of an abnormal behavior or a shift in some physiological trait that results from mutations could have the same effect in isolating a gene pool. Potential isolating mechanisms like these probably break down in nature as (or more) often as not. Yet if they do take hold, a population is subject to genetic recombination of its own without being swamped by a larger gene pool. When speciation is complete, the new population would no longer viably interbreed with the original parent population even if renewed contact were established. In other words, there are now two species instead of one; in nature, speciation by fission leads to a net increase in diversity over time. (Hybridization occurs when two putative species can interbreed because speciation is incomplete.)

Passerines birds have the ability to vocalize in complex ways that other birds cannot. Each species has a signature repertoire, a song, call, and notes that identify it as a member of that particular lineage and no other, male or female. Furthermore, some passerines, such as White-crowned sparrows (*Zonotrichia leucophrys*) consist of populations with regional dialects. A continued strengthening of a dialect under novel ecological circumstances then has the potential to become an isolating mechanism.

Voice evolution might be but one factor that drove passerine diversification. Nesting tactics and strategies could be another. Most nonpasserine arboreal birds nest in cavities like those in tree trunks and rock faces. Such sites are, over all, in short supply, and the nests themselves are vulnerable to detection by predators. Passerines on the other hand construct nests from a wealth of different materials and place them in myriad situations where concealment or inaccessibility are favored. Individual species could arise as birds exploit new opportunities in an almost endless array of nesting possibilities by fully using the three-dimensional world they live in, and the construction materials available. Nest building is not learned. It is innate. The birds get it right the first time. Natural selection, in other words, has affected the neurological wiring, giving each species the sensory and motor skills to select appropriate sites, find the required materials, and go about building the nest with only their bills and feet. Taken together, vocalization and nest building may account for much of the spectacular diversity of passerine birds.

10.5 FEATHERS AND FLIGHT

Feathers are obviously essential for flight. They also provide insulation, give shape and contour to the body, function in display, and serve in other less conspicuous functions. Feathers are arranged along epidermal tracts called *pterylae* that are separated by patches of bare skin. They do not emerge randomly over the body. Exceptions occur in some birds like ratites and penguins where feathers are more uniformly distributed. Feathers themselves are made up primarily of beta keratin, a protein similar in structure to that found in the scales of lizards and snakes. A typical feather is anchored in a follicle at the base of the central axis (quill) called the *calamus*. The axis continues outward as the *rachis*, which supports the feather *vanes* on either side of it. Vanes are constructed of minute, interlocking hooked barbs and barbules. Barbs near the base are not hooked therefore softer and more filamentous. *Flight feathers* include the *remigies* on the wing, and the *rectricies*, or tail feathers that attach to the *pygostle* (the caudal appendage consisting of a few fused vertebrae). Flight feathers (once employed as writing instruments) are a type of *contour feather*. Other contour feathers contribute to the overall form of the body and provide insulation. At the base of the tail is the large *uropygeal* gland that produces sebaceous, oily secretions for waterproofing that a bird transfers to its contour feathers with its bill.

Semiplumes lie beneath contour feathers. The rachis is not as stiff so semiplumes provide additional insulation and help fill out the body profile. Down feathers are of various types although entirely plumaceous—the rachis is short or absent—and they serve exclusively for insulation. The insulating feathers of hatchlings, called *natal down*, precede the development of other types of feather. Powder down is simply keratin that has broken down into granules and scattered over the body to provide additional waterproofing. *Bristles* are short feathers with a stiff rachis and usually no barbs. Bristles aid in keeping debris away from the eyes. In some birds, like poor-wills and flycatchers, bristles around the mouth function as tactile sense organs. *Filoplumes* are fine, hair-like feathers with a few short barbules at the tip of an otherwise naked rachus. Generally filoplumes are not exposed. Nerve endings in the follicles relay information about the status of the other feathers to help keep them properly aligned. In some birds, filoplumes appear over the contour feathers as a display.

Molt: Like integumentary features of other vertebrates, feathers become worn and tattered and must be replaced. Molting is the process of replacing old feathers with new ones. Birds of prey molt gradually, replacing feathers here and there as necessary. Most other birds complete their annual molt in late summer or early fall before migration. Others undergo partial molts in the late winter or early spring, and still others, especially shorebirds, complete their molt after winter migration when they've arrived at their breeding grounds. Molting results in fresh plumage and birds are at their showy best at this time. Many shorebirds and aquatic species like grebes are in bright breeding plumage for only a few weeks. During the winter months, these birds can be rather drab and field identification is a challenge, so be aware of seasonal changes in plumage when you are in the field.

Avian Flight: Flying is the preferred method of travel by animals that are not aquatic. That case is made easily when we include not only birds and bats but also the world's millions of insect species. Insects have not sacrificed any appendages to create wings, as have vertebrates; instead, insect wings derive from modified extensions of the exoskeleton.

Flying requires lift and a means for attaining forward motion, or thrust, to oppose the forces of gravity (mass) and drag (air is dense, not frictionless). Birds are variable in numerous ways, yet all must obey certain principles of physics if they are to fly, regardless of how they do it. Consider mass. The muscle power required for flight increases by a factor of 2.25 for every doubling of body weight. A simple means of keeping body mass low among birds is the ultralight pneumatic (hollow-boned) skeleton. [Penguins are an exception; the bones of these extraordinary divers are solid to reduce buoyancy.] Still, large birds require larger wings, yet big wings cannot beat as fast as small ones so greater force is necessary to achieve minimum wing beat frequency. At some point, however, muscle mass becomes too great a burden and flapping flight is no longer possible. Thus, 25 lbs. (ca. 12 kg) is about the upper limit in body weight for flying birds. Mass and energy further explain why most birds are small. They can take off nearly vertically and are highly maneuverable. Large birds such as vultures can only take off from the ground at a shallow angle and require a runway to do so.

The wing of a bird is an airfoil designed to achieve a desired reaction to the air around it. The forelimb forms the wing frame. Flight feathers (remigies) complete the airfoil. The flight feathers closest to the body, called *secondaries*, are located on the forearm (radius and ulna). They provide lift. The primaries provide thrust and are located on the modified hand, the carpometacarpus. The most proximal wing bone, the humerus, is where sternal muscles attach that power the wing. Individual flight feathers are asymmetrical; feather vanes on one side of the central axis (the rachis) are narrower than those on the opposite side. Flapping flight involves a bewildering array of variables that practically defies analysis. Generally the down stroke produces thrust. On the recovery stroke, the feathers twist so air passes through the wing with minimal resistance; it is not solid like an airplane wing.

In a few instances, as in hummingbirds, thrust can also be achieved on the upstroke. Hummingbirds hover while feeding, their wings beating from 80 to 100 times per second, and there may be 100 or more hovering bouts per day as they dart from one nectar source to another before taking a breather. One bout usually lasts less than a minute. Around 30% of a hummingbird's body mass is made up the primary flight muscles alone, the supracoracoideus (upstroke) and the pectoralis major (downstroke). The heart beats about 1,300 times per minute while hovering versus around 500 beats while at rest. Next to hummingbirds we lumbering humans live in a slow-motion world.

As an airfoil the wing can achieve lift in one of two ways. Commonly the wing is cambered to create a pressure differential between the upper and lower surfaces. The upper surface is convex, whereas the undersurface is concave. Camber also applies to airplane wings when the wing flaps are down. Cambering decreases pressure on the upper surface because the air is passing over a greater distance in the same amount of time. If both surfaces were parallel, air pressure would be equal and no lift could occur. Cambering is deepest near the body, which is why the secondary feathers are responsible for lift. Camber decreases toward the wing tips. Lift can also be obtained by increasing the angle of attack. The principle is similar to camber but accomplished by tilting the leading edge of the wing upward to a maximum of about 15° from horizontal, the stalling angle.

Both means of achieving lift come at the cost of increasing drag. If the smooth, laminar flow of air over the wing becomes disrupted, the airflow can separate from the wing, turbulence increases, and the wing stalls. In most birds the wings are slotted (because of the tapered flight feathers) along the posterior edge to minimize such disruptions. Birds with deeply cambered wings like hawks often have a small auxiliary airfoil called the *alula* or "bastard wing" on the leading edge. These accessories help to maintain smooth airflow. In some cases, induced drag occurs near the tips of the wings. Here, eddies or vortices result from an outward flow of air from under the wing and an inward flow from the top. Lengthening the wing puts distance between the vortices and increases the surface area for smoother airflow.

Wing size, shape, and proportions correlate with specific flight modes, for example soaring, hovering, or maneuvering in close quarters. Wing loading is simply the weight of the bird divided by the area of the wing (gm/cm^2). Generally, a low ratio requires less power to sustain flight, but other considerations can come into play. For instance, small birds have lighter wing loading on average, yet a hummingbird has twice the wing loading of a swallow even though a swallow is five times heavier. The hummingbird compensates with extraordinarily high wing beat frequency. Aspect ratio is a quantitative description (ratio) of wing shape (i.e., wing length to wing width). Birds such as towhees and numerous other small passerines that dart through and around dense cover have elliptical wings with low aspect ratios. These wings are highly cambered and slotted on the outer primaries. Wing beats are rapid although flight itself is comparatively slow. Elliptical wings with a low-aspect ratio also characterize galliform birds such as quail and grouse. These largely terrestrial species rely on rapid bursts of speed over short distances.

Long, narrow wings like those on a sailplane have a high aspect ratio *and* a high lift-to-drag ratio. Wings with a high-aspect ratio typify many seagoing soaring birds such as albatross as well as some high-speed fliers like swifts and falcons. Long-distance seabirds like the albatross practice dynamic soaring on long, narrow flat wings that lack slotting. Dynamic soaring is possible only in ocean regions with persistent winds (e.g., the "roaring 40s"). A wind gradient results when the lower 15 m or so of airflow is slowed by friction from the ocean surface. The bird begins at the top of the gradient and glides downwind with increasing relative ground speed. As it nears the surface it turns back into the wind and gains altitude. At the top of the wind gradient, it turns down wind and repeats the cycle.

By contrast, static soaring requires deeply cambered (high lift) and slotted wings, such as those of vultures, eagles, and many hawks. Static soaring birds seek out and glide in air masses that rise faster than their sinking speed. Wing tips are often turned upward from the air pressure relative to the heavier body, resulting in a V-shaped profile. Watch soaring vultures in particular. Broad, deep wings reduce wing loading in static soaring birds, and the high degree of slotting responds to shifting wind currents by the individual feathers acting like mini airfoils. These birds soar slowly and cannot fly in tight spirals at high speed.

10.6 HIND LIMB LOCOMOTION

All birds are bipedal by default. The correlation between hind limb structure and lifestyle is obvious in the long legs and toes of wading birds like herons and egrets. Birds of prey use their powerful feet, or talons, for capturing prey. Some aquatic species are so specialized for swimming and diving that they are barely able to get around on dry land.

The ability to perch in trees appeared early in avian evolution, although *Archaeopteryx* was probably no more specialized for it than its closest archosaur relatives. The most specialized feet for perching like those of passerines are anisodactyl, a condition in which all four toes are free, of moderate length, and three of them extend forward and one backward. Parrots and woodpeckers are zygodactyl, meaning two toes point forward and are opposable to two others that point backward. Most land birds are anisodactyl, but there is always a compromise between specializations and bipedal posture. Some land birds have evolved cursorial (running) adaptations by lengthening the distal elements of the limb and reducing the surface area in contact with the ground. Roadrunners, for instance, are long-legged and have zygodactylous feet. Ostriches, the ultimate in cursorial birds, have only two toes. Many birds walk, one leg in front of the other as we do, like pigeons and chickens, for example. Others, notably arboreal perching birds, hop while moving about on the ground. Hopping is simply a series of jumps in which both limbs move simultaneously. Hopping is most prevalent in passerines, and for the majority, it is their only means of land-based locomotion. A few can walk as well as hop such as starlings and grackles. Among corvids, the crows and ravens walk, whereas jays hop. Birds such as woodpeckers forage by climbing tree trunks. They use not only their feet but also their stiff tail feathers for support. Nuthatches ascend and descent tree trunks without tail support, relying instead on their backward-directed toe that is longer than the others and strongly curved.

Birds of ponds and bays such as ducks are surface swimmers that employ fully to partially webbed feet for propulsion. Their foot structure, relatively short hind limbs, and generally bulky bodies are compatible with that function. It is also the reason why they "waddle" on dry land. Foot-propelled divers are the most specialized swimmers. Lobed or webbed, foot-propelled divers like grebes, loons, and cormorants have their hind limbs positioned well back on their bodies. This enables them to freely dive from the surface in search of food. The lobed foot of a grebe is able to rotate on the recovery stroke to minimize drag. Generally, foot-propelled divers cannot walk upright on land, and they never leave the water except to nest or relocate. Cormorants, however, can perch on cliff faces and man-made structures like pier pilings and even overhead power lines.

10.7 AVIAN SENSES—SIGHT, SOUND, AND SMELL

Birds are the most visually dependent of all vertebrates. Almost everything an active bird does requires precise visual discrimination: flying amid thickets, landing on branches, foraging, nesting, or securing a mate. The structure of the eye differs little from that of other vertebrates, but the eyes of birds are exceptionally large and may be equal to or exceed the weight of the brain. In many birds, the eyes are so large that they meet at the mid line of the skull and most are asymmetrical, not round like those of mammals. Some are flat whereas others are globular. The shape of the eye is primarily a consequence of accommodating large eyes in a small space, but it may also provide enhanced clarity at the periphery of the image. A thin cornea permits the shape distortion, and a ring of wafer-like bones, the scleral ossicles, surrounds the eyeball to maintain its shape. In all vertebrates extrinsic ocular muscles rotate the eyeball within the orbit. These muscles are reduced in birds, and, in owls, they are absent and the eyes are immovably fixed. To shift its gaze, then, a bird must turn its head, an action easily performed because of increased neck mobility.

In birds of prey, the eyes face forward giving them an overlapping, binocular (stereoscopic) field of view of as much as 60°–70° in owls and about 50° in hawks. Binocular vision is essential for maximizing depth of field and in assessing speed while pursuing prey. In small, ground-foraging birds like seedeaters, the eyes are laterally placed. Binocular vision is practically nil, but the field of view can cover as much as 300°. This allows them to survey a wide area for food and more easily detect possible predators. Wading birds, such herons and egrets, and those that dive after aquatic prey, like kingfishers and ospreys, have the ability to rapidly focus where light is refracted at the air–water interface. Many birds perceive ultraviolet (UV) and polarized light that we mammals cannot. UV detection of plumage reflections can be important in mate selection, and using polarized light on cloudy days is an asset with sun-compass orientation and migration (see below). Kestrels make intriguing use of UV capabilities to rapidly survey open fields for voles. Voles move along narrow trails they establish among the grassy cover, but the urine and feces they leave behind absorb light in the UV spectrum and kestrels take advantage of the signal to find their prey.

Auditory reception is second in importance to sight. Birds' sensitivity to airborne sound is about equal to that of humans despite their much smaller ears and the absence of the pinna (external ear "flaps") that mammals have. In birds, the cochlea of the inner ear has about 10 times the number of hair cells per unit length as mammals. Mammals localize the source of sounds when the brain compares the arrival time and intensity of the pulse received by each ear. This can be difficult in small vertebrates like birds because the distance between the ears is so narrow. Some evidence suggests that the pneumatic structure of the skull allows for increased sensitivity by cross-referencing vibrations between the opposing middle ears. Hearing is keenest in owls. Most owls are nocturnal or crepuscular, and they locate prey by sound. Barn owls, in fact, can capture prey in complete darkness. A facial ruff of feathers is characteristic of many owls. This feature acts as a parabolic disc that reflects a sound wave to the ear opening(s) (external auditory meatus) where it is amplified by a large tympanic membrane (eardrum) that in turn vibrates the oval window and the auditory nerve of the inner ear. The left and right ears of owls are also asymmetrical with respect to the position of the ear opening. This allows them to pinpoint the direction of the sound by acoustically triangulating on the source.

A few birds rely on echolocation, much like bats. Cave swiftlets (*Collocalia, Aerodramus*) of Southeast Asia and the tropical Pacific freely zip about in the blackness of caves by bouncing high frequency sounds off walls and nearby objects.

A sense of smell is well developed in a few birds but hardly at all in most. Vomernasal organs are absent. Hence, olfaction is the least developed of the three primary senses. Birds with good olfactory capabilities are largely limited to carrion feeders like vultures, and relatively large olfactory bulbs are present in certain ground nesting and colonial nesting species as well. The kiwis of New Zealand have nostrils located on the end of their long bills; they probe loose soil using their sense of smell to detect earthworms. Some tube-nosed seabirds such as petrels and albatross (Procellariiformes) use some olfactory cues to locate food and for orientation. After foraging far out at sea they return to their home islands from downwind to pick up the scent of their refuge.

10.8 REPRODUCTION AND DEVELOPMENT

Birds living in temperate regions typically breed in the spring and early summer. Timing is under hormonal control, which itself is activated by increasing day length. Temperature and food resources are also factors, and many birds synchronize nesting to coincide with maximum food abundance for their young. The breeding season usually ends in late summer or early fall for most species.

The number of annual broods is often just one, but that number varies by species and the mating system that they follow (see below). Nesting failure for whatever reason may terminate the breeding cycle for the rest of the year. In other cases, a replacement clutch is laid at a later date. Where several broods are produced annually they may be evenly spaced in time or overlap such that the male is attending to the fledglings of one while the female is incubating a new clutch.

Clutch size is simply the number of eggs laid and incubated at one time. The California Condor (*Gymnogyps californianus*) is a true minimalist. It lays but a single egg, and successful rearing of the hatchling takes 2 years to complete. Among small songbirds, clutch size appears to increase from the southern to the northern part of the species range. Some birds have a fixed clutch size and are said to be determinate layers. Pigeons, for example, deposit two eggs at a time. Others such as domestic fowl are indeterminate layers that lay eggs continually until a comfortable number is achieved; if any eggs are removed the female will deposit more.

Developing eggs need to be kept within a certain temperature range. Most birds incubate their eggs with their own body heat. The brood patch is an area around the belly devoid of feathers that allows for the direct transfer of heat to the eggs. Waterfowl, however, have no brood patch. They cover the eggs with their feet before settling on them. In both cases eggs are regularly turned and rotated by the incubating parent.

Prior to actual mating, courtship initiates the breeding process, which if successful leads to pair bonding between a male and female. Both auditory and visual signals may be employed in courtship behavior. Among passerine birds, for example, males sing to advertise themselves to females. The song is a species-specific vocalization different from the call and usually performed by males during the breeding season. Recent evidence reveals that passerine females may also sing, but the reason is unclear. Vocalizing is a complex, learned behavior operated by a song control center in the brain. The males are usually brightly colored or display glossy plumage at this time. Fresh, bright plumage provides visual reinforcement to singing that a female will respond to, or not, if the male is somehow deemed inferior. In many instances the male sings from a territorial perch in an area where a potential nest could be built. Thus, the song also serves to alert rival males that the area is occupied.

In many birds visual displays are the sole means of male courtship. Displays can be in the form of bright, gaudy plumage worn by the male, as for example in peafowl, to the ritualistic dances and posturing practiced by gulls and galliforms. Because females usually determine the pairings, in other words courtship was successful, it seems fair to ask if there is "truth in advertising." Some research suggests that there is. For example, female mallard ducks that mated with "more attractive" males had offspring with higher viability. In this case, viability is correlated with egg size, which is controlled by the female's physiology. The male's contribution may have been to stimulate her into investing more energy into those eggs.

10.8.1 Nests

In many birds, nest site selection and construction precede mating itself, an activity that further strengthens the courtship pair bond. Birds generally are solitary nesters; there is one nesting pair per site. However, about 98% of seabirds nest in large colonies where individual nests are little more than a wing span apart from one another. The high density of individuals in the colonies affords protection from predators.

The nest itself might be no more than a depression in the ground, the bare crotch of a tree branch, a lose pile of twigs, or elaborately woven multichambered affairs. Some nests are reused. Ospreys (*Pandion haliaetus*) engineer massive nests that are built up over the years with sticks and all manner of other materials both natural and man-made, like aluminum cans and odd pieces of rope and plastic. Many passerines build cuplike nests usually hidden from view, and some like orioles and bushtits construct hanging, pendulous homes. Nests are lined with soft material the birds collect from their surroundings. Hummingbirds often line and tighten their small cuplike nests with spider web silk (Figure 10.4).

But why bother building a nest if you can hoodwink other birds into using theirs? Cowbirds (*Molorothus* spp.) are brood parasites that do just that. The host is usually a slightly smaller bird unaware that the female cowbird has replaced one or two of her eggs with the cowbird's own. The duped hosts not only incubate the eggs but also care for the hatchlings. Cowbird hatchlings can be slightly larger and more aggressive than their host-nest mates so they have an advantage in garnering attention and receiving food brought in by the host parents. A few species, like American Robins (*Turdus migratorius*), can detect cowbird eggs and will eject them from their nest. Brood parasitism by cowbirds seems to be on the rise in North America and has been cited as a factor in the population decline of some passerine species. Cowbirds may have easier access to nests than they did in the past because of forest clearing for agriculture and urbanization.

FIGURE 10.4 Like flying economy class, these fully-fledged Anna's Hummingbirds (*Calypte anna*) have little room to move. The resourceful parents used the base of a patio light fixture to secure their tiny nest. The fledglings were gone the next day. (Courtesy of David Archibald.)

10.8.2 Mating Systems

Either one of two mating regimes are practiced among social vertebrates: monogamy or polygamy. In monogamy, male and female remain together throughout the breeding cycle. This may be for part of the season, all season when multiple broods are involved, or even for a lifetime if the pair bonding is particularly strong. Monogamy for all or only part of a season is often a function of operational sex ratios—the temporal distribution of breeding females. If all females are receptive at the same time then a single male likely cannot dominate the pool and will mate with one female at a time. But if females in the population become receptive over several weeks, then an individual male has the opportunity to mate more than once during the breeding season. Competition and sexual selection among males are strongest under these circumstances. Monogamy is presumed to be the widespread mating strategy among birds, which is logical because monogamy typically is accompanied by a high incidence of parental care. Both male and female participate in the feeding and defense of the hatchlings thereby increasing the odds for a higher survival rate.

Recent genetic studies, however, demonstrate that extra-pair matings are not uncommon among supposedly monogamous pairs. Males and females both may in engage in promiscuous activity and mate with other individuals. Hence the term, social monogamy has been introduced to describe the situation where a male and female take responsibility for a clutch of eggs but do not maintain fidelity to one another. Gametic monogamy refers to the more traditional definition: both sexes care for the brood and do not have extra-pair matings.

Why would one or both sexes of a pair engage in extra-copulations? One possibility is that either sex could increase the fitness of its offspring by mating with a genetically superior individual(s) it is not paired with. Each also avoids having all its "eggs in one basket" and risk losing the entire reproductive effort should the nest be lost to predation or some other calamity. As a bonus, the male and the female, should she deposit her eggs in the nest of the male who fertilized her, now have a dedicated pair attending to his or her offspring while they resume the responsibility of seeing to their own.

10.8.3 Polygamy

Polygamy means that an individual, male or female, has multiple partners during a particular breeding cycle. It differs from social monogamy (promiscuity) in that the individual does not participate in parental care. For instance, polygyny (*poly* means "multi"; *gyn* means "female") is where a male will mate with several females. One driver of polygynous mating is resource-limited habitat. Males stake out and defend the more favorable of otherwise scarce nesting sites. Breeding females choose among the males that may offer a greater likelihood of having more offspring than if she mated monogamously with a male in a poorer locality. Red-winged Blackbirds (*Agelaius phoeniceus*) are a familiar example. These birds nest in tall reeds of swamps and marshy areas throughout North America.

In Southern California, we can find them around our lakes and reservoirs in the spring. Look and listen for the males advertising themselves atop a reedy perch.

In an alternative expression of polygyny, the male does not monopolize a resource (nesting site) required by the female but instead demonstrates his superiority through display. Some galliform birds like the Greater Prairie-chicken (*Tympanuchus cupido*) follow this strategy. Males assemble in aggregates known as *leks*. Each male in the lek has a small territory in which he struts, postures, and otherwise puts on his best courtship display. The audience of observing females selects one lucky male to mate with based on his performance.

Polyandry (*andr* means "male") is a clever polygamy tactic that favors the female instead of the male. She is the one with multiple partners. This scheme is adopted by birds like the Spotted Sandpiper (*Actitis macularia*) who have a need for multiple, rapid clutches. Food is abundant, but the fledgling success rate is low because the simple grass-lined nest is a mere depression on the ground where predation can be high. Males establish territories that the female visits. She remains with him until the clutch is deposited then moves on to another male leaving him to incubate the eggs. She may return to the first male and breed again if the clutch has been destroyed. Thus, a replacement clutch is the resource that the female controls. The energetic cost to the female is lessened by an abundance of food available to her, the fact that she lays only a few small eggs at a time, and her lack of incubation responsibilities. Multiple matings increase the likelihood that all her eggs will be fertilized. Secondly, the fitness of her offspring could be increased if at least one of her partners possesses some superior quality of its own.

10.8.4 Hatchlings

Baby birds emerge from their shells displaying a spectrum of physical maturity and independence that can be attributable to differences in yolk mass and is specific to a particular lineage. Typical of hatchling passerines are the baby House Finches (*Carpodacus mexicanus*) being reared in the nest in your backyard. These altricial young are naked, blind, and utterly helpless at birth. They require continual warmth and feeding and, only after several days, do they sit up and become aware. In about 2 weeks, they have acquired essentially full plumage and neuromuscular control of their wings. At this point they leave the nest but remain in the vicinity, where they are still fed for a few more weeks until the parents gradually neglect them, and they learn to fend for themselves. Even then they may stay together as a family until the fall. At the other extreme are the precocial young of birds like galliforms (e.g., chickens and quail) and anseriforms (ducks, geese) that hatch already covered in down. Although still small they are alert and able leave the nest immediately with a fair degree of independence. They imprint on their parents and follow them about learning how to find food. The most precocial hatchlings are probably the mound-nesting megapodes (Galliformes: Megapodiidae) of the Tropical Pacific and Indo-Australian region. These birds emerge from the nest ready to fly. Hence there is no female parental investment after the eggs have been deposited. The male maintains the mound.

Between these two extremes are variations in hatchling independence that ornithologists recognize as semi-precocial and semi-altricial. The former include chicks of gulls, terns, auks, and goatsuckers whose hatchlings have their eyes open, are covered in down, and are able to walk but don't stray far from the nest. Semi-altricial birds include herons, bitterns, and birds of prey. The hatchlings are covered in down, have their eyes open (except for owls), but are unable to leave the nest for a considerable time.

10.9 FEEDING AND FORAGING

Birds are endotherms and thus have steep energy requirements. Most of their active time is spent in search of food. They rely on their primary senses to locate food items and use their bills to seize it or tear it apart.

Birds are not the only toothless vertebrates with beaks. Turtles have beaks and so did several lineages of extinct Archosaurs, for instance pterosaurs. The avian bill consists of a bony upper maxilla + premaxilla (referred to, incorrectly, as the *upper mandible* in the bird watching lexicon) and a "lower" mandible (homologous with our lower jaw). Both are covered in a keratinized sheath. Birds within the egg develop an "egg tooth" externally on the upper jaw. The egg tooth is epidermal, not bony, and assists in cracking the shell. It is resorbed or cast off after hatching.

Bill shapes widely vary among birds because shape is correlated with diet. Hawks and owls capture and kill prey with their talons and their curved raptorial bill is an obvious adaptation for tearing flesh. Other carnivorous birds like herons and egrets use their long bills as spears whereas gulls, roadrunners, and some corvids stab at their prey. Note that corvids like crows and jays have an unremarkable bill shape that reflects their generalized diets (Figure 10.5). These birds feed on almost anything that is convenient. Shrikes have slightly hooked bills that they use to capture insects, small mice, and other birds. Shrikes are famous for impaling their prey on thorns, spines, and even barbed wire for later consumption (Figure 10.6). *Granivores* such as finches, towhees, sparrows, and cardinals (and grosbeaks to the extreme) have conical bills for cracking seeds. Many seedeaters forage on the ground beneath shrubs. Towhees, for example, will hop and scratch the soil surface to uncover seeds. Others like buntings feed directly on the grain stalks of tall grasses. Warblers flit about in trees using their thin needle-like bills to glean insects from leaves. The bill of a pelican is a dip net with which they capture prey from a plunge dive. Swifts and nightjars are aerial feeders with small weak bills and a wide gape. Dabbling ducks strain food from water through transverse grooves on their bills.

A moveable tongue is an amniote innovation, but the tongue of most birds is small, more or less triangular, and not especially manipulative, so it is not typically employed for feeding or drinking in most cases (Figure 10.7). Taste buds are few or absent, although an array of tactile sensors is present so that a bird can discriminate among food items based on feel between the tongue and the palate. The long thin bills and agile tongues of hummingbirds are exceptional

FIGURE 10.5 Bill shape correlates with the type of preferred food. In the case of this Western Scrub Jay (*Aphelocoma californica*), the bill has no obvious specialization. It is a generalized bill that reflects the omnivorous diet of jays, crows, and ravens (Corvidae).

FIGURE 10.6 This White-crowned Sparrow (*Zonotrichia leucophrys*) met its end from a Loggerhead Shrike (*Lanius ludovicianus*) who impaled it on a mesquite thorn for later retrieval. Shrikes, sometimes called "butcher-birds," are notorious for this behavior. White-crowned sparrows are abundant winter residents in Southern California. The Loggerhead Shrike is a year-round but uncommon resident in open country habitats.

FIGURE 10.7 Humans drink water by creating suction with their lips and facial muscles. Because birds lack this capability, they take in water by scooping a mouthful then tipping the head back and letting it run into the throat, like this Acorn Woodpecker (*Melanerpes formicivorus*) is doing. Pigeons are among the few birds that can drink with their bills immersed.

and adapted for extracting nectar from tubular flowers. (Hummingbirds will on occasion also consume small insects.) Woodpeckers also have an exaggerated tongue that they use to probe for insects and larvae once they've bored a hole into woody tissue. When retracted, the tongue bones and muscles actually wrap around the back of the head internally. Acorn Woodpeckers (*Melanerpes formicivorous*), on the other hand, cache acorns in the holes they drill in trees and utility poles.

In all vertebrates, the esophagus conducts food from the mouth to the stomach. Part of the esophagus in birds is modified into a storage sac called the *crop*. The food can be swallowed later or regurgitated to feed nestlings. Many birds, especially seedeaters, have altered part of the stomach into a gizzard. In the absence of teeth, this muscular organ, which may contain pebbles and grit, pulverizes husks and seed coats.

Wading shorebirds that forage around tidal mud flats are an excellent illustration of the correlation between morphology and behavior and how multiple species and hundreds of individuals can feed in the same area at the same time. Visit any of Southern California's bay estuaries in the winter months. Notice that the smaller birds, for example some sandpipers (*Calidris* spp.), have comparatively short legs and bills and therefore forage along the shallowest parts of the flats. In deeper water are the willets (*Catoptrophorus semipalmatus*), whimbrels, and curlews (*Numenius* spp.), and deeper still are the very long legged and long billed Black-necked Stilts (*Himantopus mexicanus*) and American Avocets (*Recurvirostra americana*).

As foraging endotherms, birds have been subjected to considerable ecological speculation. One question is whether individuals take food items in proportion to their abundance without discriminating, or does a bird forage by optimizing (balancing) food availability against energy expended versus energy gained? The assumption here is that a continual abundance of a stable food supply is not likely and therefore some sort of optimization strategy is probably in play. But assuming it to be true creates an immediate bias if it could actually be tested at all, which is next to impossible to do empirically. Measuring the energy content of a food item and the amount expended to find and consume it, for example, simply cannot be done in the field. However elegant the idea appears, generalizations would be fraught with exceptions even if optimization could be demonstrated at some level in a handful of species. Tuck the idea away while you are in the field and see what insights you can gain by simply considering body morphology, bill shape, and the foraging tactics of the species you see.

10.10 MIGRATION

St. Joseph's Day, March 19, marks the return of Cliff Swallows to Mission San Juan Capistrano in southern Orange County. The birds' spring arrival is celebrated with festivals, and famously in song, as the mass of the population joins the early arrivals over the next couple of weeks. These birds have just traveled 7,500 mi (12,000 km) from Argentina in about a month's time. They will spend the next few weeks repairing and rebuilding anew their mud nests beneath the eaves of the mission and then set about rearing young until they depart once again for the Southern Hemisphere in late August. Cliff Swallows (*Petrochelidon pyrrhonota*) breed across North America, and in Southern California the endless construction of sheltered vertical concrete surfaces (bridges, freeway overpasses, buildings, etc.) created nesting sites where they once did not exist. Sadly, however, human engendered opportunity has also led to a fairly steep decline in Cliff Swallow numbers in recent decades. Maintenance crews and annoyed property owners have removed the colonies of mud nests, and the open landscapes they need to forage for insects have largely been paved over.

I was curious as to how they were faring at the mission, so my wife and I drove to Capistrano at the end of April in 2015. It was a cool, beautiful spring day. Once inside the mission grounds, we began scanning for nests in all the likely places. Not a one was found. In fact, there were no swallows even flying about overhead. Rather dumbfounded I sought out a staff member. It turns out that beginning in the early 2000s mission preservationists had them all removed. The symbol of the old church and historic encampment (built in 1776) is no more. Presently in its place are a few "nesting reenactment exhibits" consisting of synthetically cast swallow nests placed strategically in a couple of areas. The hope, apparently, is to entice the birds to return and nest once again, or the catch phrase might have to be amended to the past tense: "when the swallows *came* back to Capistrano."

Migration is the annual movement of a population to a fixed location, usually by the same route. Migration is common in both marine (e.g., some fish, sea turtles, and whales) and terrestrial (amphibians, birds, and mammals) vertebrates. Physical barriers are a challenge to terrestrial, nonvolant species, so migratory routes and distances are dictated by topography. Birds are not so constrained. Some, like the cliff swallows, make yearly migrations of extraordinary distances between the northern and Southern Hemispheres. The Short-tailed Shearwater (*Puffinus tenuirostris*) travels 18,650 mi (30,000 km) round trip from its breeding area in southern Australia to its wintering region in the north Pacific. Their migration route circumnavigates the Pacific Ocean clockwise to take advantage of prevailing winds.

In the 1920s and 1930s, Frederick Lincoln, an ornithologist with the U.S. Biological Survey, began looking for patterns in the migration routes of North American birds. Using records of banded individuals, Lincoln identified four primary migration routes, or flyways: Atlantic, Mississippi, Plains (central), and Pacific. The Pacific flyway is responsible for the influx of most of the wintering shorebirds that arrive from northern latitudes to Southern California beginning in the late fall. A number of passerines follow this route as well, like wintering sparrows and the spring and fall populations of colorful warblers. Sometimes flocks of a species arrive over a period of several weeks, in other cases almost within a few days of the same date each year. In Southern California, White-crowned Sparrows show up in mid to late September with dependable, clock-like consistency.

Nowadays some migration routes have been confirmed by outfitting birds with tiny transmitters that relay their coordinates by way of GPS signals.

Many migrating species are consistent with the flyways, but they are not restricted to them or to any one in particular. Nor is a straight north-south round trip necessary. Some birds are reluctant or unable to cross large, inhospitable features like deserts or massive expanses of water. Instead of a straight line, they take a longer indirect route to reach their destination. Soaring birds such as hawks and eagles generally migrate along narrow pathways beside mountain ranges where updrafts provide the necessary wind currents. Swainson's Hawks amass in spectacular numbers this way in the Borrego Valley, primarily in the spring when they return from South America and head to open-country breeding areas farther north and east. Some birds fly a loop migration, like the Short-tailed Shearwater mentioned previously. The American Golden Plover (*Pluvialis dominica*) is another one. This bird breeds across northern Alaska and Canada. It flies southeast along Hudson Bay and the Maritime Provinces then across the Atlantic to its wintering stations in South America. In the spring, it returns primarily over land via Central America, Mexico, and the Mississippi Valley, a route with more suitable ecological conditions at that time of year. Then there are those montane species that remain more or less in the same geographic area throughout the year and simply migrate vertically, moving to lower elevations in the winter and return to higher elevation in the summer to breed.

Whether a species follows a narrow path or a broad front chiefly depends on the geographic extent of its breeding range. Broad fronts will be taken by birds whose breeding habitat spans multiple degrees of longitude, for example the boreal forests of southern Canada and the northern United States favored by many passerines.

Around 40% of Northern Hemisphere birds migrate. Migration requires two attributes: stamina and an ability to navigate. Both of these requisites are fairly well understood, but *why* birds migrate, or do not, is multifaceted and draws on a complex number of variables, not all of which are applicable to every species. In general, migration is a response to seasonal variation in resources and typically correlates with breeding. Resource variation is a function of seasonality and climate, which in turn are determined by latitude. The majority of migration routes trend north and south for this reason. Consider, moreover, that the latitudinal zonation of terrestrial biomes changes over time. During the last (Wisconsinan) glacial period, biomes were compressed toward the equator and did not have the same dimensions and character that they do today. Most modern migratory routes therefore were established following retreat of that glacial episode from 18,000 to 12,000 years ago.

Northern Hemisphere birds that migrate to southern latitudes for the winter do so to avoid temperature stress or because food supplies may be insufficient during the coldest months. For many birds, more favorable conditions are met in a variety of habitats within warm temperate, tropical, and subtropical latitudes. For others, however, available wintering grounds might be limited because of historical or ecological circumstances. The rarity of some North American bird species may indeed be on account of limited wintering habitat that can only accommodate small populations.

Birds return to their summer residences in regions where structural habitat meets their nesting requirements. Gonadal development occurs at this time. In late summer, following breeding, gonads regress and birds begin their molt. The longer summer days mean more time for foraging and rearing of young. Snow Geese (in the millions) and many shorebirds summer in the harsh, high latitude tundra of the arctic. Although the breeding season is short, the days are long and food resources are plentiful. Some intriguing evidence also indicates that predation risk is lowered in the open high arctic, where, paradoxically, predation by arctic fox, snowy owls, and jaegers can be intense. Crucial to this dynamic are several species of lemmings (*Lemmus*, *Dicrostonyx*) whose population densities are cyclical over a 3- to 4-year period. When lemming densities are high, predation on nesting shorebirds (e.g., *Calidris* spp.) is reduced, but predation increases when lemmings are experiencing off years. Where lemmings are essentially absent altogether, the predation pressure on nesting shorebirds would be acute every year and viable breeding populations probably could not be sustained. Hence, many of the arctic-breeding shorebirds are found only within the geographic range of the lemmings.

10.10.1 The Migratory Journey

The urge to migrate is triggered by both photoperiod (changing day length) and internal biological clocks. Before migration, birds prepare for their journey by storing up energy as fat reserves to as much as 50% of nonfat body mass, a condition called *zugdisposition*. For example, the four-gram Ruby-throated Hummingbird (*Archilochus colubris*) of eastern North America (a close relative of our western Black-chinned Hummingbird [*A. alexandri*]) flies nonstop for 20 h (approximately 500 mi ([800 km]) across the Gulf of Mexico to Central America. In preparation it will put on 2 g of extra fat, which is enough for 24–26 h of sustained flight. Hence there is an energy reserve to aid in recovery and for contingencies.

The Ruby-throated hummingbird, however, is pretty exceptional. Most birds do not fly nonstop to their destinations. Many, especially passerines, migrate at night, and then refuel during the day before continuing. Waterfowl also break up their migration and depend on landmark recognition during daylight hours to identify specific stopovers while on route: estuaries, wetlands, and other coastal and interior habitats. Sad to say, many of these have been lost to urbanization and agriculture. The loss of stopover habitat has occurred worldwide and affects many species and populations during a stressful period of their life cycle. Fortunately in California, man-made lakes and reservoirs throughout the state have, in part, offset the loss of historically significant interior wetlands. The accidental creation of the Salton Sea (pg. 113) is Southern California's signature example.

10.10.2 Orientation and Navigation

Successful migration requires that a bird knows where it is starting from, the direction it wishes to go, and an ability to stay on course, sometimes over many miles off apparently featureless terrain. Yet the vast majority of individuals reach their destination. And they can do so with astonishing precision, often returning to the same park and even tree from one year to the next. Some species have much better homing instincts than others. Certain breeds of pigeons are renowned for that ability, whereas some sparrows, by contrast, fail to locate their nest after being experimentally displaced only a few kilometers.

Long-distance migrants usually navigate by celestial cues. Diurnal migrants make use of sun-compass orientation, at least on clear days. Because the sun changes its position in the sky throughout the day a bird, therefore, must be able to "tell time." The same holds for birds that migrate at night navigating by the position of stars. To demonstrate reliance on celestial navigation, experimenters conditioned day-migrating birds to a 6-hour shift (one quarter of a day) in their photoperiod. When released, the conditioned birds flew 90° off course, but on cloudy days a photoperiod shift had little effect on their orientation and most reached their destination without incident. Apparently, other means of navigation were substituted for the sun compass on cloudy days. One option is a sense of magnetic north that can also be used as a compass. Other birds make use of polarized and UV light, and a few, like the petrels and albatross noted previously, can orient by detecting distinctive odors. Remarkably, too, many birds imprint on low frequency sound waves that are below what humans can detect. The myriad features of the Earth's terrain create low-frequency signatures that a bird learns to identify. Of course, a bird wintering in South America cannot detect its far away summer home by infra sound directly, but it does learn the sounds of geographic sections along the way and breaks up the trip accordingly. Because birds migrate in flocks, the younger birds may learn the wave points from their more experienced elders.

Numerous details are yet to be learned about avian migration. Clearly, though, birds possess a redundancy of hierarchies and cues to navigate between summer and winter residences.

10.11 BIRD WATCHING 101

Legions of local bird-watching organizations and regional Audubon chapters attest to the popularity of identifying and counting birds as a recreational pastime. Further evidence? See the abundance of field guides at your bookstore. Southern California is a truly enviable setting for this activity. Consider the available habitats—sea shores, bays, and estuaries to valleys, foothills, mountains, and deserts—and it is not surprising that around 550 species of residents and migrants have been recorded from our region.

Beginning bird watchers confront a daunting challenge. No shortcuts are available to becoming a competent identifier in the field. Like all crafts, proficiency is gained through the time put in to get good at it, which of course is where the fun is. The first thing we usually do is buy one of those field guides assuming that the book alone will, like a cookbook, provide the ingredients and techniques. Field guides, and many fine ones are out there, are in reality effective only after you have learned to identify a variety of species. It's sort of a catch-22. You have to learn *how* to identify birds before you can learn who they are. Flipping through pages while at the same time trying to keep your binoculars trained on a bird before it takes off will only lead to mind-snapping frustration. So what is the beginner to do?

Become familiar with your field guide by all means. But get started by going on field trips with knowledgeable observers. Visit sites where a reasonable diversity of birds can be expected. Go in the morning hours when many species of birds are most active. For now, avoid the individual species of wintering shorebirds, sparrows, or migrant warblers. These can be tough to identify. When you return home, study in your field guide the birds you saw that day. Here is that point where your field guide can significantly elevate success. You've come to realize that color, for example, is rarely the first criterion in identifying a bird. Instead, by following the old adage of "general impression, shape and size," you learn first to discern what type of bird it is that you are seeing. Hawks and ducks, say, are simple enough at that level, but not so for many others. So you need to learn the different groups, and once you have that familiarity the identification process can be cultivated. Each species account in the field guide highlights the subtle differences that distinguish a bird from those with which it would most likely be confused. These include such characteristics as bill shape, seasonal plumage, body markings like wing bars, eye rings, facial stripes, and finally any peculiar behavioral tics, for instance dipping its tail or flicking its wings while perched.

Learn the birds in your backyard, your neighborhood park, or nearby canyon. Use these species to develop fixed reference points to take with you on later excursions. Also pay attention to voice if a bird is audible. Learning the call notes and songs of common species is another useful identification tool and an added enjoyment to the experience. And do not overlook geographic distribution. Ask yourself (and consult your field guide's range maps) if this species reasonably could be expected to occur here, in this habitat, and at this time of year.

Before long you will at least be able to look at a new bird and say what it is not and thus narrow your possibilities as to what it could be. If you can say no more than it is a sparrow or woodpecker, then that's okay. Eventually you will refine your search image with practice and be able to whittle down the possibilities to one or two species.

Accumulating a lifetime list of birds is fashionable among many bird watchers. The life list can evoke a memorable field trip or be spent as currency for good-hearted bragging rights at the next get-together. But let bird watching take you beyond stamp collecting. Use it to probe and extend your interests in all realms of natural history.

11 Mammals

Across all vertebrate lineages mammals stand out for their adaptability. They are endothermic and at birth are nourished by milk secreted by the mother's mammary glands. Mammary glands and hair, unique mammalian characteristics, have negligible chance of preserving in the fossil record, and neither does endothermy, although it can be inferred. By examining the fossil record, however, it is possible to extrapolate function from form and make plausible estimates about when and how mammals came to be. Mammals have an upright posture that is reflected in the bones of the limbs and girdles. The mandible is composed of a single element, the dentary, and forms a unique joint with the base of the skull. Teeth exhibit a division of labor; some like canines are used for piercing, whereas molars are for cutting or grinding. Mammals have three tiny bones in the middle ear (an evolutionary consequence of their unique jaw articulation), whereas other tetrapods have only one. These and other characteristics and their precursors are to varying degrees available in the fossil record.

11.1 SYNAPSIDA

The Cenozoic is often called as the "Age of Mammals." The living orders of placental mammals make their appearance at that time and radiated in spectacular fashion. Mammalian history, however, extends back to the beginning of their synapsid ancestry. The Synapsida, recall, have a single temporal opening on either side of the skull. Mammalia are the most derived, but they were preceded by two major groups of nonmammalian synapsids that initiated the structural and physiological transformations that came to define the Mammalia. This history extends back more than 300 million years to the Carboniferous Period in the latter part of the Paleozoic. Collectively, the nonmammalian synapsids are sometimes called the "mammal-like reptiles," an unfortunate and misleading term because it implies a relationship with reptiles as we know them today and that evolution follows a smooth, linear trajectory. "Pelycosaurs" were the earlier of the two groups and were followed by the "therapsids." Both of these groups are paraphyletic (unnatural) taxa whose relationships are still being sorted out.

Pelycosaurs appear shortly after the earliest amniotes in the late Carboniferous. They diversified to the extent that in the Permian Pelycosaurs accounted for more than half of the known amniote genera. The majority of them were from about the size of a cat to a large dog, although *Dimetrodon* may have reached 4 m, including the tail. Pelycosaurs displayed a generalized tetrapod design including a somewhat lizard-like sprawling posture. A few were herbivores, but most were carnivorous. At least in the early Permian they seemed to have been denizens of equatorial zones and exploited aquatic ecosystems for food. Perhaps the most distinctive were *Dimetrodon* and a few others that sported a prominent sail down the back composed of skin supported by extended neural spines from the vertebrae. This distinctive feature is doubtless the reason that they have been erroneously called "dinosaurs" in children's literature. Pelycosaurs declined rapidly at the end of the Permian and became extinct by the beginning of the Triassic.

The Permo-Triassic boundary was marked by a global mass extinction event triggered by volcanic activity. Ash clouds and acid rain led to dramatic climatic cooling and even polar ice. About half of all marine animals disappeared at this time, and many terrestrial ones vanished as well, perhaps as many as 90%. The therapsids were one of the groups of land vertebrates that survived the Permo-Triassic upheaval. They arose in the Permian and prospered into the Triassic. Like pelycosaurs, therapsids included both herbivores and carnivores. Some of the herbivores were large and massive, about the size of a cow; others were small, almost rodent-like. A few of the carnivores were as large as a big dog and rather menacing looking, whereas others were smaller, nearer that of a rabbit, and may have been insectivorous.

Becoming a Mammal. Cynodonts were one of several therapsid lineages that flourished during the Triassic. Cynodonts are of particular interest because they exhibited some of the morphological tendencies that were fully expressed later in early mammals, such as a more upright posture, specialization in teeth and jaws, and a small body size. The more trenchant characteristics are described below.

- The temporal fenestra was exaggerated and the canine teeth and associated maxillary bones were enlarged. The enlarged temporal opening and other feature on the side of the skull would suggest that cynodonts had robust jaw muscles, and thus more food was being consumed and processed orally per day. Additionally, there was a trend toward reduction of the lower jaw to a single bone. In pelycosaurs (and most all other vertebrates), the *dentary* (the tooth-bearing bone) makes up the anterior half of the lower jaw. The posterior end is composed of several elements including an articular component that forms the joint with the skull. In cynodonts and a few other therapsids, the postdentary bones are reduced to the point that in mammals only the dentary remains and forms its own articulation with the *squamosal* bone (part of the temporal) of the skull. The *dentary/squamosal joint* creates a powerful bite force because of its more anterior position. It is one of the hallmarks of mammalian evolution. Fossil evidence shows that two of the ancestral postdentary bones, the articular and the quadrate, were not lost but were instead incorporated into the middle ear as the malleus and incus, respectively. Along with the stapes (present in all tetrapods), these three ear ossicles transmit sound impulses from the eardrum to the inner ear.

- The teeth of pelycosaurs, like most other nonmammalian vertebrates, were homodont, and tooth replacement was more or less continuous throughout life (polyphyodont). Homodont teeth are undifferentiated from one another. That is, the teeth in the front of the jaw look just like those at the back. Within therapsids, teeth became increasingly heterodont, or specialized from front to back for particular masticatory functions, for example, nipping, slicing, and grinding. Continuous tooth replacement was suppressed. Mammals have at most two sets; our baby teeth and our permanent teeth. Molars are not replaced at all but erupt later in life. Molars also have precise occlusions between the upper and lower cusps, which make it possible to thoroughly process food. Only mammals masticate their food. In humans and herbivores, like cows, the articulating condyle on the lower jaw is elevated above the tooth row such that the jaws move in a rotary fashion. These mammals can only chew on one side of the mouth at a time. In carnivores the condyle is level with the tooth row so that the jaws close like a pair of scissors.
- A secondary palate arose consisting of both bone (anteriorly) and later soft tissue (posteriorly). A primary palate is present in bony fish and tetrapods, including pelycosaurs, and consists of a medial series of fused ventral skull bones overlying the mouth. In tetrapods and their ancestors, the nasal passage reaches the mouth through a pair of openings in the primary palate, the internal nares or choanae. The secondary palate (also present in crocodilians) created two separate passageways, one dedicated to the movement of air and the other for the passage of food. The animal can breathe and chew at the same time.
- Therapsids underwent profound changes to the vertebral column that accompanied modifications to the limbs and girdles. Both sets of changes correspond with an upright posture and an active lifestyle in which locomotion is no longer a form of lateral undulation. The limbs now move to and fro, parallel to the vertebral column. Weight reduction of the pelvic girdle, the addition of more vertebrae to the sacrum, and the reduction of ribs on the lumbar vertebrae are in keeping with the hindquarters providing most of the power stroke. In essence, mammals are built with rear-wheel drive.

The mammal-like features found among cynodonts and other therapsids suggest an active lifestyle that required greater food intake to maintain. Endothermy, then, may have evolved concomitantly as a series of steps. High metabolic rates and small body size among cynodonts, for example, would also favor insulation. Hair is known possibly as early as 160 million years ago and definitely by 120 million years ago.

Therapsids waned in the late Triassic as the earliest mammals began to appear. Mammals emerged in a landscape dominated by dinosaurs, saurishcians in particular, and were the only therapsid line to prosper through the rest of the Mesozoic and survive the mass extinction of dinosaurs at the end of the Cretaceous 65 million years ago. In other words, mammals are a kind of cynodont, which are a type of therapsid, which are a group of synapsids.

11.2 MESOZOIC MAMMALS

The first mammals were smaller than even the smallest cynodonts, about the size of a shrew, though we cannot be certain if they occupied a new adaptive zone. The skull indicates a proportionately larger brain. Postcanine teeth had divided roots. These early mammals were taxonomically diverse, but all shared a more or less similar design. They were too small for an herbivorous diet that required the processing of fibrous material and were more likely insectivores and omnivores. Mammals did not diversify into larger forms with varied diets until the Cenozoic. It is possible that dinosaurs may have prevented mammalian radiation into broader adaptive zones during the Mesozoic, yet it seems unlikely that they competed directly.

Mammals appeared to have experienced two major Mesozoic radiations. Through the Jurassic into the early Cretaceous were the shrew-like forms that included morganucodonts, triconodonts, docodonts, and symmetrodonts. Recent fossil discoveries of Jurassic mammals from China, however, demonstrate that a greater diversity of body types existed. Some were built for digging and others for climbing and swimming. Perhaps the most dramatic fossil is the terrier-sized *Repenomamus*, an evident predator that was preserved with a baby dinosaur in its stomach.

The second radiation began in the late Jurassic/early Cretaceous and coincided with the appearance and diversification of flowering plants (angiosperms) and new predators such as snakes. This group of mammals was a more derived assemblage that included the rodent-like multituberculates that became extinct in the early Cenozoic and the first therians. Multituberculates are named for their broad, multicuspate molars specialized for grinding. Some multituberculates were squirrel-like climbers; others were more terrestrial. Their demise may have been hastened by the appearance of rodents and rodent-like primates in the early Cenozoic.

Evidence suggests that there was a Laurasian (Northern Hemisphere) radiation known as *boreosphenidans* that were the ancestors of multituberculates and therians, and a southern diversification of a group called *australsphenidans* from which we find the origin of monotremes. The earliest eutherians including the ancestors of the living placental mammals arose at least 100 million years ago.

11.3 LACTATION AND SUCKLING

Histologically (at the tissue level), mammary glands are similar to the sebaceous glands that secrete oily products (sebum) near the hair follicle to condition and water proof the fur. Sebaceous glands are absent from the palms and soles of the feet but are present around the vagina, penis, and next to the mammary nipples where they lubricate the skin. Their specialization into mammary tissue may have had its start as pheromone producers functioning to help the infant imprint on its mother and keep it from straying. The addition of other organic products would only benefit. For example, aside from its nutritional value, milk is vital in helping to establish the immune response in all infant mammals. It seems particularly important in controlling inflammatory responses. An increase in production coupled with added proteins and antimicrobial agents would further enhance its value.

Lactation also has an adaptive advantage in that it decouples the reproductive effort from the seasonal availability of food. Birds time their reproduction to coincide with maximum food availability for their chicks. Female mammals, on the other hand, can store food as fat and later convert it to milk and the stress associated with viviparity is lessened.

Suckling is a uniquely mammalian feature and a structural consequence of a primary and secondary palate along with the appearance of diphydonty. The reduction of tooth replacement to two sets would only happen if mammals were fed milk early in their development before the first teeth erupt. The jaw would continue to grow, and permanent teeth would appear only after the jaw achieved near adult size. Suckling also depends on extensive facial muscles that assist in forming a seal involving the mouth in association with the tongue and secondary pallet. The suckling individual can swallow while at the same time breathe through its nose. In adult humans, the secondary seal posteriorly involving the epiglottis has been lost because the larynx shifts to a more ventral position in early childhood. (Humans are thus more likely to choke on their food.)

Mammals are the only vertebrates capable of making facial expressions, although they vary in that ability. Horses, for example, use their lips when nipping grass and can also manipulate their face in several ways. Cows use more of their tongue to feed and therefore are mostly expressionless beyond their cavernous bovine stare. Small mammals like rodents likewise do not reveal much emotion through their facial muscles. Interestingly, the elaborate facial expressions of primates are the same in other mammals under similar emotional circumstances, even though they are derived independently. Facially, a snarling dog looks not all that different from an anger-infused human being.

11.4 HAIR

Mammals have high metabolic rates, and hair provides an insulating cover of fur, except for cetaceans, armadillos, and a few other species. Hair is an epidermal derivative. The base of an individual hair, the root, is located within in a hair follicle that is imbedded in the dermis. The nonliving hair shaft is a continuation of the root that projects above the skin. The growth of a single hair eventually slows and the hair dies. Stem cells in the dermal papilla produce a new follicle; the old hair shaft falls out and is replaced by a new one. The hair growth cycle is intrinsic and cutting it does not effect it. Hair also grows in a distinctive pattern or grain, and force applied against the grain is resisted. Petting your dog or cat in this manner is met with annoyance, hence the expression of rubbing someone the wrong way.

Fur, or pelage, is a thick covering of hair. Guard hairs are the thick, coarse hairs apparent on the body surface. The underfur lies beneath the guard hairs and consists of finer, shorter hairs. Both function as insulators. In marine mammals like cetaceans (whales, dolphins), the underfur is absent and only a scattering of guard cells remains. Insulation comes from large stores of fat. The sea otter (*Enhydra lutris*) off the Pacific coast has an especially dense, plush pelage. Indeed, its value to the fur industry (and along with that of the fur seals) almost led to the sea otter's extinction at the beginning of the twentieth century (see beyond).

Apart from skin impressions of a few Mesozoic fossils, hair itself is too soft and pliable to preserve well in the fossil record. Obviously, hair became a vital insulator coincident with the evolution of endothermy, but its earliest role and appearance are not fully resolved. One explanation is that hair arose as filamentous tufts between the hinges of scales and may have had a sensory or tactile function. Even in modern mammals, nerve endings at the base of some hair follicles are sensitive to mechanical vibrations. The whiskers of carnivores and rodents are evident examples of the tactile sensitivity hair can provide. In early mammals, natural selection may have favored longer hair shafts as an insulating mechanism, which became its primary function while still retaining a sensory adaptation secondarily.

11.5 LIVING MAMMALS AND THE RISE OF PLACENTALS

As we have seen, present-day mammals were preceded by a fascinating diversity of Mesozoic forms. The three major living clades reflect differences in body plans and reproductive pathways: *Prototheria*, or monotremes are egg layers, whereas the *Theria* give birth to live young and include the Metatheria (marsupials) and the Eutheria (placental mammals).

11.5.1 Monotremes

Monotremes were never particularly diverse. They have hair and suckle their young like other mammals but retain several characteristics of their generalized amniote ancestry. For example, their name is a reference to the *cloaca*, a single body opening shared by the excretory and reproductive tracts (*mono* means "one"; *treme* means "hole"). Besides laying eggs, they lack nipples (lactation occurs across milk fields) and external ears. Adult monotremes are toothless, and males produce thread-like sperm cells similar to those of birds.

The platypus (*Ornithorhynchus anatinus*) with its distinct leathery bill is perhaps the best known of the monotremes. This is a semiaquatic mammal of eastern Australia and Tasmania that uses its bill to detect electromagnetic signals emitted by its prey. The male platypus has a spur on each hind leg that discharges a poison to thwart its adversaries. Venomous mammals are rare and otherwise known only among a few shrews that produce toxic saliva.

The other living monotremes are the two species of echidna. The short-nosed Echidna (*Tachyglossus aculeatus*) is a specialist of ants and termites of Australia (including Tasmania) and southern and eastern New Guinea. The long-nosed Echidna (*Zaglossus bruijni*) is found only in New Guinea.

11.5.2 Marsupials

Marsupials give birth to live young at a late embryonic state. The offspring haul themselves into the mother's pouch, or marsupium, where the mammary glands are located, and there they complete their development. At this stage, the hind limbs are not formed. Claws on the manus (hand) are used to pull themselves into the pouch and thus precluded the evolution of hooves. The pouch is absent in male marsupials, and in some species, females may also be pouchless.

Current thought is that extant marsupials might be divisible into four lineages, but minimally there is a fundamental division between the marsupials of the New World (Ameridelphia) and those of (mainly) Australia (Australidelphia). Most marsupials today live in Australia, but their origin apparently took place in North America from where they dispersed in two directions. In the Eocene, they spread to Europe and North Africa where they eventually became extinct. The other wave spread to South America, Antarctica, and Australia before these continents had fully separated. The Australian radiation is a case study in evolution. The 195 or so species of marsupials there today are a diverse assemblage of insectivores, carnivores, herbivores, and omnivores that include such familiar forms as the bandicoots, koala, wombats, wallabies, kangaroos, and the endangered Tasmanian devil. The only placental mammals present are bats from about the early Eocene and rodents whose ancestors reached Australia about 4 million years ago. Today Australian rodents are represented by 27 species, the largest of which is the Water Rat (*Hydromys chrysogaster*) that is about the size of a rabbit (up to ca. 1.3 kg).

Though not as diverse in terms of species, the New World marsupial radiation is nonetheless impressive. Many of the approximately 85 species belong to the Didelphomorpha, a group that includes a diverse array of opossums. Most are small- (20 g) to medium-sized (6 kg) arboreal or semiarboreal omnivores. An even greater diversity of didelphomorhs existed in the Tertiary, several of which were ecological equivalents of some placentals like the Plio-Pleistocene saber-tooth analogue, *Thylacosmilus*. Today, the only living marsupial in the United States is the Virginia Opossum (*Didelphis virginianus*).

11.5.3 Placental Mammals

Placental mammals are the largest, most diverse, and widespread of any group of living mammals (about 4,500 species). Placentals are distinguished in part by the placenta, a vascular organ that connects the embryo with the female uterus. Although a short-lived, placenta does form in some marsupials, and similar nutritional-respiratory support is also known in certain fish, amphibians, and reptiles; reproduction in all placental mammals is entirely placental-based.

Placental mammals had their start in the late Mesozoic but saw much of their diversification and adaptive radiation take place against the backdrop of drifting continents, Cretaceous extinctions, shifting vegetation, and climatic swings during the past 65 million years. The end of the Paleozoic joined all of the earth's landmasses into the super continent Pangaea. Pangaea spit into northern (*Laurasia*) and southern (*Gondwana*) halves in the middle Mesozoic, and by the latter part of the Cretaceous these two landmasses were rifting into smaller continent separated by incipient oceans. The Atlantic was just emerging, so most continents were still connected by land bridges. The climate was mild even in polar regions.

Fragmentation and rotation of the continents continued in the early Cenozoic. Ancestral placental mammals moved into Africa and the New World, and the isolated stocks were the genesis of the groups that evolved later on separate continents. They originated in Asia and used land-bridge connections between the continents to reach the New World. Placental diversification appears to have happened rapidly as there are few clues that help us create the morphological trajectories that characterize the living orders. The infusion of molecular data in recent years has produced some interesting results that do not emerge from phylogenies based on morphology alone. Currently, four major groupings are recognized: the Afrotheria, Xenartha, Euarchontoglires, and the Laurasiatheria, with the latter two groups comprising the more inclusive Boreoeutheria.

Long suspected on morphological grounds, the Afrotheria is an endemic grouping of African mammals, like elephants, dugongs, manatees, hyraxes, and aardvark but also several smaller forms, such as tenrecs, elephant shrews, and golden moles. The South American anteaters (Cingulata) along with the sloths and armadillos comprise the Xenartha. This group is currently neotropical but was widespread in North America until the end of the Pleistocene, for example the giant ground sloths (see beyond). Afrotheria and Xenartha may be sister lineages, and currently they cluster at the base of the eutherian phylogenetic tree (Figure 11.1). In both instances these endemic groupings speak to a long period of isolation that fits with continental drift in the Cenozoic.

Southern California's mammal fauna is listed in Table 11.1 and discussed below.

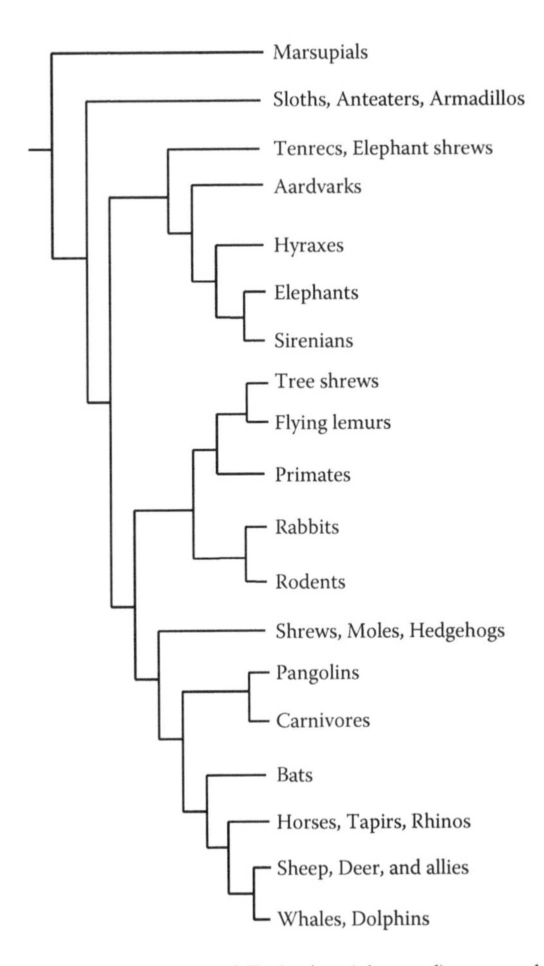

FIGURE 11.1 A phylogenetic tree of Eutherian (placental) mammals, with marsupials as their sister group. Placental mammals began diversifying in the early Cenozoic. Southern California has living representatives of primates (humans), rabbits, rodents, shrews, moles, carnivores, bats, horses (introduced), sheep, deer, whales, and dolphins (artiodactyls) (Table 11.1). (From O'Leary, M.A. et al., *Science*, 339, 662–667, 2013.)

TABLE 11.1
Mammals of Southern California

Marsupialia: Marsupials
  Didelphidae: Opossums
    Didelphis virginiana: Virginia Opossum (Int.)
Lipotyphla: Shrews and Moles
  Soricidae: Shrews
    Notiosorex crawfordi: Desert Shrew
    Sorex monticolus: Montane Shrew
    Sorex ornatus: Ornate Shrew
    Sorex trowbridgii: Trowbridge's Shrew
  Talpidae: Moles
    Scapanus latimanus: Broad-footed Mole
Chiroptera: Bats
  Molossidae: Free-tailed bats
    Eumops perotis: Greater Bonneted Bat
    Nyctinomops femorosaccus: Pocketed Free-tailed Bat
    Nyctinomops macrotis: Big Free-tailed Bat
    Tadarida brasiliensis: Mexican Free-tailed Bat
  Phyllostomidae: Leaf-nosed bats
    Choeronycteris mexicana: Mexican Long-tongued Bat
    Macrotus californicus: California Leaf-nosed Bat
  Vespertilionidae: Vesper bats
    Antrozous pallidus: Pallid Bat
    Corynorhinus townsendii: Townsend's Long-eared Bat
    Eptesicus fuscus: Big Brown Bat
    Euderma maculatum: Spotted Bat
    Lasiurus blossevillii: Western Red Bat
    Lasiurus cinereus: Hoary Bat
    Lasiurus xanthinus: Western Yellow Bat
    Myotis californicus: California Myotis
    Myotis ciliolabrum: Western Small-footed Myotis
    Myotis evotis: Long-eared Myotis
    Myotis lucifugus: Little Brown Myotis
    Myotis thysanodes: Fringed Myotis
    Myotis velifer: Cave Myotis
    Myotis volans: Long-legged Myotis
    Myotis yumanensis: Yuma Myotis
    Pipistrellus hesperus: Western Pipistrelle
Lagomorpha: Pikas and rabbits
  Leporidae: Rabbits and hares
    Lepus californicus: Black-tailed Jackrabbit
    Sylvilagus audubonii: Audubon's Cottontail
    Sylvilagus bachmani: Brush Rabbit

(*Continued*)

TABLE 11.1 (*Continued*)
Mammals of Southern California

Rodentia: Rodents
- Sciuridae: Squirrels and Chipmunks
- Ground squirrels
  - *Ammospermophilus leucurus*: White-tailed Antelope Ground Squirrel
  - *Ammospermophilus nelsoni*: Nelson's Antelope Ground Squirrel
  - *Spermophilus beecheyi*: California Ground Squirrel
  - *Spermophilus lateralis*: Golden-mantled Ground Squirrel
  - *Spermophilus mohavensis*: Mojave Ground Squirrel
  - *Spermophilus tereticaudus*: Round-tailed Ground Squirrel
  - *Spermophilus variegatus*: Rock Squirrel
- Tree squirrels
  - *Sciurus carolinensis*: Eastern Gray Squirrel (Int.)
  - *Sciurus griseus*: Western Gray Squirrel
  - *Sciurus niger*: Eastern Fox Squirrel (Int.)
- Flying squirrels
  - *Glaucomys sabrinus*: Northern Flying Squirrel
- Chipmunks
  - *Tamias merriami*: Merriam's Chipmunk
  - *Tamias obscurus*: California Chipmunk
  - *Tamias panamintinus*: Panamint Chipmunk
  - *Tamias speciosus*: Lodgepole Chipmunk
- Geomyidae: Pocket gophers
  - *Thomomys bottae*: Botta's Pocket Gopher
- Heteromyidae: Kangaroo rats and pocket mice
- Kangaroo Rats (Dipodomyinae)
  - *Dipodomys agilis*: Agile Kangaroo Rat
  - *Dipodomys deserti*: Desert Kangaroo Rat
  - *Dipodomys heermanni*: Heerman's Kangaroo Rat
  - *Dipodomys ingens*: Giant Kangaroo Rat
  - *Dipodomys merriami*: Merriam's Kangaroo Rat
  - *Dipodomys microps*: Chisel-toothed Kangaroo Rat
  - *Dipodomys panamintinus*: Panamint Kangaroo Rat
  - *Dipodomys simulans*: Dulzura Kangaroo Rat
  - *Dipodomys stephensi*: Stephen's Kangaroo Rat
- Pocket mice (Perognathinae)
  - Spiny pocket mice
    - *Chaetodipus californicus*: California Pocket Mouse
    - *Chaetodipus fallax*: San Diego Pocket Mouse
    - *Chaetodipus spinatus*: Spiny Pocket Mouse
  - Smooth pocket mice
    - *Chaetodipus baileyi*: Bailey's Pocket Mouse
    - *Chaetodipus formosus*: Long-tailed Pocket Mouse

(*Continued*)

TABLE 11.1 (*Continued*)
Mammals of Southern California

Chaetodipus penicillatus: Desert Pocket Mouse
Perognathus alticolus: White-eared Pocket Mouse
Perognathus longimembris: Little Pocket Mouse
Cricitidae: Wood rats, Deermice, Voles, and Allies
Wood rats
Neotoma albigula: White-throated Wood Rat
Neotoma macrotus: Big-eared Wood Rat
Neotoma lepida: Desert Wood Rat
Grasshopper mice
Onychomys torridus: Southern Grasshopper Mouse
Deermice
Peromyscus boylii: Brush Deermouse
Peromyscus californicus: California Deermouse
Peromyscus crinitus: Canyon Deermouse
Peromyscus eremicus: Cactus Deermouse
Peromyscus fraterculus: Northern Baja Deermouse
Peromyscus maniculatus: North America Deermouse
Peromyscus truei: Piñon Deermouse
Harvest mice
Reithrodontomys megalotis: Western Harvest Mouse
Voles and Muskrats (Arvicolinae)
Microtus californicus: California Vole
Microtus longicaudus: Long-tailed Vole
Ondatra zibethicus: Muskrat
Cotton rats
Sigmodon hispidus: Hispid Cotton Rat
Old world rats and mice (Int.)
Mus musculus: House Mouse
Rattus norvegicus: Brown or Norway Rat
Rattus rattus: Black or Roof Rat
Erethizontidae: Porcupine
Erethizon dorsatum: Porcupine
Castoridae: Beavers
Castor canadensis: American Beaver
Carnivora: Carnivores
Canidae: Dogs, foxes, and Allies
Canis latrans: Coyote
Urocyon cinereoargenteus: Gray Fox
Urocyon littoralis: Island Gray Fox
Vulpes macrotus: Kit Fox
Felidae: Cats
Felis catus: Feral house cat
Lynx rufus: Bobcat
Puma concolor: Mountain Lion (Puma)

(*Continued*)

TABLE 11.1 (*Continued*)
Mammals of Southern California

- Mustelidae: Weasels and allies
  - *Enhydra lutris*: Sea Otter
  - *Mephitis mephitis*: Stripped Skunk
  - *Mustela frenata*: Long-tailed Weasel
  - *Spilogale gracilis*: Western Spotted Skunk
  - *Taxidea taxus*: Badger
- Procyonidae: Raccoons and Ringtails
  - *Bassariscus astutus*: Ringtail
  - *Procyon lotor*: Northern Raccoon
- Ursidae: Bears
  - *Ursus americanus*: Black bear
- Otariidae: Sea lions
  - *Arctocephalus townsendi*: Guadalupe Fur Seal
  - *Callorhinus ursinus*: Northern Fur Seal
  - *Eumetopias jubatus*: Steller's Sea Lion
  - *Zalophus californianus*: California Sea Lion
- Phocidae: Seals
  - *Mirounga angustirostris*: Northern Elephant Seal
  - *Phoca vitulina*: Harbor Seal

Perissodactyla: Odd-toed ungulates
- Equidae: Horses
  - *Equus asinus*: Feral burro (Int.)
  - *Equus caballus*: Feral horse (Int.)

Artiodactyla: Even-toed ungulates
- Bovidae: Sheep and goats
  - *Ovis canadensis californiana*: Bighorn Sheep
- Cervidae: Deer
  - *Odocoileus hemionus*: Mule Deer
- Suidae: Pigs
  - *Sus scrofa*: Wild boar and Feral hog (Int.)

Cetacea: Whales, dolphins, and porpoises
- Mysticeti: Baleen whales
  - *Balaenoptera acutorostrata*: Northern Minke Whale
  - *Balaenoptera borealis*: Sei Whale
  - *Balaenoptera musculus*: Blue Whale
  - *Balaenoptera physalis*: Fin Whale
  - *Eschrichtius robustus*: Grey Whale
  - *Eubalaena glacialis*: Northern Right Whale
  - *Megaptera novaeangliae*: Humpback Whale
- Odontoceti: Toothed whales
  - *Delphinus capensis*: Long-beaked Common Dolphin
  - *Delphinus delphis*: Short-beaked Common Dolphin
  - *Globicephalus maccrorhynchus*: Short-finned Pilot Whale

(*Continued*)

TABLE 11.1 (*Continued*)
Mammals of Southern California

Grampus griseus: Risso's Dolphin
Lagenorhynchus obliquidens: Pacific White-sided Dolphin
Orcinus orca: Killer Whale
Phocoenoides dalli: Dall's Porpoise
Physeter catodon: Sperm Whale
Pseudorca crassidens: False Killer Whale
Tursiops truncatus: Bottle-nosed Dolphin
Historically extirpated species
Carnivora
Canis lupus: Grey Wolf
Panthera onca: Jaguar
Ursus arctos: Grizzly Bear
Artiodactyla
Antilocapra americana: Pronghorn

11.6 MARSUPIALS

11.6.1 Didelphidae

The Virginia opossum (*Didelphis virginiana*) is indigenous from Argentina to the eastern half of the United States. It was first introduced in California around the Bay Area in 1910 and is now found throughout the Pacific states, generally below 3,000 ft in areas that are neither too dry nor severely cold. It is fairly common in Southern California lowlands. This scruffy, cat-sized marsupial has large canine teeth but lacks the chisel-like incisors of rodents. The long thick tail is prehensile and the hind foot has an opposable "thumb." The opossum is an omnivorous scavenger fairly common in agricultural regions and has adapted to some urban areas as well. It is harmless but can be a nuisance when it digs around houses and foundations. It is known for its catatonic behavior ("playing possum") when confronted with danger, a defensive response that serves it well.

The name *Didelphis* means "two wombs," a reference to the complete separation of the uterus into separate uterine horns. Correspondingly, the penis is divided at the apex. Like all marsupials, young are born after a brief gestation (ca. 12–14 days in this case) and the embryo-like newborns complete their development in the female's pouch.

11.7 LIPOTYPHLA—SHREWS AND MOLES

Until recently, these small mammals along with several Old World relatives were classified in the Order Insectivora, a taxon now abandoned because of the phylogenetic realignment of some African members. Shrews are the smallest mammals in North America; many weigh no more than a couple of grams and are at the structural and physiological limit for how small a mammal can be. Shrews in our area are around 4 in. (100 mm) including the tail. Moles are slightly larger. Shrews and moles have long, tapered snouts and fine, keenly pointed teeth. The sharp cutting molars easily penetrate the exoskeletons of their insect prey. Shrews in particular are fearless little beasts that will attack animals much bigger than themselves and devour them on the spot. Their primary diet consists of arthropods, but they will also eat mice. They differ from moles by the presence of small but clearly visible eyes and evident external ears (pinnae). The teeth of some species are tainted with a reddish pigment. The eye opening is minute or grown over by skin. Shrews favor damp undergrowth or stream sides, whereas moles live in underground tunnels.

11.7.1 Soricidae—Shrews

Of the 14 species of shrews found in California, four occur in our area. All but one species belong to the widespread genus *Sorex*, which are known as the long-tailed shrews. The *Ornate Shrew* (*S. ornatus*) is common (if such can be said of these secretive little animals) in Southern California west of the deserts. It occurs along most of the Coast Ranges and much of the Central Valley south into northern Baja. A small population also exists on Santa Catalina Island. The Montane Shrew (*S. monticolus*) ranges from Northern Mexico north through the Rocky Mountains all the way to Alaska. In California, it is mainly a denizen of the Sierra Nevada, but outlying populations occur in the San Bernardino and San Gabriel Mountains. Trowbridge's Shrew (*S. trowbridgii*) inhabits coniferous and mixed forests along the Pacific coast from Washington state south through the California Coast Ranges to Santa Barbara County.

The Desert Shrew (*Notiosorex crawfordi*) is distinct from our other shrews by its short tail, and for a shrew, fairly large ears. It is an arid land species as its common name implies. This shrew is capable of meeting its water requirements through its diet of insects and other arthropods. The Desert Shrew occurs in coastal sage scrub and chaparral in Southern California as well as both of our deserts. The species ranges east to western Oklahoma and northern parts of Mexico.

Most shrews have an unusual life cycle, more like that of an annual flowering plant. Young are born in the spring or early summer but do not reach sexual maturity until late the next winter. Usually there is only a single litter. After the young are weaned, the adults die so that by the end of the summer the population is composed of an entirely new generation.

11.7.2 Talpidae—Moles

Moles occur across the Northern Hemisphere and spend practically their entire life underground. And are they well built for it. The forelimbs are stout and powerful. Hands are broad and look like spades. The hind limbs and pelvic girdle are designed to anchor the body as hands and forelimbs push soil up and forward.

Rather than lying flat, the fur sticks straight out so the animal is not hindered while moving either forward or backward in the tunnel. External ears are absent, and eyes are covered over to keep out soil particles. Moles feed mostly on arthropods and earthworms and prefer loose sandy soils. They build mounds like gophers, but with no evidence of a plugged hole, and the soil mound itself is clumpy. They will also construct horizontal tunnels near the surface.

Four species of talpid moles occur along the Pacific coast, but only one of these is found in Southern California, the Broad-footed Mole (*Scapanus latimanus*). This species inhabits valleys and mountain meadows from south-central Oregon south throughout much of California and into northern Baja.

11.8 CHIROPTERA—BATS

Bats occur worldwide (except polar regions) and at around 1300 species is the second-most diverse order of mammals next to rodents. One group, the megachiropterans (Pteropidae), is confined to the Old World tropics and islands of the southwest Pacific. Informally called "flying foxes," they are a fruit-eating group of mammals encompassing some 166 species. As the name implies these are large bats—some have wing spans more than 5 ft (1.5 m)—and are visually oriented. All other bats are microchiropterans, the small ones we typically think of that account for 82% of the order. From here on "bats" refer to microchiropterans.

Some are solitary, whereas other species roost in colonies. Most feed on insects although some dine on fruit and nectar. Bats have figured in folklore in numerous ways, as omens of malevolence (vampire movies) or as symbols of decrepitude ("she has bats in her belfry"). Then there is the view that they carry blood-sucking bugs that can be transferred to humans. Not true. Bats do have ectoparasites of their own, but they are not the kinds that are interested in humans. In this respect, bats are entirely harmless, and to the contrary, they perform an invaluable service to humankind by the astonishing number of insects they consume. Bats are known to carry rabies, although they are not the only mammalian vector of the virus.

The earliest bats are known from about 50 million years ago with their unique ability for powered flight already evident. The wing is constructed from digits two to five; the thumb is a tiny vestige and does not participate. A thin membrane covering the bony scaffolding extends to the legs and tail as the *interfemoral membrane* where it forms a basket called the *uropatagium*. In most bats, a small bone called the *calcar* projects from the back of the foot and keeps the interfemoral membrane deployed when the bat is flying. The hind limbs are rotated in such a way that the knees bend backward instead of forward. This allows the uropatagium to hold a captured insect so it can be consumed in flight.

Bats are by and large nocturnal even though their eyes, although capable of seeing just fine, are small and of little use in distance orientation. Still, bats can avoid objects in total darkness, a fact that has been known since the eighteenth century. In the 1950s we learned that bats would, however, fly into objects if their ears were covered. This discovery led to the realization that bats depend on echolocation for collision avoidance and for locating insect prey. They do it by producing ultrasound (frequencies above 20 kHz) in the larynx that is transmitted through the mouth or nose and focused by the complex folds of skin on the face. The sound waves bounce off objects that in turn are captured by the ears. The nervous system interprets the returning signal with remarkable speed and precision, filtering out random noise while focusing on the relevant stimulus.

Bats are most likely to be seen at dusk when their coarse fluttering flight is recognizable against a soft pastel sky in summer evenings. Many bats begin their sorties by first skimming the surface of a pond or lake to take a drink of water. After foraging, they return to their night roosts to sleep. Some will make another run at dawn. An insectivorous diet means that physiological adjustments must be made during winter months when insect prey is scarce. Some species avoid the shortage by migrating to southern latitudes. Others stay put and instead slip into periods of dormancy.

Bats have few natural predators. Owls prey on them occasionally, and others may fall victim to snakes that lurk near cave entrances waiting for them to emerge or that seek them out directly as they hang upside down sleeping. The California Lyre Snake is pretty adept at this tactic. West Indian boid snakes (*Epicrates* spp.) that inhabit the cave-riddled and brutally unforgiving terrain of limestone forests also employ this strategy.

In the past decade, a more vicious threat has appeared in North America in the form of a lethal fungus (*Pseudogymnoascus destructans*) that has decimated bat populations in the eastern half of the United States. The fungus infects bats during hibernation by forming a ring of white fuzz around the snout, which has given it the name "white-nose syndrome." The fungus's affect seems to be energy depletion and dehydration, and it can cause some bats to leave their shelter when it is too cold to do so. The mortality has been staggering since the syndrome was discovered in 2008. An estimated 6 million bats have been wiped out in 28 states and 5 Canadian provinces. And it doesn't appear to be ebbing. In March 2016, white-nose syndrome turned up in Washington state. This plague is eerily familiar—recall our discussion of the pathogenic chytrid fungus that is collapsing amphibian populations worldwide.

Forty-six species of chiropterans occur across the United States and Canada. Southern California boasts a rich bat fauna itself, harboring 22 of the 25 species that inhabit the Pacific States (Table 11.1).

11.8.1 Molossidae—Free-Tailed Bats

Species in this family have one-half or more of the tail extending beyond the interfemoral membrane, hence "free-tail." In other bats, the tail projects little if at all beyond the edge of the membrane. Large ears are also a prominent feature. Molossids are mostly a family of tropical bats, and four of the five species reaching the United States include Southern California in their range. Those in our area typically breed in late winter or early spring and a single young is born in June or July. Nursery colonies are known in at least one species, the Mexican Free-tailed Bat. A lactating female allows any young to nurse from her.

The Mastiff or Greater Bonneted Bat (*Eumops perotis*) is the largest bat in North America at a total length of 6–7½ in. (160–190 mm). They occupy caves and building usually in open areas near tall cliffs and forage for strong-flying insects at fairly high altitudes (600–700 m). The Mastiff Bat is an inhabitant of the Southwest from Southern California to east Texas, and then discontinuously through Mexico south to Argentina. Mating takes place in early spring and one to two offspring are born from late June into early September.

The Mexican (or Brazilian) Free-tailed Bat (*Tadarida brasiliensis*) is the most common member of the family and the most common free-tailed bat in California. It occurs coast to coast across the southern United States and through the West Indies to northern South America. This species is known to gather in enormous colonies around buildings, cave, and mine shafts. These bats breed in late winter and gestation lasts around 100 days with birth taking place in nursery colonies in late June and July.

The Pocketed Free-tailed Bat (*Nyctinomops femorosaccus*) takes its name from the hollows or "pockets" present in the interfemoral membrane. It is otherwise similar in appearance to the Mexican Free-tail Bat. This bat is known throughout Southern California but is most common in the Colorado Desert. It otherwise ranges through southern Arizona to Texas and south to Mexico including Baja California. One offspring is born in late June or early July; nursing may continue into September. Another species of *Nyctinomops* is the Big Free-tailed Bat (*N. macrotis*). This one has its large ears extending beyond the tip of the snout. It is an uncommon resident of piñon-juniper habitats around cliffs and ranges east to Kansas and south into South America. Breeding activity is like that of the Pocketed Free-tailed Bat.

11.8.2 Phyllostomidae—Leaf-Nosed Bats

Phyllostomids (formerly Phyllostomatidae) are a diverse assemblage of mostly Middle and South American bats but with two species represented in Southern California. The family name derives from the projection of skin on the snout—a "leaf nose." Some phyllostomids like the Mexican Long-tongued Bat (*Choeronycteris mexicana*) feed on fruit and nectar. This is the only such species in California to do so. The long snout and protrusile tongue are adaptations for extracting nectar and pollen from night-blooming plants. In California, the Mexican Long-tongued Bat only occurs in southern San Diego and Imperial Counties. It ranges across southern Arizona and south through Mexico, including Baja. A single young is born in the spring.

The California Leaf-nosed Bat (*Macrotus californicus*) is an insectivorous species restricted to drier regions of Southern California. It ranges from southern Nevada through western Arizona and most of Mexico including Baja. It is also found in the West Indies. This species has particularly long ears. Reproduction in this species is unusual for North American bats. Mating and fertilization occurs in autumn. Embryonic development proceeds slowly through the winter then accelerates in the spring resulting in the birth of a single young about 8 months after mating.

11.8.3 Vespertilionidae—Vesper or Evening Bats

This is the largest family of bats in the Northern Hemisphere and accounts for 15 species in Southern California. The species differ from one another in skeletal details, dentition, dimensions of the ears and body proportions, and habits. Most all are insectivorous. Vesper bats in our region are arranged into three groups: the giant-eared bats (*Antrozous*, *Corynorhinus*, and *Euderma*), lasiurine bats (*Lasiurus*), and those commonly referred to as "little brown bats" (*Myotis*, *Eptesicus*, and *Pipistrellus*).

In most species, mating takes place in August or September, but fertilization is delayed; sperm is stored in the upper uterus or oviduct. Ovulation occurs in the spring and two or three offspring are born in June. Generally, bats roost upside down. While giving birth, these bats face head up and the baby is dropped into the interfemoral basket. The infant bat clings to the mother while she forages. Offspring are capable of flight in about 3 weeks.

The Pallid Bat (*Antrozous pallidus*) is a medium-sized sandy or buff-colored giant-eared species found throughout most of California and much of the western United States. It ranges into Mexico and western Canada as well. Its large ears are separated at the base. Townsend's Long-eared Bat (*Corynorhinus* [formerly *Plecotus*] *townsendii*) is similar to the Pallid Bat. Another species with giant ears, it is a (mostly) cave-dwelling species of desert scrublands and piñon-juniper associations throughout much of the intermountain west, all of California (including the Channel Islands) as well as Mexico and western Canada. The Spotted Bat (*Euderma maculautm*) is the third of our giant-eared species. The three large white spots on the back our unmistakable. This is another cave-dwelling species of the intermountain west but it is uncommon and rare in Southern California's montane open coniferous forests and in deserts.

Eight species of lasiurine bats occur across the United States. Most are migratory and therefore more likely to be encountered in the fall and spring. They roost in trees and tend to be solitary. Lasiurines are colorful bats sometimes referred to as the "hairy-tailed bats" because the upper surface of the interfemoral membrane is covered by fur. The undersurface of the wing is also covered in hair along the axes of the digital bones. Ears are small. Three species occur in Southern California.

The Western Red Bat (*Lasiurus blossevillii*) was recently removed from the synonymy of the (Eastern) Red Bat (*L. borealis*). Both bats are indeed reddish-orange. The Western Red Bat is found around deciduous trees statewide in California, north to British Columbia, and south through Baja and mainland Mexico. The Western Yellow Bat (*L. xanthinus*) has, like its cousin the Western Red Bat, recently been split from its relative, the Eastern Yellow Bat (*L. ega*). This is a desert species particularly fond of palms. Its distribution includes southern Arizona to South America. The Hoary Bat (*L. cinereus*) takes its name from the dark fur frosted with white tips, rendering it distinct from any other bat in California. This is a fairly common species in wooded areas of Southern California, but it does not assemble in aggregations. The Hoary Bat ranges widely throughout North America as far north as Hudson Bay in the summer. It occurs south to Argentina and Chile. Populations of Hoary Bats are also known from the Hawaiian Islands.

Pipistrellus (one species), *Eptesicus* (one species), and *Myotis* (eight species) comprise Southern California's third group of vesper bats. These are mostly small brownish and often difficult to identify.

The Western Pipistrelle (*Pipistrellus hesperus*) is Southern California's smallest bat, about 3.25 in. (ca. 85 mm). Its black face is distinctive. Solitary or in small colonies, the Western Pipistrelle is found in numerous situations in the West, although it has a preference for deserts, grasslands, and open country near cliffs. Look for it in the early evening and notice its soft, bouncy flight.

The *Big Brown Bat* (*Eptesicus fuscus*) is one of our most common species, as it is throughout North America and into the Greater Antilles. It is considerably larger than the other brown bats in this group. It can reach nearly 5 in. (138 mm). The wing and interfemoral membranes are dark and hairless. Big Brown Bats often roost in buildings and other man-made structures.

The small bats of the genus *Myotis* are the most common chiropterans in the United States. Typical Myotis are brownish with pale underparts and are among the most difficult mammals to identify along with field mice. Distinguishing characteristics need to be considered in combination. Not surprisingly, taxonomic issues abound at the species level and many carry one or more synonyms. Thus, the nomenclature used here could change. Collectively *Myotis* have no standard common name so most species are simply referred to by the generic name along with a descriptive modifier, for example California Myotis. Some folks employ the vernacular "little brown bats" for the inclusive species but that reference traditionally has been restricted to *M. lucifugus* (the Little Brown Myotis, below).

Except for *Pipistrellus*, Myotis are smaller than the other bats in our area (total length to 3.75 in. [95 mm]). They spend the day in caves, crevices, and similarly dark refuges, as well as in trees. They emerge at dusk to forage for insects. Some are colonial, others not.

Most all *Myotis* follow a reproductive pattern similar to the vesper bats in having delayed fertilization. Mating occurs in September or October and ovulation takes place in the spring. A single offspring is born about 60 days later. These bats roost head-down except while giving birth. Then the body and head are upright so that the baby bat drops into a basket formed by the interfemoral membrane. Young Myotis are able to fly in about 3 weeks, and they are weaned shortly thereafter.

Prominent long ears distinguish two of our species, the Long-eared Myotis (*M. evotis*) and the Fringed Myotis (*M. thysanodes*). The Long-eared Myotis tends to forage after dark flying only about 4–6 ft off the ground. This species occurs widely over the western United States and southern Canada, but it avoids the deserts.

The Fringed Myotis has a delicate fringe of fine hair along the posterior margin of the interfemoral membrane. It is also a western species but includes the Mojave and Sonoran life zones in its range.

The California Myotis (*M. californicus*) is a small, short-eared bat of the western United States, Canada, and Mexico, including Baja. Like the Long-eared Myotis, it forages close to the ground. It generally inhabits lower elevation woodlands and deserts. The Western Small-footed Myotis (*M. ciliolabrum*) is similar to *M. californicus*, essentially differing in subtleties of certain proportions and pelage. This is a mostly solitary species found throughout California except for coastal coniferous woodlands. It also occurs throughout the intermountain region and south into mainland Mexico. The Western Small-footed Myotis speaks to some of the headaches that accompany *Myotis* taxonomy; at one time or another it has been assigned to *M. leibii*, *M. subulatus*, and *M. yumanensis* (below). The Long-legged Myotis (*M. volans*) has also been called the Hairy-winged Myotis because of its denser fur on the underside of the wing between the knee and elbow. Tibial length is only slightly greater than that of other Myotis. It ranges throughout the western half of the continent including all of California and Baja. The Yuma Myotis (*M. yumanensis*) is another species that ranges throughout the western States, British Columbia, and peninsular and mainland Mexico. It is most often found in canyon bottoms and low valleys especially near water. The Yuma Myotis is a colonial species of caves and abandoned buildings. The Cave Myotis (*M. velifer*) is one of the largest North American species of *Myotis* (to 4 in. [ca. 100 mm], 9–14 g). This is a colonial bat of low areas in the Southwest. In California it is restricted to caves and crevices along the Colorado River. The Little Brown Myotis (*M. lucifugus*) is quintessentially nondescript. Its geographic range encompasses most of North America including California except for the southwestern portion of the state; in our area it is confined to the eastern parts of the deserts. The Little Brown Myotis is and oft-studied subject in bat ecology.

11.9 LAGOMORPHA—RABBITS, HARES, AND PIKAS

Lagomorphs are superficially rodent-like, and these two clades are considered sister taxa, collectively paced in the group *Glires*. Two pairs of upper incisors will distinguish lagomorphs from rodents, among other features. Small incisors sit behind the larger ones in front. Rabbits (*Sylvilagus*) and Hares (*Lepus*) are familiar to us all by their enlarged hind limbs and prominent ears. Pikas much less so. Pikas are best described as cute little balls of fur. About 7 in. (180 mm) in length, they have small rounded ears and the hind limbs are only slightly longer than the forelimbs. Picas have been erroneously called conies, which is actually a species of Old World Hyracoidea. The American Pika (*Ochotona princeps*) occurs throughout the intermountain West but does not reach Southern California. Elsewhere in the state, it inhabits talus slopes and lava beds above around 5,300 ft (1,700 m) along the east side of the Cascades and Sierra Nevada, south to Tulare County.

11.9.1 Leporidae—Rabbits and Hares

Rabbits include the cottontails (*Sylvilagus* spp.) and the Pygmy Rabbit (*Brachylagus idahoensis*). Of the nine species of *Sylvilagus* in the United States, three occur in California, two of which are found in our area. [The European or Domestic Rabbit (*Oryctolagus cuniculus*) was introduced to the Farallon Islands in the mid-nineteenth century.] The Audubon's (or Desert) Cottontail (*S. audubonii*, Figure 11.2) of the western United States occurs throughout the lower two-thirds of our state into Baja and mainland Mexico. This is the default rabbit of Southern California where it is common in all manner of brushy areas of chaparral and deserts. Largely diurnal and gregarious, cottontails feed on numerous types of green plants, both shoots and leaves. This cottontail breeds from January to June and probably has two broods per year of two to six young. They may breed year around when their food supply is high quality and abundant like in irrigated areas.

Whereas Audubon's Cottontail is almost colloquial, the Brush Rabbit (*S. bachmani*) is far less familiar because of its nocturnal, largely solitary and secretive ways. They are known to climb into shrubs and to thump their foot when alarmed.

This species indeed favors dense brush. It ranges west of the Cascades and Sierra Nevada from Oregon south through Baja. Compared with Audubon's Cottontail, the Brush Rabbit is slightly smaller, has smaller less pointed ears, lacks the russet color on the nape, and is darker overall.

The four species of hares in the lower United States are known commonly as jackrabbits. Their enormous ears and long legs are unmistakable. Three other hares are restricted to boreal and arctic habitats: the Snowshoe Hare

FIGURE 11.2 An Audubon's Cottontail (*Sylvilagus audubonii*) leaps over a charging opponent during an apparent dispute. Cottontails are gregarious and common throughout Southern California in all types of brushy habitat.

(*Lepus americanus*), Alaskan Hare (*L. othus*), and Arctic Hare (*L. arcticus*). Unlike cottontails, hares are precocial at birth, born with their eyes open, ample hair, and are able to walk almost immediately. The aptly named Black-tailed Jackrabbit (*L. californicus*, Figure 11.3a and b) ranges over much of the western United States, Mexico, and Baja California in many biotic communities with open habitat. In Southern California, it is most likely to be encountered in rangelands and deserts. They breed almost any time of year with litter sizes ranging from three to four and as many as seven offspring.

FIGURE 11.3 (a,b) The enormous ears of the Black-tailed Jackrabbit (*Lepus californicus*) help dissipate heat. Powerful hind legs can propel them to 40 mi/h. The Black-tailed Jackrabbit is widely distributed in open country of the western United States, and it is the only jackrabbit species in Southern California. (Courtesy of Jason Wallace.)

11.10 RODENTIA—RODENTS

Rodents are gnawers equipped with a pair of enlarged, continuously growing upper and lower incisors covered in orange enamel. Canine teeth are absent so there is a gap between the incisors and the cheek teeth posteriorly (premolars and molars). Rodents include such familiar mammals as squirrels, mice, voles, gophers, and beavers. Rodents for the most part are herbivores though some are omnivorous and will consume a variety of food types. Many possess a suite of morphological and physiological adaptations for living in extreme conditions. For example, periods of dormancy (hibernation in the winter or aestivation in the summer) are characteristic of some species when food or water supplies are in short supply or when temperature stress is excessive. Desert species are remarkable at conserving water metabolically by excreting concentrated urine and dry feces. Many rodents live in burrows of their own construction or those abandoned by other animals. Some take refuge in rotting logs, beneath rock outcrops, or in crevices. Most rodents are excellent climbers whether or not they spend much time in trees.

Rodents comprise the largest group of mammals; worldwide there are more than 1,800 species, and in Southern California they account for almost half (54 species) of the land mammals found here. Several species are disease vectors and others are simply destructive pests. In turn, rodents themselves are a significant component in the diet of numerous vertebrate predators.

11.10.1 Sciuridae—Squirrels and Chipmunks

Squirrels and chipmunks are among the more familiar of our native mammals. Most of them are diurnal and easily recognizable. Squirrels can be small or large. They lack prominent facial markings, are often unicolor, and most have bushy tails. Chipmunks are small, delicate, and conspicuously striped over the head and body. Many sciurids vocalize by emitting sharp peeps or chips.

11.10.1.1 Ground Squirrels

Most ground squirrels are unicolor, although several are spotted, and some are striped. Usually they are found in open areas adorned with boulders and large stumps. In general they do not climb trees. Some are colonial and can be quite numerous over a small area. Seven species include Southern California in their distributional range.

At one time most species in our area were included in the genus *Citellus*, which has since been replaced by *Spermophilus* or *Ammospermophilus*. The most common, and the only species west of the mountains, is the comparatively large California Ground Squirrel (*S. beecheyi*), sometimes called Beechey's Ground Squirrel, which inhabits chaparral, fields, roadsides, and successional habitats. It has a brown, indistinctly spotted back and a grey nape (Figure 11.4). The California Ground Squirrel occurs throughout the state except for the Great Basin and the deserts. A colonial species, it can become a pest when it burrows and

FIGURE 11.4 The California ground squirrel (*Spermophilus beecheyi*) is a familiar site in open coastal sage scrub, chaparral, and successional habitats often around rocks and boulders. This colonial species constructs extensive burrow systems that can become a hazard around retaining walls and building foundations.

tunnels in unwanted places. They climb fence posts and boulders, and I have seen them up in manzanita and scrub oak as well. Their alarm call is a high-pitched peep. These ground squirrels are fairly omnivorous and will even consume carrion. The California Ground Squirrel has one litter annually of from 3 to 10 offspring, although the timing varies by geographic locality.

All but one of Southern California's other ground squirrels live north or east of the Transverse and Peninsular Ranges. The Rock Squirrel (*S. variegatus*) resembles the California Ground Squirrel, but the ranges of the two do not overlap. This species is common in canyons and around cliffs and hillsides of the intermountain West well into Mexico. However, in California, it is restricted to rocky slopes of the Providence Mountains in the eastern Mojave Desert. Rock Squirrels have one or two litters per year of five to seven offspring.

The small, unmarked sandy-colored Round-tailed Ground Squirrel (*Spermophilus tereticaudus*) in California occurs throughout the southern Great Basin and most of the Mojave and Colorado Deserts (Figure 11.5). Its range extends to western Arizona south into Sonora, Mexico. The common name derives from its sparsely haired tail. The Round-tailed Ground Squirrel has a litter of 6–12 offspring usually born in June. The similarly appearing Mojave Ground Squirrel (*S. mohavensis*) is a shy and uncommon species confined to the northern and eastern parts of the Mojave Desert. Litters of 4–6 offspring are born in the spring.

The White-tailed Antelope Ground Squirrel (*Ammospermophilus leucurus*) and Nelson's Antelope Ground Squirrel (*A. nelsoni*) are among two of six species of North American ground squirrels with a bold white stripe on either side

FIGURE 11.5 The unicolored Round-tailed Ground Squirrel (*Spermophilus tereticaudus*) blends in perfectly with the coarse sandy soil it favors. The species is common throughout the Mojave and Colorado deserts. (Courtesy of Jason Wallace.)

of the body. In this respect, they resemble chipmunks. They all have the habit of carrying their tail up and forward over their back. The White-tailed Antelope Ground Squirrel is common among Great Basin sagebrush east of the Sierra Nevada and eastward through Nevada to western Colorado and New Mexico. In California, its range continues into the Mojave and Colorado Deserts and on into Baja. Five to eight offspring are born in May. Nelson's Antelope Ground Squirrel is found only in California's Central Valley and foothills of the Coast Range from around Merced south through western Kern County. It is the only ground squirrel in the area and barely reaches Southern California around the Tehachapi foothills west toward Maricopa. Breeding takes place in mid-January and 6–10 young are born about 60 days later.

The Golden-mantled Ground Squirrel (*Spermophilus lateralis*) is another striped species, but differs from the other two (*A. leucurus* and *A. nelson*) in that the lateral stripe is bordered by obvious dark edging. This species occupies coniferous forests and sagebrush from British Columbia south through the Rocky Mountains to Arizona and New Mexico. In California it inhabits the Sierra Nevada and in Southern California as a disjunct population in the San Bernardino Mountains where it is common around campgrounds and often mistaken for a chipmunk. A single litter of four to eight young are born about 30 days after hibernation.

11.10.1.2 Tree Squirrels

Tree squirrels are forest denizens with slender bodies, large bushy tails and erect ears. They are spirited climbers and leapers yet are also active on the ground where they forage for bulbs, sprouting seeds, leafy material, and insects. Tree squirrels do not hibernate. The Western Grey Squirrel (*Sciurus griseus*) is the only indigenous tree squirrel in California. It occurs from sea level to around 4,600 ft (ca. 1,500 m) and in all of the Pacific states down into Baja. In California they are most likely to be found in association with California Black Oak (*Quercus kelloggii*) and coniferous forests along the Coast Ranges and Sierra Nevada. In Southern California, they occur in the Transverse and Peninsular Ranges. They are absent from the Central Valley and the deserts. The Western Grey Squirrel can be destructive in walnut and almond orchards, yet in some parts of the state habitat degradation has led to population declines. They have one or two litters of two to six young annually.

The Eastern Fox Squirrel (*Sciurus niger*) and Eastern Grey Squirrel (*S. carolinensis*) are non-native transplants from the eastern United States. Both have become established in a number of California cities around parks and the like (the Western Grey Squirrel is less likely to be seen in urban settings). Of the two interlopers in Southern California, the Eastern Fox Squirrel has the firmer footing, especially around Los Angeles where it was introduced in 1904. Since then it has spread to numerous localities in the greater L.A. area and south into Orange and Ventura Counties. Fox squirrels also have thrived for decades in San Diego's Balboa Park.

11.10.1.3 Flying Squirrels

Flying squirrels (*Glaucomys* spp.) are also tree squirrels in that they are arboreal. Of course they do not fly like birds or bats but glide by launching themselves from tree trunks and branches. A membrane of skin and fur called a *patagium* stretches from the arms to the legs and serves as a wing when the limbs are extended, nearly tripling the surface area. A short cartilaginous rod protruding from the wrist helps to maintain wing shape and assist in steering. The broad flattened tail functions as a rudder and stabilizer. Gliding requires tall trees with unobstructed air space where glides of 100 ft (ca. 30 m) are possible.

The two North American species of *Glaucomys* are separated geographically—one in the east (*G. volans*) and another in the north and west. The Northern Flying Squirrel (*G. sabrinus*) inhabits boreal forests across North America. In the far West, its range extends from Alaska south through the Cascades, northern Coast Ranges, and the Sierra Nevada. In Southern California, disjunct populations occur in the San Bernardino and San Jacinto Mountains. The Northern Flying Squirrel is California's only nocturnal squirrel. It associates with conifers, especially firs (*Aibes*), and feeds on berries, nuts, insects, and birds eggs. They nest in the hollows of trees and produce on litter of two to five young born in May or June.

11.10.1.4 Chipmunks

These small squirrels are easily recognized by their diminutive size and conspicuous alternating dark and light stripes that extend from the snout to the rump. The long tail may be as much as 40% of the animal's total length. Chipmunks are essentially ground-dwellers that compile underground caches of food such as nuts and seeds. They also feed readily on insects and their larvae. Twenty-two different species occur in North America and 21 of those are in the West, where most of them live in mountainous terrain. Identifying individual species in the field is a challenge because they differ from one another only in subtle variations of pattern, color, and proportions. Genital bones (*bacula*) have proven useful, but of course those characteristics are useless in the field. Geographical distribution should be considered when attempting identification. Of the 13 species in California, 4 are found in our region.

Merriam's Chipmunk (*Tamias merriami*) is the most common species in Southern California, where it inhabits chaparral foothills and pine/oak woodlands up to around 7,000 ft (ca. 2,400 m) in the Coast, Transverse, and Peninsular Ranges. It also occurs along the western slopes of the Sierras. Merriam's Chipmunk bears a litter of three to five offspring born usually in April. The California (or Dusky) Chipmunk (*T. obscurus*) was at one time regarded as a synonym of *T. merriami*. This species is diagnosable and occurs as a series of disjunct populations in dry rocky habitats of the Transverse and Peninsular Ranges into Baja. In the north of its range, it is found along the dry northern slopes on the eastern end of the San Bernardino Mountains, then across the Morongo Valley to the Little San Bernardino Mountains. Eastward, it extends into the rocky arid ranges of Joshua Tree National Monument. San Gorgonio Pass separates the northern populations from those in the San Jacinto Mountains to the south. To the east, the Coachella Valley isolates populations in the Santa Rosa Mountains in the desert. From there, the distribution leapfrogs to the Sierra Juarez and Sierra San Pedro Martir in northern and Central Baja California. In the San Bernardino and San Jacinto Mountains, the species is sympatric, or nearly so, with Merriam's Chipmunk, although the zones of overlap are narrow. In the San Bernardino Mountains, *T. obscurus* is closely associated with single leaf piñon (*Pinus monophylla*) in rocky areas whereas Merriam's Chipmunk is most often found higher up in more mesic coniferous forest. Interestingly, in the San Jacinto Mountains, Merriam's Chipmunk occupies the transitional zone *below* the California Chipmunk in habitat that may include scrub oak and manzanita. Both species breed early in late January or February of broods of three to five are born in April.

The Lodgepole Chipmunk (*T. speciosus*) is associated with Lodgepole, Jeffrey, and Yellow Pines in the Sierra Nevada and Lake Tahoe area. Outlier localities in Southern California include the San Gabriel, San Bernardino, San Jacinto Mountains (where its current status is unclear) and Mt. Pinos along the Ventura/Kern county border. Three to six young are born in July.

Our fourth and smallest species, the Panamint Chipmunk (*T. panamintinus*) ranges from 4,500 to 7,500 ft (1,500–2,500 m) along the dry piñon and juniper slopes of the eastern Sierra Nevada. From there, it occurs discontinuously in the higher mountains of the Mojave, for example, in the Panamint (where the first specimens were collected) Inyo, Kingston, Clark, Granite, Providence, and New York mountain ranges. The thin rocky soil where the Panamint Chipmunk lives is generally unsuitable for burrowing, so it shelters in rocky crevices. Three to six offspring are born in June or July.

11.10.2 Geomyidae—Pocket Gophers

Notorious as they may be, pocket gophers are splendidly suited for life below ground as small digging machines around 7–10 in. (180–250 mm) total length. Their bodies are stocky and the hands and forelimbs are robust. Eyes and ear pinnae are reduced. They even use their incisors for digging; the lips enclose the front teeth in a manner that prevents dirt from crumbling into their mouth. Whereas shrews prefer soft, damp soils, gophers will burrow in almost any friable substrate, in some cases even clay. Like their relatives the heteromyids (below), pocket gophers have fur-lined cheek pouches that they stuff with food.

Pocket gophers are rarely seen. On the occasions when a glimpse is had, the view is of a head pushing dirt out of a hole. The best evidence of their presence is the mound piled outside a plugged entrance. Tunnel systems are elaborate and are constantly being modified with new side branches as dirt is pushed into old or abandoned chambers. Gophers live solitary lives and a single individual can do considerable damage to a home garden. Gophers socialize only when it is time to mate.

Geomyids range widely across North America, from central Mexico to the boreal coniferous forests. Eighteen species are allocated among 3 genera. *Thomomys* is the only one occurring in the western states. California supports five species, one of which occurs in our region, Botta's Pocket Gopher (*T. bottae*). This species is one of the most variable of all rodents in terms of pelage. Fur color corresponds with the soil so it could be brown, reddish brown, grayish black, tan, or yellowish-brown depending on the area inhabited. Botta's Pocket Gopher ranges from Oregon to Baja and south southeast to west Texas and northern Mexico. It occurs almost anywhere within its distribution as long as the soil can be penetrated. In California, they are absent from the highest elevations of the Cascades and Sierra Nevada but are otherwise statewide in valleys, meadows, montane slopes, and along desert arroyos. Breeding takes place from late winter into summer and can be nearly year around in irrigated areas. Up to 4 litters are possible with from 2–12 young.

11.10.3 Heteromyidae—Kangaroo Rats, Kangaroo Mice, and Pocket Mice

Kangaroo Rats (*Dipodomys* spp.) are unmistakable with their large heads, dark eyes, a conspicuous white stripe on the flank, a bicolored tufted tail and, as their name implies, enormous hind feet (Figure 11.6). Forelimbs are reduced and locomotion is by saltation (jumping). Leaps of 10 ft are possible. Kangaroo rats (both sexes) possess a seasonally enlarged sebaceous gland on the back of the neck that may function in molt or reproduction. Heteromyids have fur-lined cheek pouches for caching the seeds that they collect while foraging and that are later brought back to their shelters. Insects are also part of the diet, but these are usually consumed on the spot.

Identifying the Individual species is another matter. They all pretty much look the same, differing in size, the number of toes on the hind foot (four or five), and

FIGURE 11.6 Kangaroo rats (*Dipodomys* spp.) are unmistakable in their markings and body proportions. Identifying the individual species is another matter; they all resemble one another and look much like the Merriam's Kangaroo Rat (*D. merriami*) shown here. The nine species found in Southern California differ from each other in subtleties of size, color, tail proportions and habitat preference. (Courtesy of Jason Wallace.)

subtle distinctions of the tail and color pattern. That said, there is about a fourfold difference in body mass between the smallest and largest species, that is, about 50 to almost 200 g. Geographic distribution and habitat specifics will aid in their identification. Kangaroo Rats are found only in the western United States, and of the 17 species 15 of them include California in their range, 7 of which are endemic. Unfortunately several species of California *Dipodomys* are threatened or endangered because of habitat loss brought on by agriculture, cattle grazing, and urbanization.

Practically no place in Southern California is without at least one species—chaparral, coastal sage scrub, deserts, and almost any habitat with open, sparse vegetation, and loose or sandy soil. Desert species in particular are remarkable for their ability to obtain water from the food they eat and make the most of it with textbook-efficient kidneys that excrete urine with ultra-high salt concentrations. They remain in their burrows during the day and are active at night when ambient conditions are cooler and more humid. Countercurrent airflow in the nasal passages reduces water loss by cooling exhaled humid air, such that water condenses on the nasal mucosa and is reclaimed.

In Southern California, only Merriam's Kangaroo Rat and the Desert Kangaroo rat have four toes. All other species have five. Three species (*D. agilis*, *D. simulans*, and *D. stephensi*) are found (primarily) west of the deserts. Four others (*D. merriami*, *D. deserti*, *D. microps*, and *D. panamintinus*) are occupants of the deserts and surrounding mountains, and two species (*D. ingens* and

D. heermanni) barely extend into Southern California from the north. Breeding in these species can take place from late winter through the spring months and even into summer. Multiple litters are possible during wet years, and none during dry ones. Typical litter size is from three to five offspring, but there can be as few as one and as many as six.

The Agile Kangaroo Rat (*Dipodomys agilis*) is endemic to south central and Southern California, from San Bernardino, Riverside (north of San Gorgonio Pass and Moreno Valley), and Los Angeles Counties, then north through Ventura County into central San Luis Obispo County. It ranges eastward through the Tehachapi Mountains into the southern Sierra Nevada. The Agile Kangaroo Rat is primarily a resident of montane and coastal foothills where it favors chaparral slopes on into coniferous woodland. Some populations apparently have been extirpated, but the species overall is not under immediate threat.

The Dulzura Kangaroo Rat (*D. simulans*) replaces the Agile Kangaroo Rat south of the Los Angeles basin. Until about 20 years ago, the two were regarded as the same species (*D. agilis*). The Dulzura Kangaroo Rat is a species of the Peninsula Range. In Southern California it is found in coastal sage scrub, open chaparral, and grasslands to around 3,000 ft (900 m). Its range continues well into Baja California to around Magdalena Bay where it occupies desert habitats up into coniferous forest.

Stephen's Kangaroo Rat (*D. stephensi*) is a dead ringer for the Dulzura Kangaroo Rat and is another Southern California endemic. It prefers open grassland habitats and is restricted to the San Jacinto Valley and adjacent areas of Riverside County and southwest San Bernardino County. Isolated populations occur around Warner Springs and Lake Henshaw in northwest and north central San Diego County. The historical range of this species is thus quite small and as much as 60% of it has been destroyed. Stephen's Kangaroo Rat has long been a bane to agriculture and housing development because of the desirable land it occupies. Some sites allegedly have been intentionally plowed and poisoned. It is a federally endangered species.

Merriam's Kangaroo Rat (*D. merriami*, Figure 11.6) has one of the broader distributions among species of *Dipodomys*. It occupies a variety of soil types and habitats from Nevada to west Texas and northern Mexico, including Baja. It ranges throughout the southern half of California except for the coastal zones and western slopes. Merriam's Kangaroo Rat is a small species having a total length of 7.5–11 in. (195–280 mm). This is the common kangaroo rat of the Mojave and Colorado Deserts. It can often be seen bounding across roads at night, or foraging around campgrounds after dark unperturbed by flashlight beams. The larger (12–14 in. [300–375 mm]) Desert Kangaroo Rat (*D. deserti*) is a specialist of the hottest and driest regions. It ranges throughout the Mojave and Colorado Deserts and the contiguous sandy areas of Arizona and Mexico that surround the head of the Gulf of California. Within its range the Desert Kangaroo Rat occupies dunes or sandy areas deep enough to dig burrows; ideal are the mounds that accumulate beneath creosote bushes. They avoid places with easily shifting sand. Occasionally Desert Kangaroo Rats form temporary colonies of from 6 to 12 individuals, each occupying a separate burrow system. Where they coexist with Merriam's Kangaroo Rat, the Desert Kangaroo Rat forages for larger seeds. They also engage in the interesting behavior of kicking sand at an unfamiliar object to determine if it is alive or dangerous.

Two other desert species have five toes instead of four. The Panamint Kangaroo Rat (*Dipodomys panamintinus*) ranges along the eastern Sierra from near Bishop and adjacent parts of Nevada south through the western and central Mojave to the Cajon Pass. It occupies coarse sandy or gravelly places associated with Joshua Trees, piñon-juniper stands, and scattered creosote. A disjunct subspecies (*D. p. caudatus*) occurs in the Mojave National Preserve and adjacent southwest Nevada. The slightly smaller Chisel-toothed Kangaroo Rat (*D. microps*) is largely a Great Basin species of valleys grown to saltbush from southeast Oregon, Nevada, and western Utah to northwest Arizona. In California, its range extends into the Mojave Desert in areas of piñon and juniper. The common name derives from the flat-faced lower incisors as opposed to the more rounded profile characteristic of other species.

The Giant Kangaroo Rat (*D. ingens*) is the largest species of *Dipodomys* (up to 14 in. [350 mm] total length). It is restricted to fine soils in the southwest part of the San Joaquin Valley and adjacent Coast Ranges in Merced and Kern Counties, south to the Tehachapi Mountains, and points to the west. It barely reaches our area in northern Santa Barbara County in the Cuyama Valley. This species is considered endangered by both California and federal agencies. The natural habitat of the Giant Kangaroo Rat included perennial grasses and saltbush that have since been replaced by introduced annuals. Grazing by sheep and cattle plus urbanization and industry, such as petroleum, has so fragmented the historical range that less than 2% of it remains. Small populations soldier on in isolation of one another. Heermann's Kangaroo Rat (*D. heermanni*) is another primarily northern species. This is a medium-sized, darker kangaroo rat of the Central Valley and adjacent foothills below 3,000 ft from Sacramento west to the Bay Area and south to around Bakersfield. It reaches the coast at Monterey and again at Morro Bay where its range extends into our region around Point Conception. Unlike the Giant Kangaroo Rat, this species presently is faring well throughout most of its range except for the populations from around Morro Bay south (*D. h. morroensis*) that are endangered.

Kangaroo mice (*Microdipodops*) are miniature replicas of their larger cousins. They have five toes. Neither of the two species (*M. megacephalus* and *M. pallidus*) occurs in our region. Most of their distribution is centered in the sagebrush deserts of Nevada. In California they are restricted to the extreme eastern edge of the State in Inyo and Mono Counties.

Pocket mice (*Chaetodipus* spp. and *Perognathus* spp.) are about the same size as kangaroo mice, but the forelimbs and hind limbs are nearly equal in size. They have long feet but do not jump nearly as well, and their heads and snouts are narrower. Desert species share with kangaroo rats many of the same attributes for living in arid environments. Like kangaroo rats, they also have external fur-lined cheek pouches, and feed mainly on seeds and insects. Whereas kangaroo rats might forage considerable distances into open spaces, pocket mice tend to seek food beneath the cover of shrubs. Most undergo extended periods of dormancy. Pocket mice comprise a diverse assemblage of which 10 of the 20 North American species include California in their range. Two of those are endemic.

Both genera (*Chaetodipus, Perognathus*) occur in the Southern California region. The six species of *Chaetodipus* bear greater resemblance to kangaroo rats than do the two species of *Perognathus*. The latter are smaller and the tail is minimally tufted, if at all.

The individual species of *Chaetodipus* are fairly similar externally so telling one from another can be tricky. One useful feature is the presence of stiff guard hairs on the rump and back of some "spiny" species. All others are smooth.

11.10.3.1 Spiny Pocket Mice

The California Pocket Mouse (*Chaetodipus californicus*) has stiff light-colored hairs on the otherwise dark rump, and the ears are somewhat long for a pocket mouse. It inhabits areas of sagebrush in the San Joaquin Valley, and chaparral and coastal sage scrub along the western side of the Coast Ranges from south of San Francisco Bay down into Baja. Breeding takes place from April to June. Litter size is from two to five young. This species hibernates in the winter.

The San Diego Pocket Mouse (*Chaetodipus fallax*) is similar to the California Pocket Mouse only smaller. The back and rump are brown and spiny hairs are present. A tan to light orange band runs between the darker upper parts and the whitish under parts. This species is restricted to Southern California from San Bernardino County south into Baja. It inhabits the valleys, foothills, and mountains to around 6,000 ft (2,000 m) and also ranges into the western parts of the deserts where it often associates with yucca.

The Spiny Pocket Mouse (*Chaetodipus spinatus*) has white or silvery spines on the rump that often extend onto the shoulder. A small white spot is detectable below each ear (Figure 11.7). This species ranges from the east Mojave through the western and eastern sectors of the Colorado Desert south into Baja. It prefers boulder areas and rocky slopes with sparse vegetation. The Spiny Pocket Mouse breeds from April to July and produces a litter of two or three, and rarely five, offspring.

11.10.3.2 Smooth Pocket Mice

The Long-tailed Pocket Mouse (*Chaetodipus formosus*) inhabits bouldered areas, lava beds, and rocky slopes with desert and Great Basin scrub from the eastern Sierra across Nevada to Utah and northern Arizona. It continues through the Mojave and Colorado Deserts well through Baja on the eastern side of the peninsula. Breeding takes place in the spring with a litter of five or six young. This species can be active in winter in some localities.

Bailey's Pocket Mouse (*Chaetodipus baileyi*) is a medium-sized grey pocket mouse of rocky areas in the Colorado Desert, southern Arizona, western New Mexico, much of Baja, and western Sonora, Mexico. It often associates with mesquite. Breeding takes place from June to October with one or more litters of two to five young.

FIGURE 11.7 The Spiny Pocket Mouse (*Chaetodipus spinatus*) is restricted to Southern California and Baja. A desert species, it favors rocky areas and rough terrain. Notice the stiff silvery hairs on the rump. Pocket mice, like their relatives the kangaroo rats, have fur-lined cheek pouches in which they stash seeds they collect while foraging. (Courtesy of Jason Wallace.)

A rather nondescript species, the Desert Pocket Mouse (*Chaetodipus penicillatus*) occupies sandy substrates in both of our deserts. It occurs as well in southern Nevada, southern Arizona, northwest Mexico, and the northeastern portion of the Baja peninsula.

The Little Pocket Mouse (*Perognathus longimembris*) occupies a variety of habitats from fine sand to coastal sage scrub in the southern half of the state. It is the smallest Pocket Mouse in the region at a total length of about 5½ in. (138 mm). It is found in dry areas from southeastern Oregon through Nevada and parts of Utah and Arizona. In California, its range continues into northern Baja.

Something of an enigma, the White-eared Pocket Mouse (*Perognathus alticolus*) is known only from two disjunct populations. One in the San Bernardino Mountains around Little Bear Valley has not been seen since the 1930s and may be extinct. Its taxonomic status is, therefore, uncertain, and it may not be the same species. The other population (*P. a. inexpectatus*) lives in the Tehachapi Mountains and the adjacent slopes of the San Gabriel Mountains that front the western Mojave Desert, east to Elizabeth Lake. Its lowest elevation is around 3,200 ft (1,070 m) where it dwells among stands of Joshua Trees and other high desert plants. Its upper limit is around 5,500 ft (1,830 m) into yellow pine forests. This population is uncommon to rare within its range and considered endangered.

11.10.4 Cricetidae—Wood Rats, Deermice, Voles, and Allies

This large and diverse family includes mouse to rat-size rodents. Taxonomically the family has been fraught with issues at the generic level that will likely continue. Cricetids were first recognized as a family, then as a subfamily of the Muridae, and most recently again as a separate taxon. All cricetids share common attributes of cheek tooth morphology and other cranial details but are otherwise variable in structure and habits. Most are terrestrial, some are fossorial, and others are semi-aquatic or arboreal. Currently around 130 genera and 680 species comprise the family, which is found in North America, South America, Europe, and much of Asia. Most cricetids have short birth intervals and gestation periods of no more than 3 or 4 weeks. Breeding may take place all year under favorable conditions and climates.

Cricetids occurring in the Southern California region are the woodrats (*Neotoma* spp.), grasshopper mice (*Onychomys* spp.), deermice (*Peromyscus* spp.), harvest mice (*Reithrodontomys* spp.), cotton rats (*Sigmodon* spp.), and voles (*Microtus* spp.).

Woodrats (*Neotoma* spp.) are large (12–16 in. [300–400 mm]), nocturnal, softly furred rodents with long tails and large ears. Woodrats collect twigs, sticks, cow paddies, and other debris that they assemble into a large mounded nest about a meter across and about as high. Nests are usually built on the ground often under a rock shelter, beside a boulder, or in association with cactus or agave (Figure 6.15). The abode is multichambered and occupied by a single adult or a female with pups. Nests occupied by females can become multigenerational, and all woodrat nests are sufficiently elaborate as to entice smaller animals into taking up residence. The curious behavior of amassing shiny objects like pop-tops, ammo casings, pieces of aluminum foil, and the like has earned woodrats the alternative name "packrats."

The gestation time of woodrats is from 3 to 5 weeks depending on the species, and litter size typically varies from 2 to 5 offspring that are born in the spring. Newborns have their eyes closed for the first 15 days while remaining attached to the mother's teat.

Of the 10 species of North American *Neotoma*, 3 include Southern California in their distributional range. The White-throated Woodrat (*N. albigula*) occupies a variety of habitats, often associated with mesquite, throughout the Southwest and northern and central Mexico. The western extent of its range is the eastern Mojave and Colorado Deserts including northeastern Baja. The White-throated Woodrat includes cactus pads (*Opuntia* spp.) in its diet and the dried, harden spines are allowed to accumulate around the entrance to the nest, presumably as a deterrent to predators.

Recently regarded as a subspecies of the Pacific Coast Dusky-footed Woodrat (*N. fuscipes*), the Big-eared Woodrat (*N. macrotus*) is a mostly California species. It occurs along the Sierra Nevada from around Susanville south into the Transverse Ranges of Southern California and down the Coast Ranges from Monterey Bay into northwest Baja. In our region, it is common in oak woodlands and chaparral. Nests are frequently constructed around the base of poison oak (Figure 6.15) and Toyon. Bark and foliage of Coast Live Oak, constitute nearly 80% of their diet. This type of plant material is not easy to digest, yet they have the metabolism to deal with the tannic and phenolic compounds that are present.

The Desert Woodrat (*N. lepida*, Figure 11.8) ranges throughout Nevada and portions of Utah, western Arizona, the southern half of California, Baja, and northwest Sonora, Mexico. It is not restricted to desert habitats as its name suggests. In Southern California, the Desert Woodrat is also found in coastal sage scrub and chaparral communities around boulders and rocky slopes. In some places, it can be the dominant member of the mammal community. Agave and prickly pear cactus are often used as a base for nest construction. The Desert Woodrat urinates and defecates on its nest, which causes the assembled materials to dry and harden resulting in a structure that can endure for thousands of years. The ancient nests of the Desert Woodrat can thus serve as an abridged time capsule of the local plant community at the time it was made. This species lacks the physiological adaptations for conserving water in xeric environments that are characteristic of other arid land small mammals. It compensates by feeding on leafy, succulent plants.

Grasshopper mice (*Onychomys* spp.) are small (total body length of 3½–7½ in. [90–160 mm]), unremarkable looking mice with a relatively short thick tail. Three species inhabit the western United States. What sets them apart is that they

FIGURE 11.8 Woodrats are often called packrats because their habit of collecting shiny metallic objects and other artifacts. The Desert Woodrat (*Neotoma lepida*) constructs large durable nests of plant material that are able to survive intact for thousands of years. Paleobotanists analyze them to assess the local plant community at the time the nest was built. (Courtesy of Jason Wallace.)

are fierce nocturnal predators. Grasshopper mice are sophisticated hunters that stalk their prey and then finish with a rapid close and a killing bite to the head. They devour insects, scorpions (apparently immune to the sting), and even other small mice including their own kind. Another feature is their audible howl-like vocalization, often repeated while standing on hind legs. They perform the call when facing an adversary or prior to a kill.

The three species of *Onychomys* are essentially identical to one another. The Southern Grasshopper Mouse (*O. torridus*) is the only one in our region. This mouse lives in open scrub habitats at low elevations in the desert Southwest from western Nevada and Arizona, and much of the drier open areas in the southern half of California. West of the deserts historically it was known from the Los Angeles basin south into northern Baja, where it favored mesas and valleys with open coastal sage scrub and chaparral. Low densities made it vulnerable to habitat destruction, and population declines began sometime in the early to mid-1900s. Specimen records from the past several decades have been few to none around the Los Angeles, San Bernardino, and Riverside areas as well as most of Orange County and western San Diego County. In short, its status in these places is unknown.

Deermice (*Peromyscus* spp.) are about the size of grasshopper mice and features distinguishing *Peromyscus* include large ears, big eyes, and moderate to long tails (up to 70% of head and body). The sides and back are various shades of brown while under parts are white to gray. Several species have tan to orange lateral bands. Grooves are absent from the upper incisors. Also known as white-footed mice, at least 16 species call North America home. Practically nowhere are they absent. Some areas may support as many as 4 different species. Identifying individual species requires expertise in the subtle differences in body measurements and tail coloration. Taxonomic issues are numerous. Several subspecies are being reexamined with new molecular data and might be elevated to full species status. Some already have been and others are likely to follow. Unless you are an authority, be content if you can correctly identify a mouse as a *Peromyscus*.

Deermice are very adaptable; they breed readily and feed on seeds, other plant material, lichens, and insects. They are ubiquitous and have long been favored subjects of field research into their ecological role in a local community. A few are known vectors of hantavirus. Seven species are currently recognized in Southern California.

None is more ecologically and geographically variable as the North American Deermouse (*Peromyscus maniculatus*). This is the most widely distributed rodent on the continent. It ranges from Canada to central Mexico and from the Pacific coast to Labrador on the Atlantic. In California it occurs statewide from sea level to 9,000 ft (3,000 m) in coniferous forests, grasslands, desert scrub, and chaparral. It is also found on the Channel Islands. This is the mouse most likely to invade mountain cabins in the winter where it builds a nest from any available material—scraps of cloth, paper towels, or toilet paper for instance. And it will gnaw on anything except metal and glass. In Southern California, it may breed throughout the year producing from 2–8 young per litter.

The California Deermouse (*P. californicus*) is the largest species of *Peromyscus*, ranging from 8½ to 11 in. (220–285 mm); the tail accounts for about 55% of the length. This species occurs along the Coast Ranges from San Francisco Bay south into northern Baja. It also ranges down western foothills of the Sierras to join the coastal populations in the Transverse and Peninsular Ranges. It is not found in the deserts. It favors brushy woodlands particularly where oak and California Bay are present. They are good climbers and probably forage in shrubs and small trees. The California Deer Mouse is exceptional among cricetid rodents in forming permanent pair bonds. Males participate in defending the nest and also tend to the young, which number from one to four per litter. This species was once known as the Parasitic Mouse, a vernacular still favored by some authors. That name derives from the inclination to make a home in the nest of woodrats.

A comparatively small species, the Canyon Deermouse (*P. crinitus*) occurs in the intermountain West generally blow 3,600 ft (1,800 m) from Oregon into Nevada east to western Colorado and the northern and western parts of Arizona. In California, it is found in the desert areas east of the Sierra Nevada south into northeast Baja and the adjacent stretches of northwest Sonora. It favors no particular plant associations but does prefer exposed rocky areas like canyon walls. Canyon Deermice employ many of the same physiological adaptations (e.g., concentrated urine) as Kangaroo rats and other small desert mammals. With adequately distributed rainfall, the Canyon Deer Mouse can breed throughout much of the year.

The Cactus Deermouse (*P. eremicus*, Figure 11.9) has soft silky fur with evident yellowish sides between the gray-brown back and the whitish belly. It inhabits low

FIGURE 11.9 Large ears, big eyes and long tails characterize Deermice (*Peromyscus* spp.), which are sometimes referred to as white-footed mice. Identifying individual species is a challenge best left to experts. The Cactus Deermouse (*P. eremicus*) above ranges across the Southwest, which in our regions includes the Mojave and Colorado Deserts. (Courtesy of Jason Wallace.)

sandy places with scattered vegetation throughout the Southwest, from southern Nevada to west Texas and northern Mexico. In California, it occurs in the Mojave and Colorado Deserts where it often becomes dormant during hot dry spells.

Long considered a subspecies of the Cactus Deermouse, the Northern Baja Deermouse (*P. fraterculus*) is now regarded as a species distinct from the former. This recognition also makes sense ecologically. The Cactus Deermouse is essentially a desert species whereas the Northern Baja Deermouse inhabits coastal sage scrub and chaparral. Its distribution is decidedly peninsular, stretching from the lower slopes of the Transverse Ranges south through much of the brushy scrubland of the Baja California Peninsula. In fact, the species appears to be phylogenetically closer to Eva's Desert Mouse (*P. eva*), a Baja endemic, than it is to the Cactus Deermouse.

Larger ears and a hairy tale distinguish the Piñon Deermouse (*P. truei*). It is also larger than most other *Peromyscus* and an agile climber. The species occurs throughout much of the West in mountainous terrain where piñon pine and juniper woodlands are found, although this plant association is not required. In California, it ranges throughout the state except in the highest elevations, the most arid regions, and in the Central Valley. In our area it inhabits the Transverse and Peninsular Ranges.

The Brush Deermouse (*P. boylii*) is another species that is widely distributed in the West. It is a fairly large *Peromyscus* with a long tail and an orange lateral band. It ranges from California and Oregon to western Nebraska, and from Montana south to Texas and central Mexico. In California, it is nearly statewide except for the deserts and northern part of the Central Valley. It inhabits rocky, brush covered slopes and in Southern California coastal sage scrub and chaparral.

Harvest Mice (*Reithrodontomys*) are even smaller than deermice and have a nearly hairless tail and a single groove on the upper incisor. They get their name from climbing grass stalks to gather (harvest) grain. Of the five species, two are found in California. One is the endangered Salt Marsh Harvest Mouse (*R. raviventris*) endemic to the San Francisco Bay region. The other species is the Western Harvest Mouse (*R. megalotus*) that occupies most of the western two-thirds of the United States. It is found throughout California but avoids forested areas. The Western Harvest Mouse prefers open grassy habitats such as meadows, overgrown pastures, some cultivated areas, and weedy/grassy roadsides. It builds a nest of balled up grass on the ground or occasionally a foot or so up in a bush. It breeds in the spring and sometimes again in the fall. Three to five young, sometimes up to nine, are born per litter. Offspring may breed in their birth year, a phenomenon that can lead to population spikes.

Voles and Allies (*Arvicolinae*). Voles are stocky rodents with short feet and mostly short tails. The head is broad and flat, and their small ears are nearly hidden by fur. The fur is generally long and loose and some shade of brown. Most typical voles are from about 6–8 in. (150–200 mm), a few are a scant 5 in. (130 mm). Molar teeth have characteristic flat, prismatic crowns with alternating triangles and loops. Variations in that pattern can useful in recognizing the numerous genera and species. Voles are prolific breeders although often cyclical in their abundance. Females probably live no more than 12–14 months and will have two litters in that time.

The following genera occur in California: *Arborimus* (tree voles), *Clethrionomys* (red-backed voles), *Lemimiscus* (sagebrush vole), *Microtus*, (meadow voles), and *Phenacomys* (heather voles). Voles trend toward wetter, more humid environments than what Southern California offers, so it follows that species diversity of this group is low in our region: two species of *Microtus*. *Microtus* are also known as meadow mice or field mice because of their preferred habitat. They build runways through the grass in the areas where they forage. Burrow entrances are located at intervals along the runways. *Microtus* feed almost exclusively on the vegetative parts of plants, and they are capable of inflicting considerable damage to crop plants such as alfalfa because of their often-high population densities. The California Vole (*M. californicus*) is found from Oregon to northern Baja. It inhabits most of California's lowlands and foothills primarily west of the Sierra Nevada up to about 4,500 ft (1,500 m). It is absent from Southern California's deserts except for a narrowly disjunct population around Victorville (*M. c. mohavensis*) and a remote outlier (*M. c. scirpensis*) in the eastern Mojave along a short stretch of the Amargosa River from around Tecopa to Shosone, Inyo County. This subspecies is critically endangered. In coastal regions population densities of the California Vole are stable at around 200 individuals per hectare. In more strongly seasonable habitats, densities can vary from near zero to 450 over a 2- to 5-year cycle.

The Long-tailed Vole (*M. longicaudus*) indeed has, for a vole, a long tail at a little more than 33% of the total length. This is a montane species of the West from southeast Alaska south through New Mexico where it lives in a wide variety of habitats like coniferous forests, along rivers and streams, sagebrush steppes, and also does well in disturbed areas. In California, the Long-tailed Vole occurs across the northern part of the state then south along the Cascades and Sierra Nevada. Only a disjunct population in the San Bernardino Mountains represents it in our region. The Long-tailed Vole typically does not make well-defined runways, and population densities are low, in the neighborhood of 5 to 6 (occasionally as high as 40) individuals per hectare.

Muskrats (*Ondatra zibethicus*) are aquatic Arvicolines with partially webbed hind feet and a laterally flattened tail that functions as a paddle. Muskrats are vole-like in appearance albeit much larger, growing up to 2 ft (620 mm) and weighing as much as 4 lbs. (1800 g). They range across Canada and most of the United States except for the dry regions of the Southwest and Texas. Freshwater lakes, slow-moving rivers, cattail marshes, and irrigation canals all provide appropriate habitat. Muskrats build large dome shelters from plant material. The main chamber is above water but the entrance is below. Unlike beavers, they do not construct dams. They will also dig burrows into the muddy banks. In California, muskrats occur naturally in northern parts of the state. In the early 1900s, they were introduced to the San Joaquin Valley where they now thrive in the irrigation canals and other waterways. In the Southern California region, they are only found along the Colorado River and some adjacent areas in the Imperial Valley.

They breed from late winter through spring and early summer. Newborns begin swimming in 2 weeks.

The muskrat is an important furbearer and has been introduced for that purpose throughout much of Europe, Asia, and Argentina. In many places, it is now regarded as a pest species because of its destruction to aquatic vegetation and the damage created by burrowing.

Cotton Rats (*Sigmodon*) are somewhat vole-like in appearance but closer to the size of a wood rat at around 8–14 in. (225–360 mm). The S- (sigma) shaped tooth crowns are diagnostic of the genus. Cotton rats inhabit grassy areas, and like voles, they establish runways around their nest and foraging territories. They occasionally eat insects, but the diet primarily consists of grasses and other herbaceous plants. Of the four North American species only the Hispid Cotton Rat (*S. hispidus*) reaches California and here only along the lower Colorado River and some adjacent areas of Imperial County. Elsewhere, the species ranges across the southern Gulf States west to central Arizona. The common name derives from the stiff grizzled hairs (hispid) protruding from the softer fur. Hispid Cotton Rats are active day and night and have an extended breeding period synchronous with fresh vegetation in late winter to early summer. Litter sizes run from three to seven offspring.

11.10.5 Muridae—Old World Rats and Mice

Probably the most loathed mammals on the planet, the non-native Norway Rat (*Rattus norvegicus*) and Roof Rat (*R. rattus*) are about the size of our native woodrats. They differ by their naked ears and nearly hairless tail among other features. Of the two, the Norway Rat is slightly larger and has a shorter tail, which in the Roof Rat exceeds the length of the head and body. The Roof Rat, also called the Black Rat (a color phase) or House Rat, arrived in the United States from Europe with the first colonial settlers. The Norway Rat, also called the Brown Rat, reached our shores later on, around the time of the American Revolution. (The purposely bred albinos of this species are the standard “white rat” of laboratory experiments.) Following European exploration of the New World and beyond, both species are now widespread around the world in tropical and temperate regions. They are especially common around port cities, but urban areas in general are not spared their absence. The Roof Rat is a good climber. It effortlessly scampers up buildings and trees and darts along telephone lines with ease. It feeds mostly on plant matter. The Norway Rat is an omnivore seemingly willing to sample anything it comes across and in many places it is the dominant and more destructive of the two species. It gnaws through wood, wiring, plastics, almost anything except steel and concrete. Destructive as they are to buildings, stored foods, orchards, and field corps, they can also lay waste to local populations of small native mammals and birds, particularly in insular ecosystems. Both of these rats are also vectors of plague, some forms of hepatitis, parasitic worms, and typhus.

The Norway Rat starts breeding at about 12 weeks of age and females are pretty much continuously pregnant thereafter. Litter size is from 4 to 10. The Roof

Rat is only slightly less prolific. If there is an upside, it is that neither species is particularly successful too far from human civilization. In the wild, they readily fall prey to owls, foxes, coyotes, and other carnivores.

More inconspicuous but no less odious, the House Mouse (*Mus musculus*), is another early stowaway to the United States from Europe that has become established in many urban areas around the world. The House Mouse is similar in size to our native harvest mice (*Reithrodontomys*) but differs in being a unicolor dark brownish grey and has an almost hairless tail. And unlike other native mice, they have an unpleasant odor about them. House mice infest warehouses, invade homes, gnaw through packaging to reach food stores, and foul what they don't consume. They have been trapped in fields and brushy areas and in some instances may be displacing native mice like *Peromyscus*.

11.10.6 Erethizontidae—Porcupines

The North American Porcupine (*Erethizon dorsatum*) is unmistakable. Weighing 20 lbs. (9 kg) on average, it is the largest rodent in North America after the beaver. The long stiff spines, most prevalent on the rump and thick tail, are modified hairs used for defense. Occasionally the spines are employed against rival males when vying for the attention of a breeding female.

Porcupines have poor eyesight and are by and large disarmingly tame. They have few natural enemies. The Fisher (*Martes pennanti*) is the one predator that seems to have success. Martins, mountain lions, wolverines, and bobcats also have been known to prey on them, albeit probably out of desperation. When threatened, a porcupine backs up toward the intruder with the spines erect and the head lowered. The pose is usually sufficient to deter further advance. If a predator (or pet) makes the unwise decision to attack, the porcupine thrashes its tail and the unfortunate adversary receives a face-full of barbed quills with no means of removing them. The quills can work themselves in and if vital organs are penetrated death may result. There is no truth to the folklore that porcupines eject their spines as if throwing darts.

The porcupine diet consists of saplings, roots, stems, berries, and nuts. During winter months, they climb up into conifers and hardwoods where their diet shifts to the inner bark of the tree. An individual will often spend considerable time gnawing and feeding on a single tree. Trees sometimes die as a result of the damage, but more often the result is a stunted tree growing at odd angles. Dens are made in rock crevices, tree hollows, and abandoned burrows. Mating takes place in late autumn or early winter. The gestation period is long, about 7 months, and the female rarely gives birth to more than one pup at a time.

Erethizontidae is a South American family, and *Erethizon dorsatum* is the only North American representative. It likely arrived here after the closure of the Panamanian isthmus united the two continents around 3 million years ago. Porcupines were once common in forested areas across the United States and Canada, but nowadays habitat loss has seen them extirpated over much of the east and Midwest. In the western United States, they occupy several habitats including coniferous and hardwood forests, piñon-juniper woodland, and desert scrubland in the Great Basin. In California, porcupines are restricted mainly to coniferous forests. Where at one time they were common, in recent years they have become very rare in some places, for example in the central Sierra Nevada and points south. Reasons behind the present decline are being studied and may or may not be directly linked to the logging and timber interests in California that persecuted them for decades. Porcupines were poisoned and shot as late as the 1970s even though the damage they do to trees is overstated so far as commercial interests are concerned. The current status of porcupines in Southern California is essentially unknown. Scattered anecdotal accounts over the past 20 years suggest their continued presence in areas of the San Gabriel, San Bernardino, and Tehachapi Mountains, where historical records first documented them. Lingering reports persist of another population isolated in piñon pine forests of the Grapevine Mountains in northern Death Valley adjacent to the Nevada border. That record traces back to Alden Miller's 1946 account of bird distributions along a vertical transect from 6,500 to around 7,500 ft. In an addendum he included, without comment, porcupines in a list of mammal species confirmed at the study site. No specimen from that locality is known, nor is any current information available on the population, undoubtedly a Pleistocene relict, if it still exists.

11.10.7 Castoridae—Beavers

The single extant species of this family is the American Beaver (*Castor canadensis*), North America's largest rodent. Its size (35–65 lbs. [16–30 kg]), webbed hind feet, and flat scaly tail are hallmarks. Doubtless no other mammal was as important in opening the American West. Its valuable hides sent trappers into new territories, and in so doing, the most favorable routes through the interior to the Pacific Northwest were identified. Eastern immigrants in wagon trains were soon to follow.

Beavers are aquatic and renowned for creating ponds by building dams of tree branches and mud. Lodges are constructed within the pond. Trees are felled along a streamside and the current carries them in a way that creates an arch as the dams are being built. The arch is directed downstream or up depending on whether the dam was started from the banks or at some anchoring spot in the middle of the stream. Because beavers do not hibernate those living at higher elevations would find it difficult to locate food in the winter when the pond freezes over were it not for the sticks and saplings they stored in the mud during the summer and fall. Beavers living in the valleys, where food is more readily available during the winter months, may not construct dams at all. Instead they dig burrows into the sides of the riverbank.

The history of the beaver in California is, as in other places in the West, tied to the fur trade of the nineteenth century and its aftermath. For much of the twentieth century, the historic range of California beavers, based on the work of Joseph Grinnell and his colleagues in 1937, was assumed to be the northern tier of the state in the watersheds of the Klamath, Trinity, and McCloud Rivers, populations in the Central Valley, and those in Southern California along the lower Colorado River. Recent searches of museum records, ethnographic information, older cartographic place names and other documentation concluded that beavers probably occurred statewide historically (except the streamless low deserts) in suitable habit. This would include the coastal drainages from Del Norte to San Diego Counties. In Southern California, the evidence amply supports the argument that beaver were at one time present from San Luis Obispo to San Diego Counties. For example, logs, diaries, and correspondence of the mission padres describe beaver skins as part of the clothing worn by some Native Americans or that comprised parts of their ceremonial accouterments. Native Americans themselves, for instance the Chumash and Kumeuyaay, have their own word for beaver. Additionally a beaver (skull only) was collected along Sespe Creek in Ventura County in 1906. Grinnell had questioned its provenance but ensuing correspondence between Grinnell and the collector, John Hornung, confirmed its authenticity.

The trade in beaver pelts along the West Coast began after James Cook reported on Sea Otters (*Enhydra lutris*) in 1778. Almost immediately, ships from Boston arrived and enlisted native people to trap and trade for beaver (and otter) along coastal waterways of California. The only records kept were some of the anecdotal ones mentioned previously. Many of the populations, perhaps most so in Southern California, were likely eradicated during that campaign, well before the arrival of the American and Hudson Bay Company overland fur traders who arrived in the 1830s when more meticulous record keeping was becoming standardized. With coastal beaver populations largely extinguished, the fur traders turned their attention to the northern inland habitats.

By the early 1900s beavers had been all but extirpated throughout much of the state. With that bleak outlook, the California Department of Fish and Game ran a translocation program from 1923 to 1950. Beavers were released into three-fourths of the state's 58 counties. The results were dramatic. California's beaver population rose to around 20,000 by 1950 from a low of about 1,300 in 1940. This stock undoubtedly accounts for the presence of beaver in some of Southern California's coastal waterways today. Beavers are presently known from Meadow Creek in San Luis Obispo County, tributaries of the Santa Ynez River in Santa Barbara County, San Mateo Creek in the San Mateo Wilderness of Riverside County, and tributaries of the Santa Margarita River of Orange and San Diego Counties. They also occur in the San Bernardino Mountains, the upper reaches of the Mojave River, and the native population along the lower Colorado River has expanded into the Alamo River in Imperial County.

11.11 CARNIVORA—CARNIVORES

Small to large specialized predators, carnivores have enlarged, piercing canine teeth and laterally compressed molars for cutting flesh. The fourth premolar and lower first molar, called *carnassials*, slide passed one another in a scissor-like slicing motion to cut through sinew and muscle. Carnivores have enlarged olfactory bulbs and a corresponding keen sense of smell. Despite their predatory specializations, many carnivores are fairly omnivorous and will consume arthropods, fruits, berries, and other plant material. Even cats will eat plant matter from time to time. The earliest carnivores were an arboreal, weasel-like assemblage (Miacidae) that appeared in the early Paleocene and persisted throughout much of the tertiary.

Today the Carnivora occur worldwide in terrestrial, marine, and freshwater habitats. Seven families are found in Southern California and several species figured prominently in the California fur trade of the seventeenth and eighteenth centuries. Sea otters and fur seals in particular experienced steep declines in their numbers during that time. The grizzly bear was hunted to extinction in California, not for its hide but for its presumed ferocity, and the mountain lion was heavily persecuted well into the twentieth century.

11.11.1 Felidae—Cats

Cats are meat-eaters and superbly adapted predators. They have been called the perfect killing machine, outmatched only by humans. Meat is their staple, yet there is no reluctance to supplement the diet with fish, mollusks, fruits, insects, or carrion as necessary. Because meat prey (terrestrial vertebrates) are difficult to come by compared with other food resources, hunting and killing efficiency is finely tuned in cats. The canine teeth are long, narrow and sharp for a penetrating stab. Cheek teeth are laterally compressed for shearing. The robust jaw muscles (*temporalis* and *masseter*) produce a powerful bite. Claws are sharp and curved to capture and manipulate prey, and they can be retracted into a sheath when not in use. Cats have excellent binocular vision, next only to primates. The eyes are located anteriorly and medially that, along with a short muzzle, permit overlapping fields of view for accurately judging distances.

Cheetahs run down their prey, but most cats rely on ambush or a short chase. They depend on concealment, patience, and speed. The background color of a cat generally matches its hunting habitat, and in many species, a pattern of stripes and or spots enhances camouflage. Rapid bursts of speed are accomplished by modifications to the appendicular skeleton (arms and legs) and associated muscles so as to increase leverage output. The upper bones (humerus in the arm and femur in the leg) are reduced in length relative to the lower components (radius and ulna in the arm and tibia and fibula in the leg). In the shoulder girdle, the clavicle ("collar bone") is reduced or absent thus bringing the forelimbs in proximity to the long axis of the body and freeing the scapula (shoulder blade) so as to increase the effective length of the limb. Stride is also lengthened during a gallop because the vertebral column can flex and extend synchronously with the swinging limbs. The remarkable degree of rotation permitted between individual vertebrae is what gives a cat the ability to twist during a fall and land on its feet.

Cats are primarily twilight or nocturnal hunters (except for the cheetah) and are almost always solitary other than when mating. In fact, most cats are strongly intolerant of one another. Even the prides of social cats like lions are usually an extended family with few if any outsiders welcome.

Several groups of extinct felids are known in the fossil record, the large ones appearing coincident with the rise of ungulates in the mid-Cenozoic. Today there are around 35 living species placed among 7 genera. Most of those are species of *Felis*. The present cat fauna of temperate North America consists of the Bobcat (*L. rufus*), the Mountain Lion (*Puma concolor*), and the Canadian Lynx (*Lynx canadensis*).

Feral house cats are often mistaken for bobcats (*Lynx rufus*) mainly because they are felids of similar size. Bobcats have a short tail (hence, "bob-tailed cat"), long legs (especially for a cat), large paws, a facial ruff, and pointed ears with a short tuft (Figure 11.10a and b). All are spotted and barred on the tail and flanks

FIGURE 11.10 The Bobcat (*Lynx rufus*) is a skillful predator found in most of Southern California's biotic communities. It preys primarily on rabbits and small rodents. On average, bobcats are about a third again as large as a mature house cat. They have a short tail, longer legs, larger paws (a), and as shown in the bottom photo (b), pointed tufted ears and a facial ruff. (Courtesy of Jason Wallace.)

against a background color of yellowish to reddish brown. Bobcats weigh on average around 10–15 lbs. (4.5–7 kg), but some males can weigh 40 lbs. (ca. 18 kg) or more. "Wildcat" is the most common alternate name for *Lynx rufus* and regional variation in color and pattern has led to several others, for example "lynx cat," "spotted lynx," or simply "lynx." None these should be confused with the Canadian Lynx (*L. canadensis*). In Spanish the bobcat is called *Gato montes*.

Bobcats range from southern Canada, through most of the contiguous United States and Mexico south to Oaxaca. They can be found in practically any habitat within this range. They occur statewide in California, from the deserts (including Death Valley) to mountain timberline. They are at home in chaparral and other brushy areas. Rocky ground and boulders are often favored retreats, but if unavailable, they will take refuge in trees. Bobcats are generally absent from intensively cultivated areas like the Central Valley and from dense urban areas devoid of habitat. Bobcats can be bold and sometimes observed at close range. They do not always shy from humans at initial contact, and show no aversion to buildings, as demonstrated by Figure 11.10b. Although they prefer to hunt at dusk and later at night or near dawn, they may be abroad by day. I had one amble the entire length of our patio on a fall afternoon.

Home range varies widely among individuals; on average it is a little more than 6 mi^2 (16 km^2) for females and about 15 mi^2 (40 km^2) for males. Rabbits, ground squirrels, gophers, and other small rodents comprise a significant part of their diet, which may also include reptiles and, rarely, ground-nesting birds like quail. They pose no threat to game bird populations as some people contend. Bobcats are known to raid poultry farms, and to prey on lambs, but the damage they do in these instances is considerably less than that borne by coyotes and dogs. Eastern bobcats are significant predators of White-tailed deer (fawns mostly). These bobcats are larger (males can weigh as much as 50 lbs.) than their western counterparts and the White-tailed Deer is smaller than our Mule Deer.

Bobcats do not hibernate during the winter. From 1 to 6 kits are born in the spring and summer after 50-day gestation. If a mating is unsuccessful with the first estrus period, another estrus will follow.

Because of its silky pelage, bobcats were regularly trapped in California and elsewhere as a furbearer even though the pelt was of poor quality and not especially valuable, at least historically. Pelt prices soared to near $600 in 2013 because of demand from China, Russia, and Europe; 65,000 were exported from the United States that year. By 2015 prices had dropped to $300. As a species, bobcats face no immediate threats beyond the usual destruction of habitat that most vertebrates face. In Southern California, automobile traffic takes a toll, and some foothill communities have applied rodenticides to eradicate "vermin" that has led to the unintentional death of bobcats.

Mountain Lion (*Puma concolor*) is the most common name for this large cat, which may also go by cougar, panther, or puma. The average adult male is 6 to 8 ft in total length (1.8–2.5 m) and from 110 to 180 lbs. (49–84 kg). Females are about 30%–40% smaller. The long cylindrical tail has a black tip and the ears are rounded and not pointed and tufted like those of the bobcat. Despite the obvious size difference, more than 80% of reported mountain lion sightings in California turn out to be feral house cats. Less than one-tenth of 1% of people in California will ever see one alive in the wild. Tracks, scats, and tell-tale claw marks on trees and stumps are the most likely signs that they have been in the area.

The mountain lion is the most widely distributed mammal in the New World, occurring from northwestern Canada to Patagonia at the tip of South America. In the United States, it has been extirpated over most of the eastern half of the country and the Florida population is rare and endangered. Today the mountain lion is essentially an animal of the 14 western states, and as its name implies, favors mountainous terrain. In California about 46% of the state has suitable habitat such as forests, chaparral, tall grass, and rocky cliffs—essentially any place that offers sufficient cover (and prey) from near sea level to around 11,000 ft (ca. 3,400 m). Mountain lions do not occur in the open deserts or for the most part in the Central Valley. They were once known along the Colorado River from northern Riverside County to the delta, but they are now likely extinct in that area. In California, mountain lions are probably most common along the west side of the Sierra Nevada from around 2,000 to 5,000 ft (ca. 930–1,800 m). Densities are highest where deer are abundant. Where deer are less concentrated, as in Southern California, mountain lions are less common. In these latter cases a single lion may occupy a territory of 200 mi^2 or more.

Mountain lions are solitary, stealthy hunters that prey heavily on deer. They are excellent at avoiding detection, and with such a powerful body, a deer is dispatched in an instant. They typically do there hunting at night. Where deer are abundant, a lion may feed on a kill until it is sated and then moves on. Where prey density is lower, the lion might cover the carcass with dirt and sticks and then return to it for 3 or 4 more nights until it is consumed. Concealing the kill varies with locality, season, prey abundance, and habits of the individual lion.

A female mountain lion breeds every other year, bearing a litter of around two or three kittens on average. Kittens are spotted for the first 6 months of life. They may stay with the mother for up to a year before they start foraging on their own.

Because of their secretive and wide-ranging behavior, the number of mountain lions currently in California is difficult to ascertain with any precision or for that matter even within tolerable limits. Dividing the number of square miles of habitat by a general density estimate (for instance, no. lions /100 mi^2, say) will give a result but not one with high confidence. In the mid-2000s the California Department of Fish and Wildlife estimated between 4,000 and 6,000. More recent census projections place the statewide population at around 3,000.

In Southern California, mountain lions occur in most of our montane habitats, and some roam surprisingly close to our metropolitan areas much like Black Bears do (see below). Biologists with the National Park Service have radio-collared more than 50 mountain lions in recent years in and around the Santa Monica National Recreation Area. Much has been learned from these efforts about the daily movements of mountain lions, the distances they travel, and their survivability. Freeways (and urbanization) are probably the biggest threat to mountain lions in Southern California. Motor vehicle traffic accounts for a significant amount of mountain lion mortality. Freeways also are dispersal barrier. Gene flow among populations is restricted and viability drops because of inbreeding. The population in the Santa Ana Mountains, for example, is effectively isolated from the Peninsular Ranges to the east by the I-15. In the Santa Monica Mountains, the Pacific Ocean on one side and the 101 Freeway on the other isolate the population in that area. Other sources of mortality include rat poison, starvation, abandonment of offspring, and territorial fights with other mountain lions.

No other mammal in California has occasioned so much misinformed folklore, prejudice, and exaggeration than mountain lions. They have been loathed and feared since western Europeans first settled in California to the point that from 1907 to 1963 the state paid a bounty on mountain lions to contracted hunters. During those years, based on the number of submitted bounty claims, 12,460 mountain lions were killed. After 1963 when the bounty was rescinded, the status of the mountain lion changed from predator to big game, which meant that trophy hunting was still permitted. In 1972 a moratorium on mountain lion hunting was signed by then Governor Ronald Reagan that lasted until 1986 when the state legislature failed to renew it. However, lobbying by various citizens groups maintained the protection until 1990 when Proposition 117 was passed as the *California Wildlife Protection Act.* With its passage, mountain lion hunting was banned statewide despite the forceful opposition of disgruntled hunters and stockmen. Proposition 117 also mandated that a minimum of $30 million was to be spent annually on wildlife habitat acquisition and restoration. Subsequent amendments allowed for the killing of mountain lions under special depredation permits that identified specific individuals as a threat to public safety or to Big Horn Sheep.

Mountain lions move in and out of California to adjacent states, where hunting is permitted and stockmen are a powerful advocacy for removing them from their area even though the death of livestock is almost always attributable to a single individual and not from a population; the monetary loss overall is minimal. Mountain lions also are often blamed for missing pets, but the evidence is slim at best. It is more likely that small dogs and cats fall prey to coyotes and Great Horned Owls. Mountain lions are valuable predators that maintain deer populations at a level commensurate with what the habitat can support. They can be dangerous to humans but such encounters are rare. From 1890 to 1910, there were 3 attacks in California each of which was fatal, albeit 2 of the deaths were from rabies. No attacks were recorded again until 1986. Since then, Californians have experienced 15 mountain lion attacks of which 5 were fatal. The last was in 2007. Eight of the attacks and 3 of the deaths occurred in Southern California. Tragic as those incidents are, they underscore our region's large human population, its proximity to mountain lion habitat, and our affinity for recreational activities in our mountains and foothills.

Until the middle of the nineteenth century, mountains lions were not the only big cat in California. Unfortunately, along with the Grizzly Bear we must include the Jaguar (*Panthera onca*) as another large mammal that was historically extirpated from the state. This huge spotted cat, the largest in the Western Hemisphere, is third in size among the felids behind the African lion and Asian Tiger. Mature males can weigh more than 300 lbs. (145 kg). Jaguars once occurred from the U.S. Southwest to southern Argentina. They have been extirpated throughout much of this range because of hunting and habitat destruction. Jaguars are mostly a tropical rainforest species yet do well in evergreen woodland and other habitats where a food supply is available. In the 1800s, they occurred across the Southwest from California to west Texas.

Jaguars were resident in Southern California and Baja but probably in low numbers. The Kumeyaay Indian name for them is *tu' kwat*, and two records from near Monterey in 1814 suggested that they roamed the Coast Ranges. The only description of a breeding den in the United States is that of an adult pair with two kittens found near Tejon Pass in the Tehachapi Mountains in 1837. The last Jaguar known in California was shot near Palm Springs in 1860 when it attacked a Cahuila Indian who was clad in deer antlers and a skin.

Over the past couple of decades, Jaguars have reappeared in parts of southern Arizona and western New Mexico, no doubt having wandered up from Sonora, Mexico, where the northernmost population exists today.

11.11.2 Canidae—Dogs, Coyote, and Foxes

Canids are well furred, have busy tails, long limbs, lithe bodies, a long muzzle, and erect ears. They are alert and intelligent and possessed of an acute sense of hearing and smell. Several species have exceptional endurance. Their association with humans runs deep. Dogs were the first mammals to be domesticated, and canids as a group are remarkably adaptable in their behavior and ecological plasticity. They can develop complex social systems or live solitary lives. Worldwide, nondomestic canids number around 35 species ranging from the arctic to the tropics. Three genera and 4 species are found in Southern California.

The most familiar and widespread is the Coyote (*Canis latrans*), a name that comes from the Aztec word *coyotl* and translates as "barking dog." Coyotes (Figure 11.11) have a slender build, long legs, large erect ears, and a bushy black-tipped tail that is depressed while the animal is running. The coarse fur is some blend of buff, rufous, brown and grey with pale underparts. Adult males can weigh up to 45 lbs. (20 kg) although the average is closer to 30–35 lbs. (13–16 kg), females somewhat less. Coyotes are among the fasted mammals in North America and can achieve speeds up to 40 mi/h. They are mostly crepuscular and nocturnal although diurnal activity is not uncommon, especially during winter months. Their familiar high-pitched chaotic howl heard from a distance can sound like three or four coyotes in a chorus. It seems improbable that a single coyote could produce such varied sounds and intonations; yet it can.

In the first half of the twentieth century as many as 19 subspecies of *C. latrans* were recognized (3 in California). Nowadays, a better understanding of Coyote biology has made subspecific designations limited in value and trinomials have been largely abandoned. Coyote distribution was originally centered in the grasslands of the western United States. Since the middle of the nineteenth century,

FIGURE 11.11 Coyotes (*Canis latrans*) are smart, opportunistic predators and scavengers found in most any terrestrial habitat in Southern California, including urban canyons. Notice the slender build, long legs, long snout, and erect ears. The bushy tail has a black tip. (Courtesy of Jason Wallace.)

it has extended to Alaska, east to Quebec and Nova Scotia and south through Mexico to Middle America. Coyotes inhabit forest edges, deserts, open chaparral, agricultural areas, and urban settings. In Southern California, it is not uncommon to see a coyote on a residential street that has wandered up from a nearby canyon. As people cleared land, expanded agriculture, and in the eastern United States extirpated its larger competitor, the Grey Wolf (*C. lupus*), coyotes exploited the new opportunities even while they endured persecution by trappers and hunters.

Coyotes are opportunistic predators and scavengers. Mice, ground squirrels, and rabbits make up a considerable part of a diet that may also include amphibians, reptiles, birds, invertebrates, berries, fruits both wild and cultivated, grasses, and carrion. Small prey like rodents is captured by ambush; larger mammals are run down in relay. They have a reputation for preying on domestic stock, in particular chickens and sheep, but that proclivity is probably exacerbated by careless husbandry practices. They have also been known to attack and eat small dogs. Attacks on small human children are extremely rare.

Dens are excavated or modified from burrows dug by other animals and are found near rock ledges, in brush, embankments, or in open country. Like wolves, a mated pair may return to the same nest site year after year. Thus, they are generally monogamous and pair bonds can last for years though not necessarily for life. Courtship takes place in December and early January, and mating from January to March. Gestation is 60 days with an average litter size of 4 to 7 or as few as 2 and as many as 12. Males participate in rearing the young by bringing food to the female and guarding the den. Pups are fully grown in about 9 months. All 3 species of North America *Canis* (*latrans*, *lupus*, and *familiaris*) will interbreed.

Coyote–dog pairings result in a "coydog," but behavioral differences and physical barriers usually intervene beyond the first generation.

Coyotes will aggregate in packs of from 3 to 8 individuals that occupy and defend a territory. The fundamental social unit is the mated pair, who is joined by adult and adolescent relatives. Pack members come and go annually while the mated pair persists. Some coyotes are solitary residents of a particular area with no mate or pack association. Others are nomadic transients that have no site fidelity at all.

The coyote's larger relative, the Grey Wolf (*Canis lupus*) once ranged over most of North America including California where early travelers frequently wrote about wolves they had seen. The documentation, however, is rife with misinformation, and coyotes and their skins frequently were misidentified as wolves. Grey wolves roamed at least the northern quarter of California and well down the Sierra Nevada into the early 1800s. By the time the Gold Rush was underway in the 1850s, cattlemen and settlers saw to their quick extermination. As a result, next to nothing is known about the natural history of the Grey Wolf in our state. The first museum specimen (Museum of Vertebrate Zoology, University of California at Berkeley) was not obtained until 1922. The skin and skull were taken from a wolf trapped years earlier in Lassen County. That same year (1922) a wolf was trapped alive in the Providence Mountains near Lanfair in the eastern Mojave Desert. The trapper had intended to return the wolf to Long Beach for exhibition, but the animal did not survive the arduous 15-h drive. It, too, became a museum specimen. There is no way of knowing if this particular wolf was a resident or a nomad that had wandered in from Nevada where the Grey Wolf was gone by 1923. In either case, like the Grizzly Bear, it marks a sad closing chapter to the life of a large, impressive mammal from the state and from our region.

11.11.2.1 Foxes

The Grey Fox (*Urocyon cinereoargenteus*) is common throughout Southern California, recognizable by its small size (about 6–15 lbs.), grey back, reddish-brown forelimbs and underparts, and black-tipped tail (Figure 11.12). It is sometimes called the "red fox" because of its color, but this is an unfortunate appellation since it is unrelated to the true Red Fox (*Vulpes vulpes*) that occurs across much of North America except for the Southwest. In California, the Red Fox is native to the higher elevations of the Sierra Nevada and the Cascades. A second population in the Central Valley, the stock of which probably originated from one of the Great Plains states, was likely the result of escapees or an intentional release from a fur farm.

The Grey Fox is found throughout the United States except for the northern Rocky Mountain region, upper Great Basin, and Washington state. It ranges south through Mexico to northern South America. In Southern California, the Grey Fox inhabits coastal sage scrub, urban canyons, chaparral, pinion juniper woodland, agricultural areas, meadows, and rocky desert. It is the only canid (with the possible exception of the Corsac Fox [*Vulpes corsac*] of Asia) that readily climbs

FIGURE 11.12 The Grey Fox (*Urocyon cinereoargenteus*) is a small canid typically weighing less than 15 lbs. with luxurious silvery and reddish-brown fur. It is widespread in Southern California although not seen as often as the coyote.

FIGURE 11.13 The Grey fox is one of only two canid species that climb trees. This fellow is feasting on the fruits of a date palm at the California Desert Studies Center. (Courtesy of Jason Wallace.)

trees. Trees are sought out for foraging (Figure 11.13) and as a refuge if the fox is being pursued. The Grey Fox is an omnivore with a diet that varies by season and locality. Mice and other small rodents are significant, and insects can also figure prominently along with birds, nuts, berries, other fruits, and carrion.

The Grey Fox is primarily crepuscular and nocturnal and prefers to remain concealed during the day. It will, however, venture out during daylight hours to visit water stations during the hot, dry months of the year. Females produce a single litter annually of four pups on average. The breeding season varies geographically. In the West, Grey Foxes mate in the late winter and gestation lasts about 60 days. Males may assist in rearing and occasionally two females will rear their litters in the same den. Pups begin foraging independently at 4 months. Dens are used year round and include such retreats as rock shelters, crevices, hollow logs, and brush piles.

Grey Foxes fall prey to coyotes, mountain lions, bobcats, and golden eagles. Up until the middle of the twentieth century, they were routinely harvested by the fur industry. During a 6-year period in the 1920s, 18,800 were trapped in California alone. Nowadays, strict state and federal regulations have all but eliminated the practice of harvesting wild foxes. Today the fur industry relies on farm raised animals.

The second and only other species of *Urocyon* is the Island Grey Fox (*U. littoralis*) endemic to the six largest Channel Islands off the Southern California Coast (San Miguel, Santa Rosa, Santa Cruz, San Nicholas, San Clemente, and Santa Catalina). They are absent from the smallest islands Anacapa and Santa Barbara, which lack fresh water and other essential resources. The Island Grey Fox is similar in most aspects of pattern, color, and life history to the Grey Fox on the mainland except for being about one-half to two-thirds their size. They likely reached the northern islands (San Miguel, Santa Rosa, Santa Cruz) during the late Pleistocene when the islands formed a single landmass that was about 4 miles from the mainland. Less clear is when and how foxes reached the southern Channel Islands. On San Clemente, at least, they have been present since prehistoric times. Colonizing Native Americans may have brought them along.

With no native predators or competitors, the Island Grey Fox has few restrictions on habitat utilization. For food they rely primarily on insects (beetles and grasshoppers), fruit when seasonally available, and occasionally birds. Rodent densities are relatively low on the islands and therefore are a limited part of the diet.

In 1999, an outbreak of canine distemper on Santa Catalina Island caused the population to plummet from 1350 foxes in the 1990s to 100 by 2002. The virus may have been carried to the island by dogs or possibly raccoons (who find it easy to stowaway on yachts). At that point, a recovery program began that included moving the remaining healthy foxes to the west end of the island from the contaminated east end (the two are separated by an isthmus). Vaccinations and captive breeding with subsequent release brought dramatic results. By 2015 the Santa Catalina population stood at 1,750.

Around the same time that the Catalina foxes were in steep decline, Island Grey Foxes on San Miguel, Santa Rosa, and Santa Cruz were also being decimated. In these cases, the decline came about by the adverse presence of feral pigs, goats, cats, habitat loss, and most importantly a novel predator, the Golden Eagle (*Aguila chrysaetos*). Golden Eagles had been infrequent visitors to the islands until the decline of the Bald Eagle (*Haliaeetus leucocephalus*) from DDT exposure in the 1960s. Golden Eagles became fairly common on the islands in the 1990s, and by the end of the decade, they were nesting. The foxes were ill-prepared to defend themselves and population levels sunk to 15 each on San Miguel and Santa Rosa by 2000, and to 62 individuals on Santa Cruz in 2002. Intense and sustained recovery efforts brought results similar to those on Santa Catalina. Golden Eagles were relocated to the mainland and feral pigs were eliminated. By 2015, the populations had rebounded remarkably on all three islands. That accomplishment stands as the fastest recovery of any mammal listed under the federal Endangered Species Act. Nonetheless, ongoing monitoring of the Island Grey Fox will be necessary to ensure its survival on the Channel Islands.

The Kit Fox (*Vulpes macrotis*) and its close relative the Swift Fox (*V. velox*) are the smallest North American canids. Total length is around 32 in. (80 cm) and they weigh from 3–6 lbs. (1.4–2.7 kg). The Kit Fox is a nocturnal, arid adapted species with light colored yellow-gray or buff-gray pelage, a bushy black-tipped tail, slender build, and large ears. The limbs are long and the soles of the paws are adorned with tufts of fur. "Kit" refers to the small size of the animal. It is a term applied to the young of most any fur-bearing mammal.

The Swift Fox is a resident of the central Great Plains whereas the Kit Fox is a species of western scrublands and deserts mostly in areas with level sandy terrain. Rabbits and rodents are the principle food items. Occasionally they also feed on insects, birds, lizards, and cactus fruit. As with many desert mammals, they do not require constant access to water.

The Kit Fox ranges from southeast Oregon through Nevada, parts of Utah and western Colorado and south into Arizona, New Mexico, and west Texas then on into central Mexico. In California, Kit foxes are confined to the southern half of the state in three populations. The desert cohort is sustaining itself, but the San Joaquin Valley foxes are imperiled. The third population known from west of the Transvers Range is probably extinct. The desert population ranges through the Mojave and Colorado Deserts south through the Baja peninsula. In the Colorado Desert, they can be found west to Palm Springs in Riverside County. In the Mojave, they have been found west to Victorville.

In the Central Valley, the isolated San Joaquin Kit Fox (*V. m. mutica*) once ranged as far north as Stanislaus County on the east side of the valley and to Tracy in San Joaquin County on the west. Its southern extent was near the mouth of Tejon Canyon in Kern County. By the 1930s, the San Joaquin Kit Fox was already restricted to the dry plains of the southern and western parts of the valley. Intensive agriculture destroyed most of their habitat, and poison-baited traps intended for coyotes also contributed to their decline. The San Joaquin Kit Fox is now listed as federally endangered.

The third presumably extinct population in Southern California was once known from the Perris and San Jacinto valley areas and northwestward along the base of the San Gabriel Mountains into San Fernando Valley. The last specimen was obtained near Moreno in 1903.

Kit Fox burrows are used year around. In flat sandy locations there may be 8 to 10 burrows per hectare although they are not all occupied simultaneously. Nursing burrows are more widely separated from one another. In the Mojave Desert Kit Foxes will also den along high rocky, juniper-covered hills. Kit foxes are monogamous and the mating pair stays together throughout the year. Mating takes place from December to February and 4 to 6 young on average are born following a gestation period of 49–55 days. The family unit stays together until October when the pups disperse into new territories.

11.11.3 Ursidae—Bears

Bears are unmistakable large heavy-bodied carnivores with strong stout limbs and a short nearly imperceptible tail. They have broad paws and feet and a plantigrade posture (the entire foot makes contact with the ground, Figure 11.14). Claws are non-retractable. And despite their size, some bears are good climbers. The carnassials are reduced in favor of broad crushing molars. Bears are largely nocturnal although they may move about almost any time of day. They often travel in pairs during the mating season but are otherwise mostly solitary except for a female and her cubs. Four genera and eight species are recognized. North America is home to three species—the Black Bear (*Ursus americanus*), Grizzly Bear (*U. arctos*), and the Polar Bear (*U. maritimus*).

FIGURE 11.14 Shown here is the imprint of the right forepaw of a Black Bear (*Ursus americanus*). The dime at the bottom of the impression is for scale. Black Bears have a plantigrade posture, meaning the entire foot contacts the ground. In Southern California, Black Bears are largely confined to the Transverse Ranges, a few of which have found their way into the San Jacinto Mountains and other parts of the Peninsular Ranges. Most of the regional population derives from bears released in the San Gabriel and San Bernardino Mountains from Yosemite National Park in 1933.

The Black Bear is the most widely distributed of the three North American species and is the only bear now found in California. It ranges from Alaska throughout much of Canada, the mountainous regions of the United States, in similar habitats in the New England states, south through Appalachia and in woods and swamplands of the Southeast. It also ranges into the Sierra Madre of northern Mexico.

Black Bears in good health weigh on average from 200 to 300 lbs. (ca. 90–140 kg). In some instances, males can approach 500 lbs. (220 kg). Fur color varies geographically. Eastern Black Bears are black, whereas western populations trend toward brown or dull reddish brown (cinnamon). Coastal populations from Alaska to British Columbia are blond. Black Bears are consummate opportunists, and the most omnivorous of all the carnivores. They feed on lots of plant matter like nuts, berries, grasses, and acorns. They are fond of underground fungi as well as beetles and their larvae. Occasionally they eat mice, ground squirrels, fish, and rarely larger mammals such as fawns, sheep and pigs. Black Bears are notorious for raiding municipal dumps and residential trash bins in mountain communities. They may also visit orchards and garden plots. Defaced trees that result from a bear rearing on its hind legs and scratching at the bark as high as it can reach often identify Black Bear trails.

In California, Black Bears (*Ursus americanus californiensis*) are most abundant north and west of the Sierra Nevada. Here they are found in all forested areas down to sea level where they are known to wander onto beaches. Density estimates are as high as 1–2.5 bears per square mile. In the Sierras they range from Plumas County south to Kern County and from around 2,500 ft (750 m) to near timberline. Occasionally they descend into the Central Valley following river courses.

Prior to the extinction of the Grizzly Bear in California in the early 1900s (see below), Black Bears were absent from the Coast Ranges south of San Francisco and all of Southern California south of Kern County. Following the Grizzly's demise Black Bears expanded their range southward, beginning with a westward extension at the southern end of the San Joaquin Valley along the Tehachapi Mountains. In 1925 they were reported in Santa Barbara and Ventura Counties. In 1933, the California Fish and Game Commission relocated 28 Black Bears to Southern California from Yosemite National Park (where they had become overly numerous). One died of injuries incident to its capture and transport. Eleven were released near Crystal Lake in the San Gabriel Mountains northeast of Los Angeles. The others were freed in the San Bernardino Mountains around Santa Ana Canyon. The introduced populations have thrived since their release and were probably joined by individuals that dispersed from Santa Barbara and Ventura Counties. At some point, a few crossed the San Gorgonio Pass along the San Andreas Fault Zone (the I-10 corridor) and took up residence in the San Jacinto Mountains, which then gave them access to the rest of the Peninsular Ranges. Their present numbers there are poorly known. The earliest sighting in San Diego County was in 1994 around Mt. Palomar.

By the mid-2000s individual Black Bears began showing up around the edge of foothill communities adjacent to the Transverse Ranges where sightings have continued to the present. Recent (since 2012) reports include the Santa Monica Mountains, Sylmar, Pasadena, Monrovia, Rancho Cucamonga, and east to Yucaipa and Yucca Valley. They've been seen up in trees, in swimming pools, and wandering the streets. Most of these bears have been juveniles around 1 to 2 years of age.

Black Bears mate in the summer and females breed only every other year because of physiological demands. Implantation is delayed for several months such that one or two cubs are born in midwinter when the mother has retreated to a den for the season. The young nurse on milk converted from fat storage accumulated during the previous summer and fall. The cubs follow their mother about when she emerges in the spring and stay with her through the rest of the year, even denning with her that winter. The next spring they are turned out to fend for themselves. These are the bears most likely to wander into foothill communities.

As humans build closer to the wildland interface encounters with large mammals can be expected. In California, the situation is exacerbated by the 5-year drought that began in 2012. Abetted by beetle infestations, massive tree die-offs resulted. Food becomes scarce and Black Bears will search wherever they can to find it. They are protected in national parks, and in California by strict hunting laws as well. Statewide, the Black Bear population in 2012 was estimated at 25,000–30,000, an impressive increase since 1982 when the estimate was 10,000–15,000. As of 2015, population size for Southern California is probably no more than a few hundred.

The mighty Grizzly Bear (*Ursus arctos*) once roamed California concerned with nothing more than its next meal. In North America they ranged from northern Mexico across the western United States north to the Arctic Circle. Today they are restricted to Alaska and western Canada; a few hundred still reside in northwest Washington, Idaho, Montana, and Wyoming. Grizzlies are various shades of brown; they have a prominently humped shoulder and exceptionally long claws. Healthy males can weigh more than 1,300 lbs. (600 kg).

Like other bears, grizzlies are omnivores. They use their claws to excavate rodent burrows, dig for tubers, rip apart decaying logs, and ransack beehives. They eat fruits, berries, grasses, and insect larvae and readily consume carrion, fish, and other small vertebrates.

Native Americans living in California recognized their potential ferocity but also understood that the Grizzly Bear posed little direct threat to them, and they coexisted without incident. The onslaught and extirpation of grizzlies from California began with the arrival of western Europeans. Spanish rancheros shot those that threatened their cattle and at times captured them to stage combats with bulls in a gruesome spectacle. In 1828, Jedediah Smith became the first American to reach California by an overland route. American settlers followed. The Grizzly Bear was already being persecuted elsewhere, and the immigrants continued the practice by shooting or poisoning them with honey or meat laced with strychnine. The occasional hog or sheep that fell prey to a grizzly and a primal fear for their own safety were sufficient reasons to eliminate them. At one point, some landowners were offering $500 bounties—an exorbitant sum for the time. By around 1900, the grizzly was all but gone from the state. In the Yosemite region, for example, one of the last killed was in 1887. Grizzlies persisted longer in Southern California. They were still common around Fort Tejon in 1895 as they were in the San Gabriel foothills. In the Santa Ana (Trabuco) Mountains, they survived well into the early 1900s. The last California Grizzly Bear in captivity was caught in the San Gabriel Mountains in 1889 and exhibited in Golden Gate Park in San Francisco where it lived until 1911. The last grizzly killed in Southern California was a 254 lb. female shot in October 1916 in lower Tujunga Canyon near Sunland in Los Angeles County. The bear had been ravaging a vineyard. The last reliable sighting of a Grizzly Bear in California was made in 1924 near Sequoia Nation Park.

The nineteenth-century accounts of the Grizzly Bear in California included scant information beyond the hunting and shooting. In other words, not much natural history can be gleaned from these often-anecdotal reports. The Grizzly Bear's legacy in California survives with its image on the state flag and in numerous place names throughout the state. In Southern California, for instance, we have Oso Flaco (lean bear) Lake in northern Santa Barbara County, Big Bear Lake and Bear Valley in the San Bernardino Mountains, Bear Valley Road between Victorville and Hesperia, Oso Parkway in Orange County, and Little Bear Valley in San Diego County.

11.11.4 Procyonidae—Raccoons and Ringtails

This New World family of about 15 species includes not only the raccoons and ringtails but also their Neotropical relatives the kingajous and coatis. (The White-nosed Coati [*Nasua narica*] reaches southern Arizona and extreme south Texas.) Most procyonids have alternating light and dark banding on the tail and prominent facial markings are present in some species. The muzzle is pointed and the hind foot posture is plantigrade; the entire foot is placed flat on the ground. The hands are mobile and are used to secure and manipulate food. Procyonids are relatives of bears and pandas. Lacking in procyonids are the carnassial teeth of most other carnivores. All are good climbers and most usually den in trees.

The Northern Raccoon (*Procyon lotor*) is easily recognized by its dark mask and ringed tail. Adult males average about 13 lbs. (6 kg) and have a total length of around 3 ft (90 cm, the tail accounting for about one-third). The raccoon is an adaptable carnivore that ranges over most of North America south through Central America. It is nocturnal and in California occurs statewide except for the low deserts and the higher reaches of the mountains. It is most common on the slopes, foothills, and valleys. Raccoons are curious and easily adapt to human presence. In urban areas they may occupy storm drains, for example.

Raccoons are generally associated with water and are good swimmers. However, they can be found in woodland areas far from it and in marshes where trees are absent. They will eat almost anything and generally feed on whatever is available including fruits, berries, small mammals, crayfish, frogs, fish, birds, and their eggs. They scavenge in garbage bins and dumps and invade orchards and cornfields.

Captive animals provided with water wash their food before consuming it and rub it with their paws if water is unavailable. (The species name *lotor* means "washer").

Raccoons were once an important component of the California fur industry and were a regular quarry of horse and hound hunts. Unless it was shot on sight, a raccoon survived by seeking refuge up in a tree and by their ferocious response to dogs.

Raccoons breed in late winter. Females den inside hollow logs and give birth to 3 to 5 young after a 65-day gestation period. Offspring remain with their mother until the end of summer. They are sociable mammals and family groups may remain together through the winter.

The Ringtail (*Bassariscus astutus*) has a slender build and long banded bushy tail. Total length is from 24 to 32 in. (60–80 cm) and an adult weighs, on average, a little more than 2 lbs. (1 kg). They lack the facial mask of raccoons and have instead huge dark eyes offset by white rings (Figure 11.15). Ringtails range across the western and south central United States from southwestern Oregon to Colorado, Kansas, Texas, and on into Central America. In California and Baja, they are found in semiarid oak woodland, pinon-juniper woodland, coniferous forest, chaparral, brushy deserts, and around cliffs and rocky areas at low to mid-elevations often along watercourses. They are known to take refuge in mountain cabins. Ringtails are absent from most of the Central Valley and the extreme parts of the deserts and generally shun urban areas. Like the raccoon, the Ringtail is

FIGURE 11.15 The photogenic Ringtail (*Bassariscus astutus*) is a secretive nocturnal relative of the raccoon (Procyonidae). In Southern California, Ringtails occupy a variety of woodland habitats often in association with rocky areas, canyon bottoms, and water-courses. (Courtesy of David Wyatt.)

nocturnal. It does not, however, forage in water. The diet consists of small rodents, berries, and fruits. Dens are placed among large boulders near canyon bottoms and mating takes place in late winter. Three to four young are born in May and June. Ringtails are still trapped legally in Arizona, New Mexico, Colorado, and Texas. However, the fur is thin and not very durable.

11.11.5 Mustelidae—Weasels, Skunks, Badgers, and Otters

Mustelids are specialized small to medium-sized carnivores with a short blunt face, long canines, and well-developed shearing and crushing cheek teeth. Eyes are small, and the ears are rounded. The body form is long and slender in some species, for example, weasels and minks, stout like the Badger, or generalized as in skunks. A pugnacious disposition is well earned in the case of the formidable Wolverine and the feisty Badger. Anal scent glands are variously developed among the species. Using the secretions for defense is best known in skunks, but all mustelids have a discernable musky odor. Reproduction often includes delayed implantation (embryonic diapause) in which development ceases after mating and fertilization. Following a prolonged quiescent period implantation takes place and cell division resumes.

Seventy species in 25 genera are found worldwide except on Antarctica, Australia, and most oceanic islands. Ten genera and 17 species inhabit North America, and 11 species in 8 genera occur in California predominately in wooded and mountainous areas in the northern half of the state. Southern California is home to 5 species.

The American Badger (*Taxidea taxus*) is stoutly built for a semi-fossorial lifestyle. It has a low, flat profile, powerful forelimbs with long claws, and long shaggy fur on the back and sides (Figure 11.16). The tail is relatively short, about 5 in. (13 cm), and the black and white striped face is distinctive. At 10–25 lbs. (4.5–11.5 kg), the badger is also one of the larger mustelids species. If cornered or tormented a badger will become quite belligerent and assume a fearless even aggressive attitude. Its common name was brought to the United States from Great Britain where "badger" is also applied to a distant relative, the European Badger (*Meles meles*).

The American Badger occurs from southern Canada, the central and western regions of the United States, and well into Mexico. They inhabit open areas with loose soil where they dig up rodent burrows in search of prey like pocket gophers and ground squirrels. Rangeland, meadows, marshes, and deserts are all suitable. Badgers are solitary and tend to be nomadic. After occupying for a time an excavated rodent burrow, a badger moves on to another area.

Badgers were common and familiar to California's early settlers in farm country. Nowadays they are often difficult to find. As early as the 1930s, the Badger had been reduced in numbers over almost all of its range in California. In our area, most of their original habitat was lost to urbanization. Badgers now occur in areas mostly unaffected by human activity. They were a minor player in the fur trade except for a few years in the 1920s–1930s when the caprice of fashion brought a spike in demand for their pelts.

FIGURE 11.16 The American badger (*Taxidea taxus*) ranges across much of the United States, Canada, and Mexico. Badgers prefer open areas with loose soil. They use their powerful forelimbs and long claws to dig up rodent burrows in search of such prey as gophers and ground squirrels. In Southern California, the badger has lost much of its original habitat to urbanization and is now difficult to find. This one was visiting a water source in the eastern Mojave Desert. (Courtesy of Jason Wallace.)

Badgers den inside extensive burrow systems with conspicuous elliptical entrances. Mating takes place in late summer but implantation is delayed until January of February. One to four young are born in March or April.

Of four North American species of *Mustela*, the Long-tailed Weasel (*Mustela frenata*) is the only one to be found in Southern California. All species share the characteristic elongate body and short limbs. The Least Weasel (*M. nivalis*) is North America's smallest carnivore. The Ermine (*M. erminea*) is similar to it although larger, and the Black-footed Ferret (*M. nigripes*) is a masked weasel now extinct in the wild; several formerly occupied areas are presently under careful management with reintroduced populations bred from captive stock.

The Long-tailed Weasel occurs from southern Canada south through most of the United States and on into South America to Bolivia. It is absent from the arid Southwest including our local deserts. The Long-tailed Weasel occupies all manner of habitats within its range and may be seen almost any time of day or night and in any season. Males are the larger of the sexes and may be up to 18 in. (450 mm) in total length; the black-tipped tail can be nearly half that length or more. The pelage is dark brown and will molt to white in populations living at higher latitudes.

The Long-tailed Weasel is an agile, adept, and fearless hunter. Although it prefers rats and mice, it will take almost any type of prey. For instance, a weasel easily climbs trees to seek out nesting birds and their eggs. They kill by seizing the back of the neck then pierce the softer parts of the skull. With this technique, a weasel is able to subdue prey larger than itself, like rabbits. Long-tailed Weasels are known to kill domestic fowl, but more often than not they are after the mice and rats that infest chicken coops.

The Long-tailed Weasel mates in mid-summer and, like other mustelids, implantation is delayed until winter. From four to eight young are born in June.

The Striped Skunk (*Mephistis mephistis*) is familiar to most everyone, even to folks who know it only as the amorous cartoon character Pepé Le Pew. The black pelage has two dorsal stripes that converge in a "V" on the neck. Stripes are highly variable in length and shape, however, as is the amount of white on the tail. In any case, the bold pattern serves to advertise a skunk's noxiousness; skunks have few natural enemies and show no fear. Humans are the primary cause of early mortality through road kills, along with poison, shooting, trapping, and other activities associated with predator and rodent control. Skunks plod about with complete nonchalance and are easily approachable. Be aware, though, that they can expel the contents of their scent glands with accuracy to around 12 ft (ca. 4 m). Before spraying, a skunk usually raises its tail and stomps its front feet. Too, skunks are a common source of rabies. In bygone days they were often referred to as "civet cats" yet they are neither a true civet (Viveridae) nor a cat (Felidae). *Polecat* is a term used by pioneers who confused them with Old World foul-smelling mustelids.

The Striped skunk occurs everywhere in North America from southern Canada to northern Mexico. It is absent only from parts of the Mojave and Colorado Deserts. These nocturnal feeders are omnivores that seek out insects, their larvae, earthworms, small vertebrates, and plant matter. Unfenced gardens are inviting, as is pet food left on the back porch.

The Striped Skunk mates in late winter and unlike many other mustelids there is no delay in implantation. Four to seven young are born in May or June. If you're lucky you may happen upon a mother skunk on a summer evening with her offspring following along behind her in single file. A cute image for sure, but don't approach them too closely.

Less commonly encountered, the Western Spotted Skunk (*Spilogale gracilis*) is nonetheless widespread in the western United States from southern British Columbia to central Mexico and Baja. It inhabits brushy, rocky areas and is an omnivore that forages by day or night. It occurs throughout California including the Channel Islands of Santa Rosa and Santa Cruz. It seems only to avoid the high mountains. The glossy black fur with white spots all over is unmistakable. The Eastern Spotted Skunk (*S. putorius*) is very similar and until recently both taxa were considered a single species.

The Western Spotted Skunk is known for its acrobatic handstand where the legs and tail are held high above the body, a posture it assumes before spraying. It dens in burrows dug by other mammals. Mating takes place in the fall or winter and unlike the Striped Skunk implantation is delayed. Two to six offspring are born in June.

The Sea Otter (*Enhydra lutris*) is the largest of the living mustelids—males average 65 lbs. (30 kg) and females 45 lbs. (20 kg). As the smallest marine mammal, it is also the only one with dense luxurious fur, flipper-like hind feet and a flattened tail. Sea otters have no blubber and must depend on dense underfur to trap air for insulation. The outer guard hairs are an aid in waterproofing. Sea otters inhabit near-shore waters among kelp forests with rocky or smooth bottoms. Here they dive for clams, mussels, crabs, abalone, and other mollusks. They also have a particular fondness for sea urchins. The sea otter diet is reflected in broad crushing molars, and they are one of a few tool-using mammals. Floating on its back with a shellfish and stone "anvil" on its chest, the otter hammers the shell with another rock until it cracks. The otter then retrieves the soft parts for a meal. When not foraging, which a sea otter must do regularly to maintain a high metabolic rate, it spends its down time grooming.

Historically the Sea Otter ranged along a narrow band from northern Japan, along the Kamchatka Peninsula and Commander Islands, east across the Aleutian Islands to Alaska, then down the west coast of North America to central Baja. Sea otters were immediately prized for their thick velvety coat as soon as they were first discovered by Russian explorers in the mid-eighteenth century. Magnificently lustrous and durable, it is the finest fur in the world. It was also the most expensive, being coveted by royalty and those with the means to purchase the products. Russian ships systematically harvested tens of thousands of sea otters from along Kamchatka, east along the Aleutians, and eventually to the Alaskan coast south to British Columbia.

Farther south along the California coast, Spanish explorations from 1784 to 1789 set out from San Blas, Mexico, to barter for sea otter hides with Native American tribes. At that time, sea otters doubtless could be found along the entire length of the coast wherever dense kelp forests grew. They were especially abundant around the Farallon Islands, in San Francisco Bay, around Monterey, San Luis Obispo, Santa Barbara, and the Channel Islands. In exchange for pelts the Spaniards traded iron scraps, knives, beads, abalone shell, and cloth. The plan was to return the skins to Mexico for tanning and then ship them to China in exchange for quicksilver that the Spanish needed for their mining operations. The 6-year effort, however, yielded only modest returns. For the first couple of years between 1100 and 1700, skins were obtained annually. Thereafter the take was only a few hundred per year. The Spanish endeavor was unsuccessful in part because of the poor quality of the pelts and the indifference on the part of local tribes to develop the enterprise. It remained for later hunters from other countries to decimate the California sea otter populations. For the next 35 years, the harvesting of sea otters (and fur seals) constituted the only important industry along the California coast. Few ships visited that did not come for the purpose of trading in furs.

Sailing up from the south, the illustrious Captain James Cook was commanded to seek new sources of fur and seal oil after the elephant and fur seals were wiped out from the British held Falkland Islands in the south Atlantic. Russian and American interests followed Cook. They along with the British set up trading operations in the Pacific Northwest. In 1812 Russia extended its Alaskan enterprise south to San Francisco Bay. Within 5 years Russian ships had taken some 50,000 skins and at least 5,000 each year thereafter. At first regarded as inexhaustible, by 1820 the sea otter's demise from the Aleutians to the Pacific Northwest was becoming evident. In 1824, Russia relinquished to the United States that part of the Oregon territory that included most of British Columbia, which subsequently (1846) was ceded to Britain. When the harvesting of fur-bearing mammals played itself out, Russia opted to sell Alaska to the United States in 1867. Russia had already exported Alaskan fur for around six times what it paid the United States for the territory. Nonetheless, sufficient numbers of sea otters remained that the hunting continued by U.S. interests. Between 1881 and 1890, the Alaskan Company took in 47,500 skins. By 1900, the total had dropped to 127, at which point the U.S. fur trade did not resist the enactment of an international treaty in 1911 that protected all sea otters north of the 30th parallel, essentially its entire historic range. Even the sale of sea otter fur was outlawed. The last pelt was legally sold that year in London for around $2000. Harvesting of sea otters continued sporadically in California for another decade, despite the ban.

By then, the sea otter had essentially disappeared from along the California coast. Thus, it was to the elation of naturalists everywhere that in 1938 sea otters were spotted off Carmel. A few years later a small colony of 50 sea otters had reestablished itself around Monterrey Bay. It grew and became the nucleus of the California sea otter population of today. They now range from Monterrey to Santa Barbara, with occasional sightings farther north and south; in 2016 the population was estimated at around 3,270. Sea otters continue to face threats from fisheries interaction, predation by sharks and orcas, contamination, and disease.

Before its decimation by the fur trade, California sea otters often hauled out onshore or visited river deltas. Nowadays they remain at sea and clamber on to coastal rocks only during severe storms. They occupy a home range of a few kilometers and do not migrate. Mating behavior is not well understood, other than males are weakly territorial and appear to be polygynous; males have several female partners. There is no breeding season per se but births peak between January and March in California. A single pup is born at sea in the kelp beds after a 6- to 9-month gestation, a period which includes delayed implantation. The precocial newborn is fully furred and active at birth. It is usually independent at 6–8 months but may continue nursing until fully grown.

11.11.6 Otariidae and Phocidae—Sea Lions and Seals

These two families plus the walrus (Odobenidae) are marine mammals known collectively as *pinnipeds*. Pinniped translates as "fin-foot" meaning that the fore and hind limbs are developed into true flippers, not the flipper-like hind feet of sea otters. Flippers replace the tail as the force of locomotion in swimming. Pinnipeds spend most of their time at sea foraging and feeding but will congregate on land or ice floes, especially during the breeding season. Locomotion on land is laborious. In the water, however, pinnipeds are swift and graceful. Their large eyes are adapted for seeing underwater, and despite having very small external ear flaps (pinnae), or none at all, their acoustic capabilities are quite keen.

Historically pinnipeds were considered to be two separate lineages (diphyletic): the otariids (sea lions, fur seals) plus walrus shared a more recent common ancestor with bears (Ursidae) whereas the phocids ("true" seals) were allied with musteloids (living mustelids plus fossil relatives). More recent evidence from molecular studies and morphological data gleaned from more than 50 fossil taxa now support pinniped monophyly. Otariids and phocids are sister taxa. Not fully resolved is the issue of which of the two outgroups is closest to them—the ursids or the musteloids. The 33 living species of pinnipeds occur worldwide and generally prefer the colder waters of the Northern and Southern Hemispheres. Here we will consider those that are or were common in the Southern California region.

Seals and sea lions are pretty easy to tell apart, even at a distance. Sea lions have small but visible external pinnae and thus are sometimes called *eared seals.* The hind limbs can rotate forward to provide a means for sitting up as well as moving about on land with an assist from an undulating vertebral column. Think of sea lions and fur seals as ungainly "walkers" as opposed to seals, which are "wrigglers." Seals lack ear pinnae and cannot rotate the hind limb forward. On land they appear as a torpedo-shaped blob. In the water, however they are the epitome of marine mammal evolution. Swimming with a side-to-side motion of their flippers seals are as acrobatic in the sea as swallows are in the air.

Pinnipeds have a thick layer of body fat that may account for as much as one quarter of their total weight. The blubber serves mainly as an energy reserve and as an assist in buoyancy. Most pinnipeds are gregarious and among otariids polygyny is the normal mating scheme—a single male mates with numerous females, as many as 40 in some species. Most phocids, by contrast, copulate in the water and female congregations are smaller. This is especially true for species that give birth on ice. The ice floe is ephemeral and breeding sites will vary from year to year. Males of some phocid species will defend underwater territories ("maritories") against rival males. Overall, however, polygyny is less pronounced in seals. Gestation among all pinnipeds typically is from 8 to 12 months, usually involves delayed implantation, and a single pup is the norm.

11.11.6.1 Otariidae

The California Sea Lion (*Zalophus californianus*) is common in many coastal areas of Southern California and also occurs in the Sea of Cortez. Sea Lions feed well off shore on squid and assorted fish. Females are the "trained seals" you see in circuses and marine theme parks. California sea lions breed from the Channel Islands south into Baja. Nonbreeding individuals will roam as far north as Vancouver, British Columbia. Like other otariids, males are much larger than females and weigh from 650 to 900 lbs. (300–400 kg).

In the breeding season, females assemble in large numbers at familiar rookeries. Males then come on shore and establish territories. Females are not bound to any single bull and may freely move from one harem to another. After the single pup is born in early summer, the female will mate again within about 2 weeks. Thus, she may be pregnant even while nursing. Pups begin venturing into the sea later in the summer. From then until next spring the mother and offspring will stay together with small groups of other sea lions and the pup may continue to nurse during this time.

Steller's Sea Lion (*Eumetopias jubatus*) is similar in appearance to the California Sea Lion but much larger. Males average around 1,200 lbs. (550 kg) and can weigh as much as 2,400 lbs. (1,090 kg). They are the largest of the otariids and occur from the Kamchatka Peninsula east across the Aleutian Islands, the Gulf of Alaska, and south along the Pacific coast to California. Steller's Sea Lion once bred on the northern Channel Islands, but they have not been seen there since the 1980s. Today, the southernmost population breeds on Año Nuevo Island off the central coast. The western populations of Steller's sea lion experienced a steep decline from the 1970s to the 1990s. Since 2000, there has been a modest increase in their numbers.

The Guadalupe Fur Seal (*Arctocephalus townsendi*), sometimes called the Southern Fur Seal, is the smallest of the sea lions along the California coast. Males average around 400 lbs. (180 kg). They are the only species of *Arctocephalus* in the Northern Hemisphere. Historically Guadalupe fur seals occurred in the Pacific waters off Baja California and were abundant around the Channel Islands. They may have ranged north as far as the Farallon Islands off of San Francisco Bay. Unfortunately, little demographic information exists because the fur trade rapidly and thoroughly decimated fur seals in the nineteenth century. They were believed extinct by 1895 until their "rediscovery" on Guadalupe Island off of Baja in 1928. In 1950, the population stood at around 200–500. By 2010 the Guadalupe Fur Seal population had rebounded spectacularly to an estimated 20,000. Breeding continues to center around Guadalupe Island and has expanded east to the San Benito Islands near Cedros. Today they also breed off Southern California on San Miguel Island and some of the other Channel Islands as well.

The Northern Fur Seal (*Callorhinus ursinus*) is somewhat larger than the Guadalupe Fur Seal with males reaching up to 600 lbs. (275 kg). It has a distinctly short head and a pointed snout. The Northern Fur Seal is a pelagic species widely distributed across the north Pacific east through the Bearing Sea to the west coast of North America. It remains at sea most of the year and comes ashore only to breed. It ranges along the California coast but typically stays from 10 to 50 mi offshore. Primary breeding sites are the Kuril and Commander Islands of Russia, and the Pribilof Islands in Alaska. The southernmost breeding locality is on San Miguel Island off the Southern California coast.

The Northern Fur Seal along with the Sea Otter were, more than any other mammal, responsible for the early exploration of the North American West Coast by the commercial fur industry. These two species accounted for most of the fur trade of the seventeenth and eighteenth century. Northern fur seals once numbered in the millions, but that vast population was reduced to around 200,000 by the early 1900s. The Northern Fur Seal Convention of 1911 by Japan, Russia, Canada (Britain), and the United States at last brought stringent regulation to the seal industry and populations have increased significantly since then.

11.11.6.2 Phocidae

The Harbor Seal (*Phoca vitulina*) is the most familiar pinniped in California. Common in our harbors and bays, they feed on a variety of crustaceans, mollusks, and fish. Occasionally they will even swim many miles up coastal rivers. Males range from 150–300 lbs. (70–140 kg). Harbor seals haul out onto rocky exposures and beaches. Their awkwardness on land is a tradeoff for their powerful swimming and diving skills; they are superbly built for a life at sea. One of the most widespread of all pinnipeds, the harbor seal occurs throughout coastal regions of the Northern Hemisphere from temperate to arctic waters. Of the five currently recognized subspecies, the Eastern Pacific Harbor Seal (*P. v. richardii*) ranges from Alaska to Baja California.

A single pup is born in the spring and mating, which takes place in the water, occurs after the pup is weaned.

The Northern Elephant Seal (*Mirounga angustirostris*) is an enormous animal. Males, displaying an outrageous proboscis, can reach 14 ft and weigh 3,000–5000 lbs. (ca. 1,500–2,300 kg). The elephant seal ranges from the Aleutian Islands to Baja and spends about 80% of its time at sea. The primary breeding localities are Southern California's Channel Islands, and on Isla Guadalupe, Benito del Este, and Cedros Island off the Pacific coast of Baja. Unlike most other phocids, the breeding system is similar to the otariids in which a single male dominates a harem. Competition among males is often fierce and bloody, and many of them bear the scars of previous combats.

The Northern Elephant Seal is another marine mammal that was nearly brought to extinction by human exploitation in the seventeenth and eighteenth centuries. In this case the prize was not fur, but the enormous amount of oil that could be extracted from their fat reserves. By around 1890 only 100 or so Northern Elephant Seals were known to exist. Left alone after the seal treaty of 1911, the population rebounded to more than 200,000 by 2010. Most of the population increase resulted from expansion in the size and number of rookeries in the Channel Islands. They now also breed at a few places along the central coast around Morro Bay, the Farallon Islands, and at Pt. Reyes.

11.12 UNGULATA—PERISSODACTYLA AND ARTIODACTYLA

These two orders comprise the hoofed mammals, or ungulates (*ungula* means "hoof"). Built for cursorial locomotion, they share similar proportions, anatomy, diets, physiology and behavior. Perisodactyls and artiodactyls have a shared ancestry through the Condylartha, a Paleocene lineage that gave rise to the large mammalian herbivores of the Cenozoic. The most obvious difference between perisodactyls and artiodactyls is the number of toes. Perissodactyls have an odd number and the axis of the foot passes through the middle (third) digit. Artiodactyls have an even number of digits and the axis passes through digits three and four.

Perissodactyls were once common and widespread but are now a group of relatively few species. And although the name means "odd-toed," the fore foot may have an even number of toes, whereas the hind foot has either one or three. Tapirs, for example, have four toes on the front foot and the hind foot has three. The three living families of perissodactyls are the Equidae (horses, asses, and zebras in Africa and desert regions of southwestern and eastern Asia), Rhinocerotidae (rhinoceroses in tropical areas of Africa and Southeast Asia), and Tapiridae (tapirs in tropical areas of Central and South America and Southeast Asia).

11.12.1 Equidae—Horses and Burros

These animals were once widely distributed in the New World but disappeared at the end of the Pleistocene 10,000–12,000 years ago. Modern horses (*Equus caballus*) as well as the burro (or donkey [*Equus asinus*] native to northeast Africa) have been domesticated for thousands of years and have been used for a variety of purposes. They were brought to North and South America by European explorers, namely the Spanish conquistadores. The wild horses of the American West are descendants of those introduced herds. Wild burros derive from stock abandoned by mining prospectors. Both had been going about their own ways for decades on private and public land until the federal government realized that they, horses in particular, are capable of doing considerable environmental damage if left unchecked. Thus, in 1971 the U.S. Congress passed the Wild Free-roaming Horse and Burro Act. The act delegated the Department of the Interior and Department of Agriculture to manage herds on public land.

As of 2014, the Bureau of Land Management (BLM) oversaw 179 designated herd management areas, each with a proscribed population size. In California, wild horses are most common in the northern part of the state in the open country of Modoc and Lassen Counties. Farther south there are populations in Inyo and Kern Counties (especially around the Naval Weapons Center at China Lake). A small herd roams Santa Cruz Island. Other herd management areas in Southern California include the Chocolate Mountains (Imperial Counties), Chemehuevi and Big Bear (San Bernardino County), Palm Canyon (Riverside County), and Black Mountains (San Luis Obispo County).

Left unmanaged, the number of free-roaming horses could triple every 6 to 8 years until adequate food and water were no longer available and horses began starving and otherwise falling into ill health. Hence, the aggregate maximum goal across all management areas is 23,000 horses. When numbers exceed an area's capacity, the BLM captures and removes the excess. As of 2012, ca. 195,000 horses had been removed from public lands. As of 2013, 45,000 horses were in captivity and about 33,000 horses roamed free in the western United States. Some of the captives are sold at auction to the public, but the demand is insufficient to put much of a dent in the numbers. This all means that tens of thousands of horses are not living the life of Hidalgo as wild mustangs in the American West, as the act of 1971 intended. Moreover, directives from congress, along with the backing of horse advocates, prohibit the euthanizing of healthy stock. The BLM, therefore, has to maintain indefinitely an ever-growing number of captives.

Feral burros are far fewer in number than wild horses. In our region, they are known from Death Valley and other parts of the Mojave and south into Imperial County. It's difficult not to respect these hardy animals for their ability to scratch out a living in some of our most inhospitable places. Feral burros will compete with Bighorn Sheep for native grasses and watering holes although the conflict is not that acute.

11.12.2 Cervidae—Deer

With around 130 species, the *Artiodactlya* are the most diverse order of large mammals. Practically all are herbivores, either grazers or browsers, and a few, like pigs, are omnivorous. Most artiodactyls are cud chewers. They have a *rumen*, which is the anterior most of the four chambers that comprise a complex digestive track. Often referred to collectively as *ruminants* they include deer, giraffes, cattle, bison, goats, sheep, antelope, and the pronghorn. Ruminants lack upper incisors, and except for elk and caribou canine teeth are absent as well.

Artiodacyls are important meat providers both as game and as livestock. They are most abundant in Africa. The three primary clades are (1) pigs and their allies, (2) camels and llamas, and (3) all other taxa such as deer and elk; sheep, cattle, and goats; antelopes; giraffes and okapi. North America is home to 13 species plus 6 exotics that were introduced for sport hunting. California is home to 5 native species, 2 of which occur in our area and 1 that has been extirpated, the Pronghorn.

The Mule Deer (*Odocoileus hemionus*) is California's most important big game mammal and is the essential prey of mountain lions. They forage extensively on succulent new shoots, grasses, and berries and thus play a role in shaping plant communities. Mule Deer range across the western United States and Canada south into Mexico including Baja. Where they overlap with their eastern relative, the White-tailed Deer (*O. virginianus*), mule deer tend to occupy drier habitats. In California, they are found in forests with open understory, meadows, fields, and chaparral throughout the state except for the Central Valley and large stretches of the Mojave. Six populations (subspecies) are recognized in California. Those in the Coast Ranges are referred to as the Black-tailed Deer (*O. h. columbianus*) because of the dark dorsal surface of the tail. Along the Colorado River in San Bernardino, Riverside, and Imperial Counties, they are called Desert, or Burro Mule Deer (*O. h. eremicus*). In Southern California, mule deer occurred nearly everywhere before European settlement but are now found largely in the south Coast Ranges, the Transverse and Peninsular Ranges, and the less-disturbed foothills. Mule Deer avoid habitat near residences where dogs run free.

Cervids are characterized by their antlers, which are found only on males except in a few instances like caribou where they adorn both sexes. Antlers are seasonal; they are shed annually. As new antlers emerge the overlying skin ("velvet") provides a vascular supply to the growing bone. The skin eventually sloughs off exposing the bare dermal bone. Antlers are dichotomously branched and tend to acquire more forks (tines) with age. In mule deer, antlers are shed in late January or February and grow anew in the summer. Mule deer are a polygynous species; males mate with multiple does. Sexual pursuit of does, called the *rut*, begins in late summer or early fall and continues into November or late January in some populations. Females begin breeding at about 1½ years of age; males about a year later. One or two fawns are born in May or June.

11.12.3 Bovidae—Sheep, Goats, and Cattle

These two-toed ungulates have "true" horns that are usually present in both sexes. Unlike antlers, the horns are neither branched nor shed. They continue to grow throughout life and consist of a bony core covered by a hardy cornified sheath derived from the integument. Horns can be coiled or spiraled and in males they are sufficiently robust so as to be able to withstand the forces of head-butting during male-to-male combat.

Native North America bovids are the American Bison (*Bison bison*), Mountain Goat (*Oreamnos americanus*), and Dall's Sheep (*Ovis dalli*). The latter two species are restricted to western Canada and parts of Alaska. (Mountain goats extend into Washington, Montana, and Idaho.) Their more southern relative, the Bighorn Sheep (*Ovis canadensis*, Figure 6.54) occurs in rugged mountainous regions of the American West, plus Canada to the north (Alberta and British Columbia) and Mexico to the south (Sonora, Baja).

Bighorn sheep have short brown pelage with a white rump patch. Males weigh from 165–300 lbs. (75–135 kg) and females from 120 to 190 lbs. (50–85 kg), The horns are impressively coiled in older rams and shorter and curved in ewes. Reproduction takes place in November and December. A single lamb is born after a gestation period of around 6 months. Rams 3 years old and older dominate the breeding even though young males can be sexually active.

Bighorn sheep have rebounded from a near total collapse in the early 1900s that was brought about by disease acquired from introduced livestock and by overhunting. By 1920, they were extirpated from most of Washington, Oregon, the Dakotas, Nebraska, and Texas. Over the years, some of these areas have been repatriated by native stock brought in from elsewhere, and most of these populations are presently stable. However, in California bighorns along the eastern Sierra (*O. c. sierra*) and in the Peninsular Range (*O. c. nelson*) are considered endangered (see discussion under Desert Vertebrates).

11.12.4 Suidae—Swine

Wild boar are present in at least 18 states in the United States and are especially common in the gulf coastal region. Texas is nearly overrun with them. These animals are simply domestic pigs (*Sus scrofa*) that have become feral. They can reach 400 lbs. or more (ca. 180 kg), are intelligent, and can be mercilessly destructive when they plow their tusked snouts through the ground in search of food. They will eat almost anything.

The domestic pig is native to North Africa and Eurasia. The Spanish and Russians first brought them to California as livestock in the early 1700s. Lax husbandry practices guaranteed that some would become feral. Then in the 1920s, the European wild boar (a putatively different subspecies) was introduced into Monterey County. This stock then bred with the already existing feral pigs such that today most of California's wild boar are a hybrid of the two. As a result, physical characteristics vary between those with long hair and snouts and short erect ears to those with short hair, blunt snouts, floppy ears and a barrel-shaped body.

Wild boar can be found in almost any habitat except for dry sandy desert. They are present statewide except for Imperial County and often travel in small herds. In Southern California, populations in Jurupa Valley and East Vale, Riverside County, have caused damage estimated in the millions of dollars. There were no restrictions on hunting them until the mid-1950s when California Fish and Game realized they could be a revenue stream and so designated them as game animals that require a license and tag to hunt.

11.12.5 Antilocapridae—Pronghorn

The Pronghorn (*Antilocapra americana*) is the only living member of this North American family that extends back 20 million years. Although people persist in calling it an antelope, it is not. The true antelopes of Africa have permanent horns that grow throughout life. The stout horns of the pronghorn consist of a short permanent bony core covered by an outer cornified sheath that is shed annually in early winter. The bony core is unbranched but the sheath itself has a smaller secondary tine (prong) pointing forward in adult males. Horns of females are smaller and the sheath is usually unbranched. Pronghorn were noted by early Spanish expeditions into the Southwest. The first use of "antelope" was by Meriwether Lewis and William Clark during their explorations of the American West in 1804–1806.

The pronghorn is a distinctive deer-size (ca. 20–55 kg) artiodactyl found only in North America. It roams grasslands and open spaces of the West from southern Canada (Alberta, Saskatchewan) to Mexico, including Baja. When startled, a Pronghorn can erupt with explosive speed, achieving 50–60 mi/h over short distances and is capable of sustaining 35 mi/h for several miles. The Pronghorn is the fastest land mammal in North America.

Pronghorns are no longer as common as they once were because of habitat fragmentation and overhunting in some places. In California, pronghorn were at one time abundant in the Central Valley from Sutter Buttes southward well into Baja California. They occurred along the coast from near Monterey to as far south as Bahia San Ignacio on the Pacific side of the Baja peninsula. On the gulf side, their range extended to well south of San Felipe. In the central peninsula they reached south to the Vizcaíno plane where the few remaining pronghorn (*A. a. peninsularis*) are found today. In Southern California, they occupied the mesas and valleys near the coast as well as the Perris plain in Riverside County. Pronghorn were common in the western Mojave and survived to the early 1930s in Los Angeles County in, appropriately, Antelope Valley. Records are also known from the Colorado Desert near, for example, Calexico and Fish Creek in Imperial County, and around Carrizo Creek and the Borrego Badlands in eastern San Diego County.

Hunting pressure, agriculture, and other forms of development eliminated most of the prime pronghorn habitat in California. They are now restricted to the grassy planes in the far northeast part of the state. Pronghorn have been reintroduced to areas around Big Pine and Bodie, and to the Carrizo Plain (San Luis Obispo County) in the 1990s, now a protected remnant of original San Joaquin Valley habitat. Similar reestablishment is being considered for the Chuckwalla Bench east of the Salton Sea, Imperial County.

11.13 CETACEA—WHALES, DOLPHINS, AND PORPOISES

Cetaceans include a few freshwater species but are otherwise entirely marine mammals. As such, we have scant opportunity to observe them unless we are at sea. Cetaceans are unmistakably built for swimming. They have streamlined fusiform bodies, a flat tail with flukes, forearms modified as flippers, and no hind limbs (vestiges of the pelvic girdle are usually present). They are essentially hairless and lack ear pinnae. Cetaceans swim by up and down undulations of the tail with near frictionless proficiency. Deep and prolonged dives are accomplished by and array of anatomical and physiological mechanisms including a reduced heart rate, circulatory shunts, tissues with high oxygen storage capacity, the ability to almost completely expel spent air from the lungs, and a tolerance for metabolic by-products like carbon dioxide and lactic acid. Bones of the skull have been reorganized as a result of the nostrils having shifted to a dorsal position with a single or double blowhole. Their sense of hearing is acute with echolocation employed for orientation. Visual acuity is also well developed.

The earliest cetaceans appear in the Eocene as descendants of terrestrial mammals and most likely share an ancestry with artiodactlys. Their closest *living* relatives might be the hippopotamuses, although this proposal remains controversial. Living cetaceans fall into two broad groups: the *Mysticeti* or baleen whales, and the *Odontoceti*, the toothed whales and dolphins.

Mysticeti develop tooth buds embryonically, but they are replaced in the adults by *baleen*. Erroneously called "whalebone," baleen is a derivative of the integument as modified bundles of hair-like strands. No bone is present. Baleen consists of a series of frayed transverse plates arrayed longitudinally on the upper sides of the mouth. These baleen skirts are used to strain krill and other planktonic organisms from water that is gulped into the mouth. Baleen whales include the largest of the cetaceans. The most familiar species to Californians is the Grey Whale (*Eschrichtius robustus*) whose annual migration brings them close enough to the coast to be observed from sea cliffs or up close in one of the many whale-watching excursions available from marinas.

Most of the great baleen whales summer in polar waters of the Northern or Southern Hemisphere where they feed on the abundance of krill and other small crustaceans. The amount of food consumed is enormous, much of which is converted to energy stores in the form of blubber. Near the end of summer, they migrate to tropical or subtropical waters to give birth. The calves grow rapidly on the mother's rich milk and by spring the young are sufficiently mature to travel back to the Arctic or Antarctic with their mother.

The Grey Whale is a remarkable study in physiological endurance and efficiency. Each year they travel 5,600 mi (9,000 km) from their summer feeding grounds in the Bering and Chukchi Seas to their primary calving and breeding area in Laguna Ojo de Liebre (Scammon's Lagoon) along the Pacific coast of Baja California from January to April. The energetic requirements of a pregnant female include basic daily energy expenditures as she gorges on arctic krill, nutrients for the developing calf, energy for the migration itself, sufficient fat reserves to maintain her insulation of blubber, and finally to exist for 8 months without eating because she consumes almost nothing during that time. If we make some basic assumptions such as 50% conversion efficiency of gross intake to biologically useful energy, she would need to take in around 2,000 lbs. (900 kg) per day to meet these needs. Four months of active feeding would consume 240,000 lbs. (108,500 kg) of food. Little wonder that nearly half of the total body mass of these whales is blubber. The migratory birthing trips are necessary because a newly born calf is ill-prepared to survive the temperatures were it born in the Arctic seas. Therefore, migrating to the warmer waters off of central Baja makes sense and is worth the energetic cost to the female.

Besides the Grey Whale, other mysticetes that pass through Southern California, albeit usually farther off the coast, are the Fin Whale (*Balaenoptera physalis*), Northern Minke Whale (*B. acutorostrata*), Sei Whale (*B. borealis*), Humpback Whale (*Megaptera novaeangliae*), and the largest animal ever known to have existed, the Blue Whale (*Balenoptera musculus*). The Northern Right Whale (*Eubalaena glacialis*) is rare. It was nearly exterminated by nineteenth-century whalers but in recent decades has made a modest come back. In the days of open boats and hand-held harpoons, they were the "right" whales to hunt because they were approachable and the body floated after being killed.

The Odontocetes are more diverse than the baleen whales at around 50 species. They include the toothed whales, dolphins, and porpoises. Odontocetes are not filter feeders like the mysticetes, and they use echolocation to find their food such as cephalopods (squid, octopus) and fish. Some odontocetes, like killer whales, have a large rounded head, whereas dolphins and beaked whales have a rostrum, variously pointed or stout, set off from the brow region. Porpoises are small dolphins with a blunt more or less rounded head. They are strictly coastal, and they often enter bays and estuaries.

Many odontocetes are cosmopolitan. Some prefer warmer waters and avoid polar regions, and others still are limited in their range to, for example, the warm Atlantic and the Caribbean Sea. The most common odontocetes that occur off the Southern California coast are the Short-beaked Common Dolphin (*Delphinus delphis*), Long-beaked Common Dolphin (*D. capensis*), Pacific White-sided Dolphin (*Lagenorhynchus obliquidens*), Bottle-nosed Dolphin (*Tursiops truncatus*),

Short-finned Pilot Whale (*Globicephalus macrorhynchus*), and Dall's Porpoise (*Phocoenoides dalli*). Less common species include Killer Whale (*Orcinus orca*), False Killer Whale (*Pseudorca crassidens*), and Risso's Dolphin (*Grampus griseus*). The endangered Sperm Whale (*Physeter catodon*) is found worldwide favoring warmer waters. Males especially will wander great distances and may occasionally pass off of Southern California.

12 The Ice Age Mammals

The Pleistocene (the ice ages, 1.8 million to 10,000 years ago) was a time of advancing and retreating glacial episodes and the concomitant lowering and rising of sea levels. It was remarkable as well for the many large mammals present in North and South America that became extinct at or near the end of the last glacial period.

The George C. Page Museum is the home of the La Brea Tar Pits in Hancock Park in Los Angeles. Many of the large mammals from North America's recent past have been recovered from these tar pits, some equaling or exceeding in size those found in Africa today. The Southern California region was a home to, among other species, several kinds of elephants, ground sloths, different horses, tapirs, camels, and formidable predator like cheetahs, saber-toothed cats, and the American race of the African lion. Also present were large vulture-like carrion-feeding birds. These large mammals were here until around 12,000 years ago, a mere blink of an eye in geologic time.

In Southern California Pleistocene fossils are not limited to La Brea. There are well-known deposits in Anza Borrego, and it is not unusual for remains of a mammoth or some other mammal from the ice ages to be uncovered during grading for a residential development or a freeway realignment. La Brea stands out not only for its taxonomic diversity but also for the depth of its chronology. The La Brea fossils span in age from around 40,000 to 8,000 years ago. Paleontologists formally recognize the past 250,000 years in North America as the Rancholabrean Land Mammal Age.

The tar pits were a natural trap. When the thick crude oil seeps to the surface (the area was once known as the Salt Creek Oil Fields), the accompanying lighter petrochemicals evaporate and leave the tar behind. The surface becomes covered in dirt and debris, or by water during the wet season, thus camouflaging the lethal hazard that awaits and animal who inadvertently ventures into it. One of the more interesting features of the tar pits is that about 90% of the mammalian species that have been recovered are carnivores and scavengers, contrary to natural mammalian demographics where herbivores dominate. The explanation for this paradox is that when a large herbivore became mired in the tar it would attract the carnivores and scavengers, who themselves became entombed in the black goo. Such an event was probably not routine. However, even once every 5 to 10 years would be sufficient to amass the wealth of bones that accumulated over 30,000 years.

Many of the genera of large mammals are now extinct such as ground sloths, mammoths, mastodons, deer, the giant short-faced bear, saber-toothed cats, a cheetah, camels, lamas, oxen, and peccary. Others persist on continents outside North America like the North American race of the African lion, horses, and tapirs. Still others survive today as the lone species of their kind, for example, several species of bison—not just one—roamed North America until around 13,000 years ago.

The Tar Pits also preserved still-living species such as black bear, grizzly bear, and mountain lion along with smaller mammals like shrews, rabbits, and many rodents. Numerous species of extant birds also became trapped including hawks, ducks, shorebirds, pigeons, and passerines. Even several species of amphibians and reptiles are preserved. In total more than 600 species of plants, invertebrates, and vertebrates have been recovered from the La Brea Tar Pits. The chronology of this remarkable historic record of *near time*—the past 40,000 years—received a windfall with the development of radiocarbon dating techniques that were developed in the 1940s by Willard Libby. Coupled with sound geological data no other dating method can match its utility and precision for establishing chronologies of the past 40,000 years. See the text box for a brief description of the method.

RADIO CARBON DATING

All living forms of life contain carbon, mainly in the stable (nonradioactive) form of ^{12}C. An unstable form of carbon, ^{14}C, is also continuously being formed in the atmosphere in low concentrations. Hence all living organisms contain both forms of carbon in those same proportions. When the organism dies carbon is no longer being replenished and the amount of ^{12}C in the dead tissues remains the same as when the plant or animal was alive. However, in the case of ^{14}C, it begins to break down (to Nitrogen-14). Like all radioactive isotopes, ^{14}C decays at a fixed rate. That rate of decay is expressed as the half-life, the time required for one-half of the "parent" isotope (^{14}C) to decay to its "daughter" isotope (^{14}N). The half-life for ^{14}C is relatively short at 5,730 years. Put another way, one-half of the original amount of ^{14}C would have decayed by this time. Thus, the ratio of $^{14}C/^{12}C$ in the fossil can be measured and used to estimate how long ago the plant or animal died. For example, if the fossil contained one fourth of the ^{14}C originally present then that amount would be equivalent to two half-lives of decay (½ of 1 + ½ of ½ = ¾) or 11,460 years. Age estimates are usually expressed with an error range (plus or minus a certain number of years). Because of the short half-life of ^{14}C, the method is most reliable up to around 40,000 years.

Radiocarbon dating offers the opportunity to raise and address numerous paleoecological topics in near time. For instance, practically all of the extinctions that occurred at the end of the Pleistocene (the late Quaternary) in North and South America were the large (>100 lbs./45 kg) mammals—the megafauna. Most of the smaller vertebrates, those less than 100 lbs. (45 kg), survived. Late Quaternary extinctions in North America saw the loss of 34 genera of large mammals and 3 genera of small ones. South America was hit even harder. That continent lost 51 genera, all large. Minus the 11 genera in common, the Western Hemisphere lost 74 genera of large mammals. Before exploring the possible causes of this catastrophic disappearance, let us consider some of those taxa that were in western North America north of Mexico in the late Pleistocene (Table 12.1).

TABLE 12.1
Late Quaternary Extinct Species of Large (<100 lbs. [45 kg]) Land Mammals of Western North America and Northern Mexico

Species	Common Name
Pilosa	
Paramylodon harlani[a]	Big-tongued Ground Sloth
Megalonyx jeffersonii[a]	Jefferson's Ground Sloth
Nothrotheriops shastensis[a]	Shasta Ground Sloth
Proboscidea	
Mammut americanum[a]	American Mastodon
Mammuthus columbi[a]	Columbian Mammoth
Mammuthus exilis	Dwarf Mammoth
Mammuthus primigenius	Wooly Mammoth
Carnivora	
Arctodus simus[a]	Giant Short-faced Bear
Canus dirus[a]	Dire Wolf
Homotherium serum[a]	Scimitar Cat
Panthera leo atrox[a]	American Lion
Miracinonyx trumani[a]	American Cheetah
Smilodon fatalis[a]	Saber-toothed Cat
Perissodactyla	
Equus conversidens[a]	Mexican horse
Equus occidentalis[a]	Western horse
Tapirus californicus[a]	Tapir
Artiodactyla	
Bison antiquus[a]	Ancient Bison
Bison latifrons[a]	Giant Bison
Camelops hesternus[a]	Western Camel
Capromeryx minor[a]	Dwarf Pronghorn
Euceratherium collinum[a]	Shrub Ox
Hemiauchenia macrocephala[a]	Long-legged Llama
Mylohyus nasutus	Long-nosed Peccary
Platygonus compressus[a]	Flat-headed Peccary

Source: Martin, P.S., *Twilight of the Mammoths—Ice Age Extinctions and the Rewilding of America*, University of California Press, Berkeley, CA, 2005.

[a] Species have been reported from the La Brea Tar Pits.

In the late Quaternary prior to their extinction, North America was home to 4 genera of ground sloths, and South America had 12 genera. None of these is with us today. Only the tree sloths of Middle and South America persist. Ground sloths were hairy, large-boned herbivores with nonretractable claws. They did not climb trees or move quickly; they probably ambled about on their knuckles in a fashion similar their living xenarthran relatives the South American giant anteater (*Myrmecophaga*). Ground sloths probably led solitary lives, other than when mating. Most ground sloths were impressively large. Ten species exceeded 1,000 lbs. (450 kg). The largest was *Megatherium americanum* of South America that weighed about 13,000 lbs. (6,000 kg), the size of an Asian elephant. The smallest North American ground sloth, and best known in the West, is the Shasta ground sloth (*Nothrotheriops shastensis*) that ranged from northern California southeast to the Texas panhandle on into the Sierra Madre Oriental of northern Mexico. It was about the size of a black bear and weighed from 300 to 1,200 lbs. (135–545 kg). Dung from the Shasta ground sloth along with mummified soft parts have been recovered from several dry caves in the Southwest.

During the Quaternary, North America had three families and five genera of proboscideans (elephants). Three survived into near time to about 13,000 years ago—*Mammuthus* (mammoths), *Mammut* (mastodons), and *Cuvieronius* (gomphotheres). Typical mammoths and mastodons stood 8–10 ft (ca. 2.5–3 m) at the shoulder and weighed from 9,000 to 13,000 lbs. (ca. 6,000 kg). They had long curving tusks, and at least some were covered in shaggy hair. The Imperial mammoth (*M. imperator*) was the largest terrestrial mammal in the North American Pleistocene at an estimated 22,000 lbs. (10,000 kg). By contrast the dwarf mammoth (*M. exilis*) of the Channel Islands (Santa Rosa, San Miguel, and Santa Cruz) stood around 5½ ft at the shoulder and weighed in the neighborhood of 1,700 lbs. (ca. 770 kg). The elephant-like gomphotheres were in South America until the end of the Pleistocene. Gomphotheres had straight tusks and an elongate lower jaw. One of those, *Cuvieronius*, ranged north from Mexico into Florida.

Of the large Quaternary carnivores, seven species survived in North America until the end of the Pleistocene. The Dire wolf (*Canus dirus*) was similar to the living Grey wolf (*Canus lupus*) but more robust. It is one of the most common fossils excavated from the La Brea Tar Pits, and rows and rows of skulls are displayed on shelves at the Paige Museum. The Giant Short-faced bear (*Arctodus simus*) was larger than any living bear, whereas the Florida cave bear (*Tremarctos floridana*) was smaller than *Arctodus* although larger than its relative living today in South America, the Spectacled bear (*Tremarctos ornatus*). The American cheetah (*Miracinonyx trumani*) and American lion (*Panthera leo atrox*) are nearly indistinguishable from the species living today in Africa and Asia (Iran).

Among other felids, saber-like canine teeth evolved independently at least four different times, including in marsupials. Two of these cats persisted until the end of the Pleistocene, the renowned Saber-tooth Cat (*Smilodon fatalis*) and the Scimitar Cat (*Homotherium serum*). Essentially there were two types of saber-tooth cats. The dirk-tooth cats like *Smilodon* had long narrow canines bearing fine serrations. The limb bones were stout and strong, almost bear-like, and they

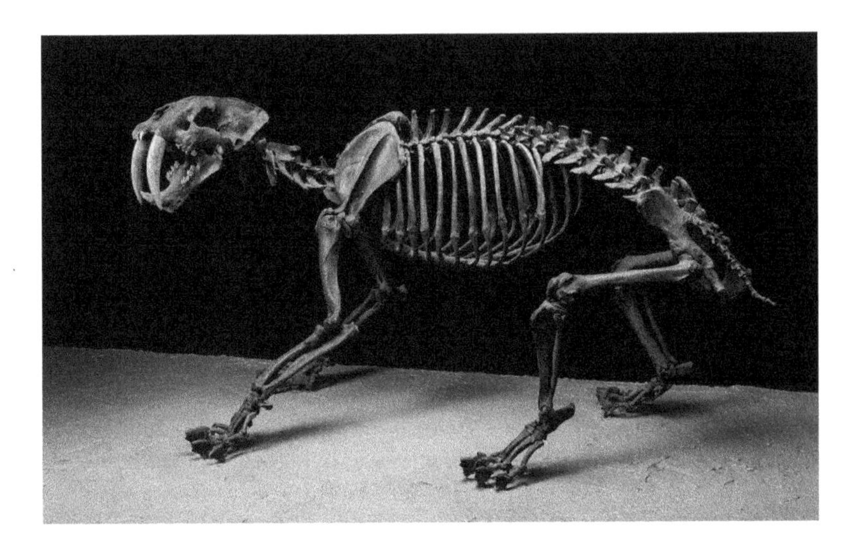

FIGURE 12.1 The Saber-toothed Cat (*Smilodon fatalis*) was a formidable predator of the late Pleistocene of western North America. Well represented in the La Brea Tar Pits, *Smilodon fatalis* (*californicus*) is the state fossil of California. The Saber-tooth was about the size of an African Lion, but with more powerful forelimbs. The exaggerated canines were probably used to subdue prey by slashing the neck and severing the trachea (wind pipe) and carotid arteries. (Courtesy of Paleobiology Collections, Smithsonian Institution, Washington, DC; Chip Clark.)

were probably ambush predators. *Smilodon* was about the size of an African lion (Figure 12.1). The canines of scimitar-toothed cats were shorter (about 4 in. as opposed to 7 in.) but broader and had coarse serrations. Over all, the skeleton was more gracile than that of the dirk-toothed cats, suggesting that they may have been pursuit predators. Otherwise, both types of saber-tooth cats shared similar features of the skull, for example a compressed facial region, small orbits (eye sockets), and a jaw gape of more than 90° (compared with about 65° in the mountain lion).

Numerous ideas have been advanced to explain how the long canines might have been used. Some of them are rather absurd such as for climbing trees or rooting about for mollusks like a walrus. Most likely the teeth were used to subdue prey. However, killing must have been performed in a precise manner because lateral forces generated by a random stabbing motion would probably cause the teeth to break. On the other hand, a controlled bite to the throat that severed the trachea and carotid arteries could have been accomplished effectively with less risk to the teeth.

Of the odd-toed ungulates (perissodactyls), only horses and tapirs survived into near time in North America. The genus *Equus* was variable and included numerous species in North and South America. Modern horses (*Equus caballus*) were right at home in North American grasslands following their reintroduction by the Spanish; in hindsight their success was predictable. Extinct tapirs are known from Florida, Kansas, Arizona, and California. Today three species of *Tapirus* exist in southern Mexico into South America and another in Malaysia.

North America still harbors a number of large native even-toed ungulates (artiodactyls) such as Mountain Goat (*Oreamnos americanus*), Bighorn (*Ovis canadensis*), Dall's Sheep (*Ovis dalli*), the American Bison (*Bison bison*), Caribou (*Rangifer tarandus*), Muskox (*Ovibos moschatus*), Elk (*Cervus elaphus*), Moose (*Alces alces*), and deer (*Odocoileus* spp.). Yet in near time there were also three genera of camels and llamas (*Camelops*, *Paleolama*, and *Hemiauchenia*), and extinct bovids like the woodland musk ox (*Bootherium*), the brush ox (*Euceratherium*), and other species of bison. Extinct cervids include the stag moose (*Cervalces*) and three deer (*Torontoceros*, *Bretzia*, and *Navahoceros*). Two extinct genera of peccary are *Mylohyus* and *Platygonus*, both of which were larger than the three living species. *Platygonus* was around 110 lbs. (50 kg) compared with the Collared Peccary (*Pecari tajacu*) living today in the Southwest at 32–55 lbs. (15–25 kg). Three genera of antilocaprids also existed where there is but one today. *Capromeryx*, *Stockoceros*, and *Tetrameryx* had V-shaped horns rather than the pronghorn of living *Antilocapra*.

The largest American rodent to go extinct was the bear-size giant beaver, *Castoroides*.

12.1 EXTINCTION MODELS

A dynamic of evolution is that species come and go through time. The fossil record demonstrates that the vast majority of species that ever lived have become extinct. Extinction can happen for any number of physical or biological reasons. A change in ocean temperatures or salinity may overwhelm the physiological capabilities of some species, or on land sudden climatic or geological shifts may eliminate or reconfigure a required habitat. But what could have caused the sudden disappearance of a staggering number of large- and medium-sized mammals 12,000 years ago leaving smaller vertebrates essentially unaffected?

This problem has vexed paleontologists ever since Thomas Jefferson received some bones and claws of a large mammal from a cave in present-day West Virginia. Jefferson did not fully grasp extinction and therefore believed that whatever the animal was others of its kind must still be alive. The bones in question were from a ground sloth named in his honor, *Megalonyx jeffersonii*. But extinct it was.

There were no meteor strikes at the end of the Pleistocene or any other disaster of global proportions. Some paleontologists have long favored a climate-based explanation. No one disputes that pulsing climates were the dominant force driving many features of the ice ages, particularly the advance and retreat of glaciers. Yet warm-cold cycles were a consistent pattern over the 1.8 million years of the Pleistocene as normal background extinction took place, perhaps including climatically sensitive species. In other words, climate change near the Pleistocene-Holocene interface was not so unique as to account for the unprecedented loss of so many large mammals at that time.

An alternative explanation is that the large mammals were ill-prepared for the ecological challenges of human arrival into North and then South America. Predation was likely the most important contributor, and some authorities propose that foreign diseases and pathogens carried by people also may have had a hand. Many species can adapt to some level of increased predation but not to the extent brought on by human hunters. In the Americas, this basic concept, known as "overkill," has been around since the 1950s. Beginning in the late 1960s, it found its voice in the work of Paul Martin at the University of Arizona.

The archaeological record in North America indicates that the earliest human inhabitants were the Clovis people, so named for their distinctive spear points that were first found near Clovis, New Mexico. Around the time Clovis artifacts show up in the archaeological record, two-thirds of the large animals disappeared along with their commensals and parasites. Some archaeologists argue that there are so few, if any, accurately dated and documented kill sites that humans could not possibly have been responsible for the extinction of animals the size of mammoths and ground sloths 12,000 years ago. Yet the fossil record rarely reveals the cause of mortality or the reason(s) for extinction. So in the face of the field evidence, the overkill model for the Americas must accommodate a massive extinction event that would involve rapid human population growth and the north (from Beringia—the ancient land now submerged below the Bering Strait) to south spread of experienced hunters encountering naïve prey. The large mammal fauna had no prior experience with humans and were without any behavioral defenses. By targeting "keystone" species, an extinction cascade would follow. For instance, large herbivores like mammoths and mastodons serve as "forest managers" that keep brush from invading the open areas preferred by grazing ungulates. When the populations of proboscideans are reduced in numbers below their ability to recover, other herbivores, and their predators, decline as well.

Some archeologists maintain that humans arrived in North America before the mass extinction of the terminal Pleistocene. Several New World archaeological sites have been cited as pre-Clovis, although most of them have been discounted by subsequent investigation (Calico Hills east of Barstow and Long Beach, for example). A few sites remain possible: Meadowcroft Rockshelter in Pennsylvania; Blue Fish Caves, Yukon, Canada; and Monte Verde, Chile. The "Clovis-first" model is also being challenged by recent archaeological sites along the West Coast of North America that argue the first humans arrived by boat beginning about 16,000 years ago. These maritime people, it is reasoned, moved quickly out of Beringia down the Pacific Coast and reached Monte Verde, Chile, about 14,500 years ago. A few tantalizing coastal archaeological sites are suggestive, but most have likely been lost to the rising sea levels that inundated shore lines since then.

Even accepting these, if humans truly occupied the interior of America thousands of years before Clovis there should be a far richer archaeological record for that time period then there is. Consider that in Australia, an environment far less hospital to sustaining the human enterprise than the Americas, and one that was inhabited far earlier, by 40,000 years ago, more than 100 verified sites are known. As in the Americas, late Quaternary extinction also hit Australia; around the time of first contact it lost 28 genera and 55 species more than 10 kg.

In addition to Australia and the Americas, wholesale extinction at first contact also characterizes deep-water oceanic islands (those not connected to a mainland continent during times of lower sea level) such as those in the Pacific. Humans sailed out from the Solomon Islands 3,500 years ago, colonizing the remote islands of Polynesia one after another. They reached Hawaii well to the north and even Easter Island 4,650 mi (7,500 km) to the east, and finally backtracking they reached New Zealand within the last 1,000 years. The impact of first contact mirrors events that occurred on the continents. In the absence of large mammals, hundreds of species and populations of birds disappeared coincident with human colonization in Oceania (Polynesia, Melanesia, and Micronesia). At first contact, people harvested the "low-hanging fruit" by seeking out the largest, most savory and easiest to find species. The large, tame, flightless birds were easy pickings and the vast colonies of seabirds were equally vulnerable. On Fiji and Tonga, giant iguanid lizards quickly vanished as well. On New Zealand, the breathtakingly swift extinction of 11 species of moas (flightless ratite birds, some larger than ostriches) coincided with human arrival. On remote islands, this then is the pattern. Once the primary resource was exhausted, people went after the smaller, less-desirable species. Indirect human impact followed from introduced alien species (e.g., pigs and rats) and deforestation for agriculture.

We might ask, "Are there any islands that were not occupied prehistorically that could serve as a control?" Indeed there are. The Galápagos in the southeast Pacific, the Commander Islands in the north Pacific, the Mascarenes in the Indian Ocean, and the Azores in the Atlantic were uninhabited prehistorically and extinction is unknown or minimal. On a similar note, extinction played out gradually on continents of human origin. In Europe and northern Asia, late Quaternary extinctions began earlier than in North America and ended later.

Taking a global perspective, the striking feature of extinction in near time is the episodic, sequential coincidence of vanishing species with the arrival of humans to places where they had never been. I want to emphasize that there is no moral implication implied in the human impact scenario of first contact and its aftermath. Nor should one be inferred. These people were simply trying to stay alive by exploiting the novel (and abundant) resources that they had in front of them. They had no awareness of outcomes. Today, humans are better aware of their environmental actions and resulting consequences but, unfortunately, that cause-and-effect relationship remains difficult for us to appreciate.

Bibliography

Hundreds of sources were consulted while composing this book. Listed here are those with broad coverage of particular topics, those with unique information, style, or of general interest, and those specifically referenced in the text.

EXPLORATION, SYSTEMATICS, AND GENERAL BACKGROUND

Archibald, J. D. 2014. *Aristotle's Ladder, Darwin's Tree: The Evolution of Visual Metaphors for Biological Order.* New York: Columbia University Press.

Baum, D. A., S. DeWitt, and S. S. Donovan. 2005. The tree-thinking challenge. *Science* 310:979–980.

Beidleman, R. C. 2006. *California's Frontier Naturalists.* Berkeley, CA: University of California Press.

Brown, J. H. and M. V. Lomolino. 1998. *Biogeography.* Sunderland, MA: Sinauer Associates.

Dana, R. H. 1909. *Two Years Before The Mast.* New York: P. F. Collier & Son.

Farquhar, F. P. 1965. *History of the Sierra Nevada.* Berkeley, CA: University of California Press.

Greene, H. W. 2013. *Tracks and Shadows—Field Biology as Art.* Berkeley, CA: University of California Press.

Marinacci, B. and R. Marinacci. 2005. *California's Spanish Place Names. What They Mean and What They Reveal.* Santa Monica, CA: Angel City Press.

Matthiessen, P. 1959. *Wildlife in America.* New York: Penguin Books.

Reisner, M. 1986. *Cadillac Desert—The American West and its Disappearing Water.* New York: Penguin Books.

Zwinger, A. 1986. *John Xántus. The Fort Tejon Letters 1857–1859.* Tucson, AZ: University of Arizona Press.

CLIMATE AND LANDFORMS

Abbott, P. L. 1999. *The Rise and Fall of San Diego: 150 Million Years of History Recorded in Sedimentary Rocks.* San Diego, CA: Sunbelt Publications.

Bailey, H. P. 1966. *The Climate of Southern California.* Berkeley, CA: University of California Press.

Brenzel, K. N. (Ed.). 2001. *Sunset Western Garden Book.* Menlo Park, CA: Sunset Publishing Corporation.

Davis, G. A. and B. C. Burchfield. 1973. Garlock fault: An intercontinental transform structure, Southern California. *Geological Society of America Bulletin* 84:1407–1422.

Lynch, D. K. 2009. *Field Guide to the San Andreas Fault.* Topanga, CA: Thule Scientific.

McPhee, J. 1989. *The Control of Nature.* London, UK: Hutchinson Radius.

McPhee, J. 1993. *Assembling California.* New York: Farrar, Straus and Giroux.

Norris, R. M. and R. W. Webb. 1990. *Geology of California* (2nd ed.). New York: John Wiley & Sons.

Sharp, R. P. and A. F. Grazner. 1993. *Geology Underfoot in Southern California.* Missoula, MT: Mountain Press Publishing Company.

PLANTS AND BIOMES

Adam, D. P. 1995. Reflection on the development of the California Pollen Record. In: *Late Quaternary Environments and Deep History: A Tribute to Paul S. Martin*, D. W. Steadman and J. I. Mead (Eds.), pp. 117–130. Hot Springs, SD: The Mammoth Site of Hot Springs South Dakota.

Axelrod, D. I. 1958. Evolution of the Madro-Tertiary Geoflora. *Botanical Review* 24:433–509.

Axlerod, D. 1978. The origin of coastal sage vegetation, Alta and Baja California. *American Journal of Botany* 65:1117–1131.

Beauchamp, R. M. 1986. *A Flora of San Diego County, California*. National City, CA: Sweetwater River Press.

Belzer, T. J. 1984. *Roadside Plants of Southern California*. Missoula, MT: Mountain Press Publishing Company.

Bowler, P. A. 2000. Ecological restoration of coastal sage scrub and its potential role in habitat conservation plans. *Environmental Management* 26:86–96.

Chase, C. P. 1982. California (coastal) chaparral. In: *Biotic Communities of the American Southwest—United States and Mexico*, D. E. Brown (Ed.), pp. 91–94. Desert Plants Special Issue 4. Superior, AZ: Boyce Thompson Southwestern Arboretum.

Cole, K. L. 1995. Equable climates, mixed assemblages, and the regression fallacy. In: *Late Quaternary Environments and Deep History: A Tribute to Paul S. Martin*, D. W. Steadman and J. I. Mead (Eds.), pp. 131–138. Hot Springs, SD: The Mammoth Site of Hot Springs South Dakota.

Davis, F. W., P. A. Stine, and D. M. Stoms. 1999. Distribution and conservation status of coastal sage scrub in southwestern California. *Journal of Vegetation Sciences* 5:743–756.

Dugan, J. E. and D. M. Hubbard. 2010. Loss of coastal strand habitat in Southern California: The role of beach grooming. *Estuaries and Coasts* 33:67–77.

Ferren, W. R., P. L. Fiedler, R. A. Leidy, K. D. Lafferty, and L. A. Mertes. 1996. Wetlands of California, part II: A method for their classification and description. *Madroño* 43:125–182.

Fultz, F. M. 1923. *The Elfin-Forest of California*. Los Angeles, CA: The Times-Mirror Press.

Griffin, J. R. 1977. Oak woodland. In: *Vegetation of California*, M. G. Barbour and J. Majors (Eds.), pp. 383–415. New York: John Wiley & Sons.

Hanes, T. L. 1977. California Chaparral. In: *Terrestrial Vegetation of California*, M. G. Barbour and J. Majors (Eds.), pp. 417–469. New York: John Wiley & Sons.

Hickman, J. C. (Ed.). 1996. *The Jepson Manual—Higher Plants of California*. Berkeley, CA: University of California Press.

Lafferty, K. D. 2005. Assessing estuarine biota in southern California. United States Department of Agriculture Forest Service General Technical Report PSW_GTR, 195:1–15.

Lightner, J. 2011. *San Diego County Native Plants*. San Diego, CA: San Diego Flora.

MacKay, P. 2003. *Mojave Desert Wildflowers*. Guilford, CT: The Globe Pequot Press.

Munz, P. A. 2004. *Introduction to California Spring Wildflowers of the Foothills, Valleys, and Coast*. Berkeley, CA: University of California Press.

Munz, P. A. and D. D. Keck. 1973. *A California Flora and Supplement*. Berkeley, CA: University of California Press.

O'Leary, F. J. 1990. California coastal sage scrub: General characteristics and considerations for biological conservation. In: *Endangered Plant Communities in Southern California*, A. A. Schoenherr (Ed.), pp. 24–41. Claremont, CA: Southern California Botanists Special Publication No. 3.

Pace, C. P. 1982. California (coastal) chaparral. In: *Biotic Communities of the American Southwest—United States and Mexico*, D. E. Brown (Ed.), p. 91. Desert Plants Special Issue 4. Superior, AZ: Boyce Thompson Southwestern Arboretum.

Pace, C. P. and D. E. Brown. 1982. California coastal scrub. In: *Biotic Communities of the American Southwest—United States and Mexico*, D. E. Brown (Ed.), pp. 86–90. Desert Plants Special Issue 4. Superior, AZ: Boyce Thompson Southwestern Arboretum.

Petrides, G. A. and O. Petrides. 1992. *A Field Guide to Western Trees*. Boston, MA: Houghton Mifflin HarCourt.

Pielou, E. C. 1991. *After the Ice Age—The Return of Life to Glaciated North America*. Chicago, IL: University of Chicago Press.

Quinn, R. D. and S. C. Keeley. 2006. *Introduction to California Chaparral*. Berkeley, CA: University of California Press.

Rogers, K. E. 2000. *The Magnificent Mesquite*. Austin, TX: University of Texas Press.

Rubinoff, D. 2001. Evaluating the California gnatcatcher as an umbrella species for conservation of southern California coastal sage scrub. *Conservation Biology* 15:1374–1383.

Schoenherr, A. A. 1992. *A Natural History of California*. Berkeley, CA: University of California Press.

Shreve, F. and I. L. Wiggins. 1964. *Vegetation and Flora of the Sonoran Desert*, 2 Vols. Palo Alto, CA: Stanford University Press.

Spaulding, W. G. 1995. Environmental change, ecosystem responses, and the Late Quaternary Development of the Mojave Desert. In: *Late Quaternary Environments and Deep History: A Tribute to Paul S. Martin*, D. W. Steadman and J. I. Mead (Eds.), pp. 139–164. Hot Springs, SD: The Mammoth Site of Hot Springs South Dakota.

Talluto, M. V. and K. Suding. 2007. Historical change in coastal sage scrub in southern California, USA in relation to fire frequency and air pollution. *Landscape Ecology* 23:803–815.

Truner, R. M. 1982. Mohave desertscrub. In: *Biotic Communities of the American Southwest—United States and Mexico*, D. E. Brown (Ed.), pp. 157–168. Desert Plants Special Issue 4. Superior, AZ: Boyce Thompson Southwestern Arboretum.

Turner, R. M. and D. E. Brown. 1982. Sonoran desertscrub. In: *Biotic Communities of the American Southwest—United States and Mexico*, D. E. Brown (Ed.), pp. 181–221. Desert Plants Special Issue 4. Superior, AZ: Boyce Thompson Southwestern Arboretum.

Westman, W. E. 1981. Diversity relations and succession in California coastal sage scrub. *Ecology* 62:439–455.

Zedler, J. B. 1982. The ecology of southern California coastal salt marshes: A community profile. *U.S. Fish and Wildlife Service, FWS/OBS-81-54*: 1–110.

VERTEBRATES—GENERAL

Feder, M. E. and G. V. Lauder (Eds.). 1986. *Predator-Prey Relationships*. Chicago, IL: University of Chicago Press.

Kardong, K. V. 2015. *Vertebrates: Comparative Anatomy, Function, Evolution*. New York: McGraw-Hill Education.

Linzey, D. W. 2012. *Vertebrate Biology*. Baltimore, MD: The Johns Hopkins University Press.

Lowe, C. H. (Ed.). 1964. *The Vertebrates of Arizona*. Tucson, AZ: University of Arizona Press.

Miller, A. H. and R. C. Stebbins. 1964. *The Lives of Desert animals in Joshua Tree National Monument.* Berkeley, CA: University of California Press.
Pough, E. H., C. M. Janis, and J. B. Heiser. 2009. *Vertebrate Life.* San Francisco, CA: Pearson Benjamin Cummings.

FISH

McGinnis, S. M. 2006. *Field Guide to Freshwater Fishes of California.* Berkeley, CA: University of California Press.

AMPHIBIANS AND REPTILES

Berry, K. H., J. L. Yee, A. A. Coble, W. M. Perry, and T. A. Shields. 2013. Multiple factors affect a population of Agassizii's Desert Tortoise (*Gopherus agassizii*) in the northwestern Mojave Desert. *Herpetological Monographs* 27:87–109.
Brown, P. R. 1997. *A Field Guide to Snakes of California.* Houston, TX: Gulf Publishing.
Duellman, W. E. and L. Trueb. 1986. *Biology of Amphibians.* Baltimore, MD: The Johns Hopkins University Press.
Ernst, C. H., J. E. Lovich, and R. W. Barbour. 1994. *Turtles of the United States and Canada.* Washington, DC: Smithsonian Institution Press.
Ernst, C. H. and G. R. Zug. 1996. *Snakes in Question—The Smithsonian Answer Book.* Washington, DC: Smithsonian Institution Press.
Estes, R. E. and G. K. Pregill (Eds.). 1998. *Phylogenetic Relationships of the Lizard Families. Essays Commemorating Charles L. Camp.* Palo Alto, CA: Stanford University Press.
Fisher, R. N. and T. J. Case. 1997. *A Field Guide to the Reptiles and Amphibians of Coastal Southern California.* San Mateo, CA: United States Geological Survey Biological Resources Division, Lazer Touch.
Frost, D. R., T. Grant, J. Faivovich et al. 2006. The amphibian tree of life. *American Museum of Natural History Bulletin* 297:1–370.
Gauthier, J. A., M. Kearney, J. Maisano, O. Rieppel, and A. B. D. Behlke. 2012. Assembling the squamate tree of life: Perspectives from the phenotype and the fossil record. *Peabody Museum of Natural History, Yale University Bulletin* 53:3–308.
Green, H. W. 1997. *Snakes, The Evolution of Mystery in Nature.* Berkeley, CA: University of California Press.
Grinnell, J. and C. L. Camp. 1917. A distributional list of the amphibians and reptiles of California. *University of California Publications in Zoology* 17:127–208.
Grismer, L. L. 2002. *Amphibians and Reptiles of Baja California, Including Its Pacific Islands and the Islands in the Sea of Cortez.* Berkeley, CA: University of California Press.
Holman, J. A. 1995. *Pleistocene Amphibians and Reptiles in North America.* New York: Oxford University Press.
Jones, L. L. C. and R. E. Lovich (Eds.). 2009. *Lizards of the American Southwest. A Photographic Field Guide.* Tucson, AZ: Rio Nuevo Publishers.
Klauber, L. M. 1972. *Rattlesnakes: Their Habits, Life Histories, and Influence on Mankind*, 2 Vols. Berkeley, CA: University of California Press.
Lemm, J. M. 2006. *Field Guide to Amphibians and Reptiles of the San Diego Region.* Berkeley, CA: University of California Press.
Lovich, J. E. and K. R. Beaman. 2007. A history of Gila Monster (*Heloderma suspectum cinctum*) records from California with comments on factors effecting their distribution. *Southern California Academy of Sciences Bulletin* 106:39–58.

Murphy, R. W., K. H. Berry, T. Edwards, A. E. Leviton, A. Lathrop, and J. D. Riedle. 2011. The dazed and confused identity of Agassiz's land tortoise, *Gopherus agassizi* (Testudines, Testudinidae) with the description of a new species, and its consequences for conservation. *ZooKeys* 113:39–71.
Pianka, E. R. and L. J. Vitt. 2003. *Lizards. Windows to the Evolution of Diversity*. Berkeley, CA: University of California Press.
Porter, K. H. 1972. *Herpetology*. Philadelphia, PA: W. B. Saunders Company.
Pough, F. H., R. M. Andrews, J. E. Cadle, M. L. Crump, A. H. Savitsky, and K. D. Wells. 1998. *Herpetology*. Upper Saddle River, NJ: Prentice Hall.
Reeder, T. W., T. M. Townsend, D. G. Mulcahy et al. 2015. Integrated analyses resolve conflicts over squamate reptile phylogeny and reveal unexpected placements of fossil taxa. *PLoS One* 10:e0118199.
Rieppel, O. 2009. How did the turtle get its shell? *Science* 325:154–155.
Sherbrooke, W. C. 2003. *Introduction to Horned Lizards of North America*. California Natural History Guides No. 64. Berkeley, CA: University of California Press.
Stebbins, R. C. 2003. *Western Reptiles and Amphibians*. Boston, MA: Houghton Mifflin.
Stebbins, R. C. and N. W. Cohen. 1995. *A Natural History of Amphibians*. Princeton, NJ: Princeton University Press.
Zug, G. R. 1993. *Herpetology—An Introductory Biology of Amphibians and Reptiles*. San Diego, CA: Academic Press.

BIRDS

Dickinson, E. C. and J. V. Remsen (Eds.). 2013. *The Howard and Moore Complete Checklist of the Birds of the World* (4th ed.), Vol. 1 (Non-passerines). Eastbourne, UK: Aves Press.
Dickinson, E. C. and L. Christidis (Eds.). 2014. *The Howard and Moore Complete Checklist of the Birds of the World* (4th ed.), Vol. 2 (Passerines). Eastbourne, UK: Aves Press.
Feduccia, A. 1999. *The Origin and Evolution of Birds* (2nd ed.). New Haven, CT: Yale University Press.
Hacket, S. J., R. T. Kimball, S. Reddy et al. 2008. A phylogenomic study of birds reveals their evolutionary history. *Science* 320:1743–1767.
Heilmann, G. 1972. *The Origin of Birds*. New York: Dover Publications. (First published by D. Appleton & Co., 1927.)
Marshall, J. T. Jr. 1957. Birds of pine-oak woodland in southern Arizona and adjacent Mexico. *Cooper Ornithological Society* 32:1–125.
Proctor, N. S. and P. K. Lynch. 1993. *Manual of Ornithology—Avian Structure and Function*. New Haven, CT: Yale University Press.
Sibley, D. A. 2003. *The Sibley Field Guide to Birds of Western North America*. New York: Alfred A. Knopf.
Steadman, D. W. 2006. *Extinction and Biogeography of Tropical Pacific Birds*. Chicago, IL: The University of Chicago Press.
Unitt, P. 2004. San Diego County Bird Atlas. *Proceedings of the San Diego Society of Natural History*, San Diego, CA, Vol. 39, pp. 1–645.
Van Tyne, J. and A. J. Berger. 1971. *Fundamentals of Ornithology*. New York: Dover Publications. (First published by John Wiley & Sons, New York, 1959.)
Wilson, R. M. 2010. Seeking Refuge. *Birds and Landscapes of the Pacific Flyway*. Seattle, WA: University of Washington Press.
Yuri, T., R. T. Kimball, J. Harshman et al. 2013. Parsimony and model-based analyses of indels in avian nuclear genes reveal congruent and incongruent phylogenetic signals. *Biology* 2:419–444.

MAMMALS

Agenbroad, L. D. 2009. *Mammuthus exilis* from the California Channel Islands: Height, mass, and geologic age. In: *7th California Islands Symposium*, C. C. Damiani and D. K. Garcelon (Eds.), Arcata, CA: Institute for Wildlife Studies.

Berta, A. 2012. *Return to the Sea: The Life and Evolutionary Times of Marine Mammals.* Berkeley, CA: University of California Press.

Grinnell, J., J. S. Dixon, and J. M. Linsdale. 1937. *Fur-bearing Mammals of California. Their Natural History, Systematic Status and Relations to Man*, 2 vols. Berkeley, CA: University of California Press.

Haynes, G. 1991. *Mammoths, Mastodons, and Elephants—Biology, Behavior and the Fossil Record.* Cambridge, UK: Cambridge University Press.

Ingles, L. G. 1965. *Mammals of the Pacific States—California, Oregon, Washington.* Stanford, CA: Stanford University Press.

Jameson, E. W. Jr. and H. J. Peeters. 2004. *California Mammals.* Berkeley, CA: University of California Press.

Kays, R. W. and D. E. Wilson. 2002. *Mammals of North America.* Princeton, NJ: Princeton University Press.

Kitchener, A. 1991. *Natural History of the Wild Cats.* Ithaca, NY: Comstock Publishing Associates.

Kurten, B. 1971. *The Age of Mammals.* New York: Columbia University Press.

Kurten, B. and E. Anderson. 1980. *Pleistocene Mammals of North America.* New York: Columbia University Press.

Lanman, C. W., K. Lundquist, H. Perryman et al. 2013. The historical range of beaver (*Castor canadensis*) in coastal California: An updated review of the evidence. *California Fish and Game* 99:193–221.

Lawlor, T. E. 1979. *Handbook to the Orders and Families of Living Mammals.* Eureka, CA: Mad River Press.

Martin, P. S. 2005. *Twilight of the Mammoths—Ice Age Extinctions and the Rewilding of America.* Berkeley, CA: University of California Press.

Martin, P. S. and W. Steadman. 1999. Prehistoric extinctions on islands and continents. In: *Extinctions in Near Time—Causes, Contexts, and Consequences*, R. D. E. MacPhee (Ed.), pp. 17–55. New York: Kluwer Academic/Plenum Publishers.

Miller, A. H. 1946. Vertebrate inhabitants of the piñon association in the Death Valley region. *Ecology* 27:54–60.

Nowak, R. M. 1999. *Walker's Mammals of the World* (6th ed.), 2 vols. Baltimore, MD: Johns Hopkins University Press.

O'Leary, M. A., J. I. Bloch, J. J. Flynn et al. 2013. The Placental mammal ancestor and the post-K-Pg radiation of placentals. *Science* 339:662–667.

Orr, R. T. 1972. *Marine Mammals of California.* Berkeley, CA: University of California Press.

Rose, K. D. and J. D. Archibald (Eds.). 2005. *The Rise of Placental Mammals.* Baltimore, MD: Johns Hopkins University Press.

Sheldon, J. W. 1992. *Wild Dogs: The Natural History of Nondomestic Canidae.* San Diego, CA: Academic Press.

Ward, P. 1997. *The Call of Distant Mammoths: Why the Ice Age Mammals Disappeared.* New York: Copernicus Books.

Wilson, D. E. and D. M. Reeder (Eds.). 2005. *Mammal Species of the World: A Taxonomic and Geographic Reference.* Washington, DC: Smithsonian Institution Press.

Index

Note: Page numbers followed by f and t refer to figures and tables respectively.

C

D

I

J

K

L

M

N

O

P

S

T